N. G. Chetaev

THEORETICAL MECHANICS

Translated from the Russian by
Irene Aleksanova

Mir Publishers Moscow
Springer-Verlag Berlin Heidelberg New York
London Paris Tokyo Hong Kong

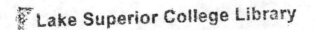

Nikolai G. Chetaev,
Emeritus Professor of Mechanics,
Moscow University

First published 1989
Revised from the 1987 Russian edition

Distribution rights for the socialist countries, India and Iran:
V/O "Mezhdunarodnaya Kniga" Moscow
For all other countries: Springer-Verlag Berlin Heidelberg
New York London Paris Tokyo Hong Kong

ISBN 3-540-51379-5 Springer-Verlag Berlin Heidelberg New York
ISBN 0-387-51379-5 Springer-Verlag New York Berlin Heidelberg

CONTENTS

FROM THE EDITORS

This course of theoretical mechanics contains a set of lectures delivered by Professor Nikolai Guryevich Chetaev. He is most famous for his essay *The Stability of Dynamic Motion* and the compilation of his research papers. Chetaev is not so well known as a teacher although for many years he delivered various courses of lectures (but always a course of theoretical mechanics) first at Kazan University and the Kazan Aviation Institute and then at Moscow University, where he was first a professor, holding the Chair of Theoretical Mechanics, and then Chairman.

Chetaev's lecturers have greatly influenced Soviet lecturers in theoretical mechanics, many of whose courses have been published since his death. Following in the footsteps of A.M. Lyapunov and N.E. Zhukovsky, Chetaev constructed his course in the classical traditions dating back to Lagrange, and creatively developed these traditions. A mathematical rigour in his exposition was combined with a large number of exquisitely selected problems which, as a rule, were significant for their applications. Being both an eminent scientist and a wonderful teacher, Chetaev often included in his course problems he was studying at the time.

In his last years, Chetaev worked on a textbook in theoretical mechanics which he intended for university-level mathematics and physics students. Although the book was not completed, the manuscript he left covers almost all the problems of the university programme, as well as a number of problems that are beyond the programme. The book is in eight chapters.

The first chapter is a concise introduction to the theory of free, sliding, and bound vectors. It contains all material on vectors necessary to study the course of theoretical mechanics.

Chapters 2, 3, 4, and 5 contain the material traditionally included in university courses. It should be noted that in kinematics the author concentrates on Euler's theorem on the instantaneous distribution of velocities in a moving rigid body and emphasizes the significance of this theorem for a correct understanding of kinematics.

Chapter 3 discusses in detail the work done by a force during virtual and true displacements and gives an elegant proof of the necessity and sufficiency of the conditions of equilibrium for a system brought about

by the principle of virtual displacements. Many subtle points concerning
the principle are illustrated by examples. Chetaev presents an original
method for defining separate reaction forces of constraints using the principle of virtual displacements, which proved very efficient, but is not usually
presented in textbooks on mechanics.

In Chapter 4, when discussing the motion of a particle in a central force
field, both the analytic method and geometric method of defining the safety ellipse are used, the latter method being almost never touched upon in
mechanics textbooks.

In Chapter 5, entitled "Dynamics of a System", the ideas of Hermann
and Euler, as developed by Lagrange, are discussed and the fruitlessness
of the arguments about the reality of D'Alembert's inertial forces is shown.
The general theorems of dynamics (without the reactions of constraints)
are derived from the Euler-Lagrange principle and are applied to the solution of some interesting problems. When deriving Lagrange's equations,
Chetaev emphasizes that they are only valid for holonomic, defining, coordinates and points out Neumann's mistake. Chetaev also gives here a
method of determining the unknown reactions using Lagrange's second-
order equations and illustrates it with numerous examples.

Chapter 6, entitled "The Dynamics of a Rigid Body", includes both
the three classical cases of integrability and some special cases (thosé of
Hess, Bobylev-Steklov, and Chaplygin).

Unfortunately, these are the only chapters Chetaev himself prepared,
and the next chapter entitled "Analytic Dynamics", to which Chetaev gave
special significance and which he included in the basic course of mechanics
for the first time, is based on lithographed lectures he delivered in Kazan.
Chapter 7 also includes a section entitled "Stability" in which Chetaev
presents the fundamental theorems on stability and instability since he fre-
quently included these themes in his lectures.

"The Theory of Attraction", which for some years Chetaev included
in the basic course, can be found in Chapter 8; again the material is based
on lithographed lectures he delivered in Kazan.

The basic course presented in the book is supplemented with abstracts
of special courses Chetaev delivered at Moscow University in 1940s and
1950s.

Chetaev delivered the special course on Poincaré's equations in 1955.
In this course he develops Poincaré's idea of using group variables in order
to write equations of motion and also considers the problem of integrability
of equations of constraints, i.e. conditions of the constraint holonomy, a
subject seldom considered in mechanics.

The special course entitled "Special Theory of Relativity" is interesting
for its unorthodox exposition and choice of material, its strict presentation,

and its geometric clarity. This course is based on the most detailed abstract written in 1938, with appropriate modifications.

The manuscript was prepared for publication by a group of Chetaev's disciples, namely, A.A. Bogoyavlensky, G.B. Efimov, V.I. Kirgetov, N.N. Kolesnikov, L.M. Markhashov, Sh.S. Nugmanova, G.K. Pozharitsky, V.V. Rumyantsev, K.E. Yakimova, A.S. Sumbatov. His wife Vera Aleksandrovna Chetaeva-Samoilova actually participated in preparing the book for publication, and his son, A.N. Chetaev, assisted her in her efforts.

The book will be of great interest to students of universities and higher schools studying theoretical mechanics and to research workers whose fields of research include mechanics, physics, and adjacent fields of science. It is of especial interest to teachers of theoretical mechanics.

V.V. Rumyantsev
K.E. Yakimova

PREFACE

I have delivered lectures in theoretical mechanics to students at the universities of Kazan and Moscow. The style of the lectures was influenced by the ideas of Professor Zeiliger and Academician Steklov, whose advice I was happy to follow—"Better less, but better".

Lectures on theoretical mechanics have been given for many years. They have been delivered by Lagrange, Euler, Poinsot, and Zhukovsky. Therefore I thought it better to use the various problems in mechanics which satisfied me than to try to be clever and put myself in the awkward position of endeavouring not to repeat clever formulations if someone else has given them before me.

I have striven to make the presentation as simple as possible.

In kinematics I deliberately omitted oblique and general curvilinear coordinates because the expressions in curvilinear coordinates necessary for dynamics can be obtained in a natural way and without any difficulty from Lagrange's equations. Further, according to Cartan, elements of tensor analysis are connected with Lagrange's equations for which covariant and contravariant defining coordinates are irrelevant.

N.G. Chetaev

CHAPTER 1

PRINCIPLES OF VECTOR THEORY

1.1 Vectors. Free Vectors

1. A vector **F** is a segment of a straight line AB (Fig. 1) which has an *origin A* and a *terminus B*. The length of the line segment AB is the *magnitude*, the *absolute value* or the *length* of the vector $\overrightarrow{AB}$, and is designated $|\overrightarrow{AB}|$. The origin A is also called the *point of action* of the vector $\overrightarrow{AB}$. The straight line along which a vector is directed is called the *line of action* of the vector. The direction of the vector $\overrightarrow{AB}$ along the line of action is denoted by an arrow placed at the end B. The direction of the line of action defines the *direction of the vector*.

Three kinds of vector are known in mechanics, namely, free, sliding, and bound vectors. A *free vector* is defined by its magnitude and the direction of its line of action. The point of its action may be arbitrary. A *sliding vector* is defined by the line of its action, its magnitude, and direction; the vector can slide freely along its line of action. A *bound vector* is defined by the point of its action, its line of action, and direction.

2. Two free vectors are considered to be *equal* if they have the same direction and magnitude.

Since the point of action of a free vector is arbitrary, we assume the origin of coordinates to be the point of action of a free vector, unless otherwise specified.

When the point of action of a free vector **F** is changed, its projections X, Y, Z on the rectangular coordinate axes remain invariant and these three projections completely define the free vector.

The length of a vector can be expressed in terms of its projections

$$F = |\mathbf{F}| = \sqrt{X^2 + Y^2 + Z^2},$$

the *direction* of the vector **F** is defined by its *direction cosines*

$$\alpha = \frac{X}{F}, \quad \beta = \frac{Y}{F}, \quad \gamma = \frac{Z}{F},$$

where α, β, γ are the cosines of the angles which the line of action of the vector **F** makes with the positive directions of the x, y, z axes.

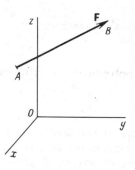

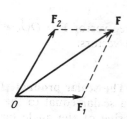

Figure 1   Figure 2

3. To multiply a free vector $\mathbf{F}$ by a positive scalar m is to construct a vector $m\mathbf{F}$ with the same direction but with length $m|\mathbf{F}|$. To multiply a vector $\mathbf{F}$ by a negative scalar $-m$ is to construct a vector $-m\mathbf{F}$ equal to $m|\mathbf{F}|$ in length but opposite in direction.

4. Vector addition. The sum of two free vectors $\mathbf{F}_1$ and $\mathbf{F}_2$ is a vector $\mathbf{F}$ which can be represented as the diagonal of a parallelogram constructed on the vectors $\mathbf{F}_1$ and $\mathbf{F}_2$ with a common origin at the point O (Fig. 2):

$$\mathbf{F} = \mathbf{F}_1 + \mathbf{F}_2.$$

The limiting cases when two vectors $\mathbf{F}_1$ and $\mathbf{F}_2$ with origins at O lie on the same straight line and the parallelogram is a straight line are treated by continuity. In other words, when $\mathbf{F}_1$ and $\mathbf{F}_2$ have the same direction their sum $\mathbf{F}_1 + \mathbf{F}_2$ is a vector equal to $|\mathbf{F}_1| + |\mathbf{F}_2|$ in magnitude and whose direction is that of $\mathbf{F}_1$ and $\mathbf{F}_2$, and when $\mathbf{F}_1$ and $\mathbf{F}_2$ are opposite in direction their sum is a vector equal to $||\mathbf{F}_1| - |\mathbf{F}_2||$ in magnitude and of the same direction as the vector with the larger magnitude.

By definition, the addition of free vectors is commutative, i.e.

$$\mathbf{F}_1 + \mathbf{F}_2 = \mathbf{F}_2 + \mathbf{F}_1.$$

The projection of the sum of vectors on an axis is equal to the sum of the projections of the summands on the same axis, i.e.

$$X = X_1 + X_2, \quad Y = Y_1 + Y_2, \quad Z = Z_1 + Z_2.$$

If the sum of two free vectors is zero, i.e. $\mathbf{F}_1 + \mathbf{F}_2 = 0$, then the length of the vector $\mathbf{F}_2 = -\mathbf{F}_1$ is equal to the length of the vector $\mathbf{F}_1$ and its direction is opposite to the direction of the vector $\mathbf{F}_1$.

The addition of several free vectors is determined by the polygon rule. Consider free vectors $\mathbf{F}_1, \mathbf{F}_2, ..., \mathbf{F}_n$. Their sum $\mathbf{F}$ must be constructed as follows. We assume the point O to be the origin and construct a polygon $OO_1O_2...O_n$ whose sides $\overrightarrow{O_{k-1}O_k}$ are equal to the vectors:

$$\overrightarrow{O_{k-1}O_k} = \mathbf{F}_k.$$

The free vector $\mathbf{F} = \overrightarrow{OO_n}$ which closes the polygon is the *geometric sum* of the vectors:

$$\mathbf{F} = \sum \mathbf{F}_k.$$

5. The scalar product. The scalar product of two free vectors $\mathbf{F}_1$ and $\mathbf{F}_2$ is a scalar equal to the product of the magnitudes of the vectors by the cosine of the angle between the positive directions of these vectors (Fig. 3), i.e.

$$\mathbf{F}_1 \cdot \mathbf{F}_2 = |\mathbf{F}_1||\mathbf{F}_2| \cos \theta.$$

But from the formula

$$R^2 = F_1^2 + F_2^2 - 2|\mathbf{F}_1||\mathbf{F}_2| \cos \theta$$

we have

$$|\mathbf{F}_1||\mathbf{F}_2| \cos \theta$$
$$= \frac{1}{2}\{X_1^2 + Y_1^2 + Z_1^2 + X_2^2 + Y_2^2 + Z_2^2$$
$$- [(X_1 - X_2)^2 + (Y_1 - Y_2)^2 + (Z_1 - Z_2)^2]\},$$

and consequently

$$\mathbf{F}_1 \cdot \mathbf{F}_2 = X_1X_2 + Y_1Y_2 + Z_1Z_2. \tag{1.1}$$

The scalar product of two vectors then reduces to the algebraic multiplication of the corresponding projections of these vectors and their addition

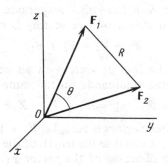

Figure 3

and is, therefore, commutative and distributive:

$$\mathbf{F}_1 \cdot \mathbf{F}_2 = \mathbf{F}_2 \cdot \mathbf{F}_1,$$
$$\mathbf{F}_1 \cdot (\mathbf{F}_2 + \mathbf{F}_3) = \mathbf{F}_1 \cdot \mathbf{F}_2 + \mathbf{F}_1 \cdot \mathbf{F}_3.$$

If the vectors are orthogonal, then their scalar product is zero.

6. The vector product. Consider two free vectors $\mathbf{F}_1$ and $\mathbf{F}_2$ (Fig. 4). We apply these vectors to the same point O. The vector product $\mathbf{F}_1 \times \mathbf{F}_2$ is defined as a free vector equal in magnitude to the area of a parallelogram constructed on $\mathbf{F}_1$ and $\mathbf{F}_2$, whose direction along the normal to the plane of the parallelogram is such that the rotation of $\mathbf{F}_1$ and $\mathbf{F}_2$ is in the positive direction. In the right-handed coordinate system, the counterclockwise rotation is assumed to be the positive rotation.

We seek the projections L, M, N of the vector

$$\mathbf{Q} = \mathbf{F}_1 \times \mathbf{F}_2$$

on the x, y, z axes, respectively, of a rectangular coordinate system. The projection on the z-axis is

$$N = |\mathbf{Q}|\cos\theta = 2\,\Pi\cos\theta,$$

where Π is the area of the triangle constructed on the vectors $\mathbf{F}_1$ and $\mathbf{F}_2$ (Fig. 5); the angle between the plane of the triangle and the xy-plane is equal to the angle θ between the normals to these planes. Assume that the vectors $\mathbf{f}_1$ and $\mathbf{f}_2$ are orthogonal components of the vectors $\mathbf{F}_1$ and $\mathbf{F}_2$, respectively, which are parallel to the xy-plane. Let Σ be the area of the triangle constructed on the vectors $\mathbf{f}_1$ and $\mathbf{f}_2$. It is equal to the projection of the area Π on the xy-plane and, therefore,

$$\Sigma = \Pi\cos\theta,$$

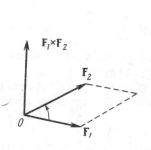

Figure 4

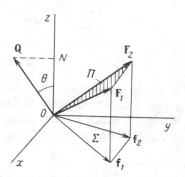

Figure 5

but

$$\Sigma = \frac{1}{2} \begin{vmatrix} X_1 & Y_1 \\ X_2 & Y_2 \end{vmatrix},$$

and, consequently,

$$N = \begin{vmatrix} X_1 & Y_1 \\ X_2 & Y_2 \end{vmatrix} = X_1 Y_2 - Y_1 X_2.$$

The other projections can be obtained from this one by means of a cyclic permutation:

$$L = Y_1 Z_2 - Z_1 Y_2, \quad M = Z_1 X_2 - X_1 Z_2.$$

If we designate the unit vectors of the coordinate axes as $\mathbf{i}$, $\mathbf{j}$, $\mathbf{k}$, then we can make a single formula out of these expressions:

$$\mathbf{Q} = L\mathbf{i} + M\mathbf{j} + N\mathbf{k} = \begin{vmatrix} \mathbf{i} & \mathbf{j} & \mathbf{k} \\ X_1 & Y_1 & Z_1 \\ X_2 & Y_2 & Z_2 \end{vmatrix}. \tag{1.2}$$

We can note that the projections of the product $\mathbf{F}_1 \times \mathbf{F}_2$ on the rectangular coordinates are equal, respectively, to the determinants derived from the matrix

$$\left\| \begin{matrix} X_1 & Y_1 & Z_1 \\ X_2 & Y_2 & Z_2 \end{matrix} \right\|.$$

A vector product changes sign when the factors are interchanged:

$$\mathbf{F}_1 \times \mathbf{F}_2 = -(\mathbf{F}_2 \times \mathbf{F}_1)$$

and is distributive:

$$\mathbf{F}_1 \times (\mathbf{F}_2 + \mathbf{F}_3) = \mathbf{F}_1 \times \mathbf{F}_2 + \mathbf{F}_1 \times \mathbf{F}_3.$$

1.2 Sliding Vectors[*]

7. By definition, we can freely displace a sliding vector along its line of action without changing its direction and magnitude.

When the vector $\mathbf{F}$ slides along its line of action, its orthogonal projections X, Y, Z on the rectangular coordinate axes remain invariant.

---

[*] The theory of sliding vectors was developed by Poinsot. Since we have to use the free vector algebra, we have to follow attentively the nature of different vectors.

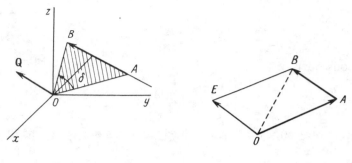

Figure 6   Figure 7

When the vector $\mathbf{F} = \overrightarrow{AB}$ (Fig. 6) slides along the line of action, not only the projections X, Y, Z, but also the following quantities are invariant: the plane passing through the fixed point O and the sliding vector $\mathbf{F}$; the area of the triangle OAB constructed on the point O and on the origin and terminus of the sliding vector $\overrightarrow{AB}$; the sense of rotation relative to the point O in the direction of the sliding vector.

All the invariants can be covered by one vector $\mathbf{Q}$ with origin at O, which lies on the straight line orthogonal to the plane of the triangle OAB and whose direction is such that the motion from A to B seems to be counterclockwise, and is equal in magnitude twice the area of the triangle OAB:

$$2\,\text{area}\triangle OAB = \delta \cdot |\overrightarrow{AB}|,$$

where δ is the distance from the line of action of the vector $\overrightarrow{AB}$ to the point O and is known as the *arm* of the sliding vector $\overrightarrow{AB}$ relative to the point O. The vector $\mathbf{Q}$, whose properties will be defined somewhat later is the *moment of the vector $\overrightarrow{AB}$ about the point O.*

We complete the triangle OAB to form a parallelogram $OABE$ (Fig. 7). It immediately follows from the figure that the moment of the vector $\overrightarrow{AB}$ about the point O is applied at O and is equal to

$$\mathbf{Q} = \overrightarrow{OA} \times \overrightarrow{OE}. \tag{1.3}$$

This formula only yields a vector $\mathbf{Q}$ without defining its nature. The vector $\overrightarrow{OE}$ is equal to the vector $\overrightarrow{AB}$ in magnitude and coincides with it in direction but is applied at the point O.

If we designate the projections of the vector $\overrightarrow{OA}$, or the coordinates of the point A, as x, y, z and the projections of the vector $\overrightarrow{AB}$ as X, Y, Z, then from (1.2) and (1.3) we find that the moment of the sliding vector

$\overrightarrow{AB}$ (X, Y, Z), applied at the point $A(x, y, z)$, about the origin O, is

$$Q = \begin{vmatrix} i & j & k \\ x & y & z \\ X & Y & Z \end{vmatrix}. \tag{1.4}$$

The projections of the moment $\mathbf{Q}$ on the coordinate axes will be designated as L, M, N. These letters also denote the moments of the sliding vector $\overrightarrow{AB}$ with respect to the x, y, z axes respectively,

$$L = yZ - zY, \quad M = zX - xZ, \quad N = xY - yX.$$

The moment $\mathbf{Q}$ is orthogonal to the direction of the vector $\overrightarrow{AB}$ and therefore

$$XL + YM + ZN = 0.$$

Consequently, only five of the six quantities X, Y, Z, L, M, N, which do not vary when the vector $\overrightarrow{AB}$ slides along the line of action, are independent, and it is sufficient to define these five independent quantities to construct the sliding vector $\overrightarrow{AB}$.

8. Addition of sliding vectors having a common origin. Two sliding vectors $\mathbf{F}_1$ and $\mathbf{F}_2$ whose lines of action meet at a point A can both be displaced, by sliding, to a position such that they would have a common origin A. Two sliding vectors $\mathbf{F}_1$ and $\mathbf{F}_2$ with a common origin A can be added by the parallelogram rule. In other words, the sum of the sliding vectors $\mathbf{F}_1 + \mathbf{F}_2$ is a sliding vector $\mathbf{F}$ applied at the point A and constructed as a diagonal of a parallelogram on the vectors $\mathbf{F}_1$ and $\mathbf{F}_2$:

$$\mathbf{F} = \mathbf{F}_1 + \mathbf{F}_2.$$

We must take the limiting cases of the parallelogram rule also by definition. If two vectors $\mathbf{F}_1$ and $\mathbf{F}_2$ lie on the same straight line and are of the same direction, then their sum is a sliding vector $\mathbf{F}$ of the same direction, lying on the same line of action and equal in magnitude to the sum of the magnitudes $|\mathbf{F}_1| + |\mathbf{F}_2|$. If two sliding vectors $\mathbf{F}_1$ and $\mathbf{F}_2$ lie on the same straight line, but are opposite in direction, and the magnitude $|\mathbf{F}_1|$ is greater than the magnitude $|\mathbf{F}_2|$, then the sum of the vectors $\mathbf{F}_1 + \mathbf{F}_2$ is a vector $\mathbf{F}$ which lies on the same line of action, has the same direction as the larger vector $\mathbf{F}_1$ and is equal in magnitude to the difference $|\mathbf{F}_1| - |\mathbf{F}_2|$.

If two sliding vectors $\mathbf{F}_1$ and $\mathbf{F}_2$ lie on the same straight line, are equal in magnitude and opposite in direction, then

$$\mathbf{F}_1 + \mathbf{F}_2 = 0.$$

The sum of two sliding vectors is zero if and only if they lie on the same straight line, are equal in magnitude, and opposite in direction.

Assume that two sliding vectors $\mathbf{F}_1$ and $\mathbf{F}_2$ are applied at the same point A. Their sum is a sliding vector

$$\mathbf{F} = \mathbf{F}_1 + \mathbf{F}_2$$

applied at the same point A.

The moment of the sum about the origin O is

$$\mathbf{Q} = \overrightarrow{OA} \times \mathbf{F} = \overrightarrow{OA} \times (\mathbf{F}_1 + \mathbf{F}_2) = \overrightarrow{OA} \times \mathbf{F}_1 + \overrightarrow{OA} \times \mathbf{F}_2 = \mathbf{Q}_1 + \mathbf{Q}_2.$$

This is the proof of *Varignon's theorem* which states that the moment of the sum of sliding and intersecting vectors is equal to the sum of the moments of the vectors.

9. Equivalence. Two systems of sliding vectors are said to be *equivalent* if we can pass from one system to the other by a number of elementary operations: (1) by sliding the vectors along their line of action, (2) by sliding intersecting sliding vectors, and (3) by adding (discarding) two sliding vectors which are equal in magnitude and lie on the same straight line but are opposite in direction.

10. Addition of parallel sliding vectors. Consider two parallel sliding vectors $\mathbf{F}_1$ and $\mathbf{F}_2$ which are of the same direction and are applied at the points O_1 and O_2 respectively (Fig. 8). We add two sliding vectors $\mathbf{a}$ and $-\mathbf{a}$ which lie on the staight line O_1O_2 and are applied at the points O_2 and O_1 respectively. We add these vectors pairwise:

$$-\mathbf{a} + \mathbf{F}_1 = \mathbf{u}_1, \quad \mathbf{a} + \mathbf{F}_2 = \mathbf{u}_2.$$

Let the sliding vectors $\mathbf{u}_1$ and $\mathbf{u}_2$ meet at a point A. Then we can add these vectors and obtain a resultant (principal) vector

$$\mathbf{F} = \mathbf{u}_1 + \mathbf{u}_2.$$

The sliding vector $\mathbf{F}$ is parallel to the vectors $\mathbf{F}_1$ and $\mathbf{F}_2$, is applied at the point A, has the same direction as the vectors $\mathbf{F}_1$ and $\mathbf{F}_2$, and the magnitude of the vector $\mathbf{F}$ is equal to the sum of the magnitudes of $\mathbf{F}_1$ and $\mathbf{F}_2$, i.e. $\mathbf{F} = |\mathbf{F}_1| + |\mathbf{F}_2|$.

The resultant (principal) vector $\mathbf{F}$ cuts the segment O_1O_2 at the point O which divides O_1O_2 into parts which are in inverse proportion to the magnitudes of the applied vectors. Since the triangles are similar, we have

$$\frac{AO}{OO_1} = \frac{|\mathbf{F}_1|}{|-\mathbf{a}|}, \quad \frac{AO}{O_2O} = \frac{|\mathbf{F}_2|}{|\mathbf{a}|},$$

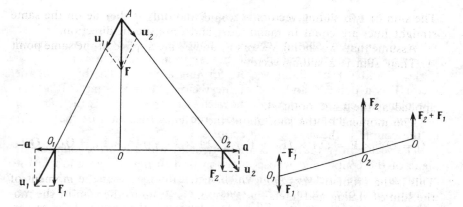

Figure 8   Figure 9

whence

$$\frac{O_1 O}{O O_2} = \frac{|\mathbf{F}_2|}{|\mathbf{F}_1|} .$$

Let us now consider the addition of parallel vectors which are opposite in direction and different in magnitude. Assume that we have two parallel opposite sliding vectors $\mathbf{F}_1$ and $\mathbf{F}_2$ applied at the points O_1 and O_2 respectively (Fig. 9). We resolve the larger vector $\mathbf{F}_2$ into two sliding vectors which have the same direction as $\mathbf{F}_2$. We make one of these vectors pass through O_1 and equal to $-\mathbf{F}_1$. Then the other vector $\mathbf{F}_2 + \mathbf{F}_1$, which is equal in magnitude to $|\mathbf{F}_2| - |\mathbf{F}_1|$ and has the same direction as the vector $\mathbf{F}_2$, larger in magnitude, will pass through the point O which, according to the aforesaid, satisfies the relation

$$\frac{O_1 O_2}{O_2 O} = \frac{|\mathbf{F}_2 + \mathbf{F}_1|}{|\mathbf{F}_1|} ,$$

or

$$\frac{O_1 O}{O_2 O} = \frac{|\mathbf{F}_2|}{|\mathbf{F}_1|} .$$

The vectors $-\mathbf{F}_1$ and $\mathbf{F}_2$, applied at O_1, cancel out. As a result we obtain only one sliding vector $\mathbf{F}_2 + \mathbf{F}_1$ applied at O which externally divides the segment $O_1 O_2$ into parts which are in inverse proportion to the given vectors.

2*

1.3 The Theory of Couples of Sliding Vectors

11. A couple of sliding vectors. A couple is a system of two parallel sliding vectors equal in magnitude but opposite in direction.

Let $\mathbf{F}$ and $-\mathbf{F}$ be sliding vectors forming a couple. The origin of the vector $\mathbf{F}$ is a point A and that of the vector $-\mathbf{F}$ is a point B (Fig. 10). The plane which passes through the vectors of the couple is the *plane of the couple* and the distance h between the lines of action of the vectors of the couple is the *arm of the couple*.

The vector $\mathbf{m}$ (we take it to originate at B) which is orthogonal to the plane of the couple and whose direction is such that the rotation of the couple is counterclockwise, and which is equal in magnitude to $m = F \cdot h$, i.e. to the area of a parallelogram constructed on the vectors of the couple, is the *moment of the couple*. We have to determine the nature of the vector $\mathbf{m}$. By construction,

$$\mathbf{m} = \overrightarrow{BA} \times \mathbf{F}. \tag{1.5}$$

12. *We can rotate a couple in its plane.* Consider a couple of sliding vectors $\mathbf{F}$ and $-\mathbf{F}$ with arm h (Fig. 11). We take a segment $CD = AB = h$ which has an arbitrary position in the plane of the couple. We shall show that this couple can be transformed into an equivalent couple with arm CD.

We draw the line of action of the vector $\mathbf{F}$ and a straight line orthogonal to the segment CD and passing through the point C. We designate the point where these straight lines intersect as P. Then we draw the line of action of the vector $-\mathbf{F}$ and a straight line through the point D at right angles to the segment CD. We designate the point where these lines intersect as S. The points where the four straight lines intersect define a rhombus for which the points P and S are opposite vertices.

At the point P we apply two sliding vectors $\mathbf{u}$ and $-\mathbf{u}$ which are equal to $|\mathbf{F}|$ in magnitude and lie on the line PC. At the point S we also apply two equal opposite sliding vectors $\mathbf{u}$ and $-\mathbf{u}$ which lie on the line SD. Then

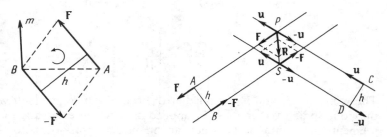

Figure 10   Figure 11

we add the sliding vectors $\mathbf{F}$ and $-\mathbf{u}$ applied at the point P together:

$$\mathbf{R} = -\mathbf{u} + \mathbf{F}.$$

In the same way we add the sliding vectors $-\mathbf{F}$ and $\mathbf{u}$ applied at the point S. We obtain

$$\mathbf{R'} = -\mathbf{F} + \mathbf{u}.$$

Since the points P and S are vertices of the rhombus and the vectors being added together are equal in magnitude and lie along the sides of this rhombus, their sums $\mathbf{R}$ and $\mathbf{R'}$ lie on the diagonal PS of the rhombus, are opposite in direction, and equal in magnitude; their sum is zero:

$$\mathbf{R} + \mathbf{R'} = 0.$$

We displace the remaining sliding vectors $\mathbf{u}(P)$ and $-\mathbf{u}(S)$ along their lines of action to the points C and D respectively. This completes the solution of the problem. The moment of the resultant couple is evidently equal to and parallel to the moment of the initial couple and has the same direction.

   13. *We can change the arm of a couple* by varying the corresponding vectors of the couple. Consider a couple of sliding vectors $\mathbf{F}$ and $-\mathbf{F}$ with arm AB (Fig. 12). We designate the midpoint of the segment AB as O. We take an arbitrary segment CD which we wish to have as the arm of the transformed couple, place it along the line AB so that its midpoint coincides with the point O. At the points C and D we apply the sliding vectors $\mathbf{u}$ and $-\mathbf{u}$ parallel to the vectors of the given couple. We find the magnitude of the vector $\mathbf{u}$ from the formula

$$\mathbf{F} \cdot AB = \mathbf{u} \cdot CD. \tag{1.6}$$

We add the vector $\mathbf{F}$ applied at A to the vector $\mathbf{u}$ applied at D. Their sum $\mathbf{R} = \mathbf{F} + \mathbf{u}$ is a sliding vector which passes through O since, by virtue of (1.6), the point O divides the segment AD in the inverse proportion to the

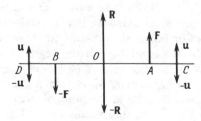

Figure 12

vectors being added:

$$\frac{OD}{AO} = \frac{CD}{AB} = \frac{F}{u}.$$

The sum of the sliding vector $-\mathbf{u}$ applied at C and the vector $-\mathbf{F}$ applied at B is a sliding vector $-\mathbf{R}$ applied at the point O. The vectors $\mathbf{R}$ and $-\mathbf{R}$ applied at O cancel out and as a result we have only a couple of sliding vectors $\mathbf{u}$ and $-\mathbf{u}$, applied at the points C and D respectively, which has the arm CD and is equivalent to the given couple. By virtue of (1.6), the moments of the initial couple and the resultant couple are parallel, equal in magnitude, and have the same direction.

14. *A couple can be displaced parallel to itself.* Consider a couple of sliding vectors $\mathbf{F}$ and $-\mathbf{F}$ with arm AB (Fig. 13) which lies in the plane π. We wish to transfer this couple to the plane π' which is parallel to π. We transfer the segment AB to the plane π' parallel to itself into a position $A'B'$. At the points A' and B' we apply the sliding vectors $\mathbf{F}$ and $-\mathbf{F}$. The sum of the vector $\mathbf{F}$ applied at A and the vector $\mathbf{F}$ applied at B' is a sliding vector $\mathbf{R} = 2\mathbf{F}$ applied at O which is the midpoint of the segment AB'. The sum of the vector $-\mathbf{F}$ applied at B and the vector $-\mathbf{F}$ applied at A' is a sliding vector $-\mathbf{R} = -2\mathbf{F}$ applied at O which is the midpoint of the segment $A'B$. The sliding vectors $\mathbf{R}$ and $-\mathbf{R}$, which are equal, opposite, and lie on the same straight line, cancel out. As a result we have two sliding vectors $\mathbf{F}$ and $-\mathbf{F}$ applied at A' and B' respectively. They form a couple whose moment is equal and parallel to the moment of the initial couple.

From the properties we have proved it follows immediately that *two couples whose moments are equal and of the same direction are equivalent.* This statement makes it possible to represent a couple of sliding vectors by the moment of the couple, to consider the moment of the couple rather than the couple itself and, as we have proved, to treat the moment of a couple as a free vector.

15. Addition of couples. Consider two couples defined by moments $\mathbf{m}_1$ and $\mathbf{m}_2$ respectively (Fig. 14). We take a point O and construct free vectors $\mathbf{m}_1$ and $\mathbf{m}_2$ with origin at O. Assume that a straight line OA lies at the intersection of planes orthogonal to the moments $\mathbf{m}_1$ and $\mathbf{m}_2$ and passing through the point O. We construct on the segment OA (taking it to be an arm) two couples of sliding vectors $(\mathbf{F}_1, -\mathbf{F}_1)$ and $(\mathbf{F}_2, -\mathbf{F}_2)$, which are equivalent to the given couples, with moments $\mathbf{m}_1$ and $\mathbf{m}_2$ respectively. Since the vectors $\mathbf{F}_1$ and $\mathbf{F}_2$ meet at the point A, we can add them together to obtain $\mathbf{F} = \mathbf{F}_1 + \mathbf{F}_2$ as a sliding vector applied at A. The sum of the sliding vectors $-\mathbf{F}_1$ and $-\mathbf{F}_2$, applied at O, is a sliding vector

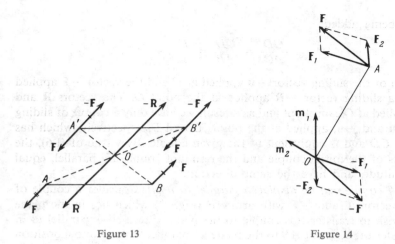

Figure 13   Figure 14

$-\mathbf{F} = -\mathbf{F}_1 - \mathbf{F}_2$ applied at O. As a result we get a couple of sliding vectors $\mathbf{F}$ and $-\mathbf{F}$ applied at A and O respectively. By Varignon's theorem, the moment of this couple

$$\mathbf{m} = \overrightarrow{OA} \times \mathbf{F} = \overrightarrow{OA} \times \mathbf{F}_1 + \overrightarrow{OA} \times \mathbf{F}_2 = \mathbf{m}_1 + \mathbf{m}_2.$$

The moment of a sum is equal to the sum of the moments of the couples being added.

1.4 Systems of Sliding Vectors

16. A resultant vector and a resultant couple. Assume that $\mathbf{F}_1$, $\mathbf{F}_2$, ..., $\mathbf{F}_n$ are sliding vectors applied at points A_1, A_2, ..., A_n respectively, and let O be the origin (Fig. 15). We add to this system a sliding vector $\mathbf{F}_i$ with origin at O, equal in magnitude to the vector $\mathbf{F}_i$ and of the same direction, and also a vector $-\mathbf{F}_i$ which is equal to it in magnitude and opposite in direction.* The vector $\mathbf{F}_i$ applied at A_i and the vector $-\mathbf{F}_i$ applied at O form a couple with moment

$$\mathbf{Q}_i = \overrightarrow{OA_i} \times \mathbf{F}_i.$$

We assume the origin of the free vector of moment $\mathbf{Q}_i$ to be at O, and then the moment of the couple $\mathbf{Q}_i$ is equal to the moment of the sliding vector $\mathbf{F}_i$, applied at A_i, about the origin O.

---

* The same notation for the vectors will not lead to ambiguity.—*Ed.*

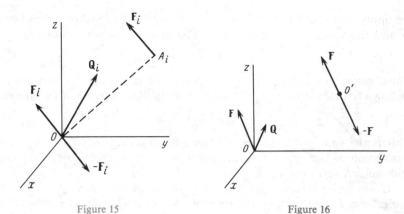

Figure 15   Figure 16

If we make similar constructions for every given vector ($i = 1, 2, ...,$ n), then, as a result, we obtain a bundle of sliding vectors $\mathbf{F}_1, \mathbf{F}_2, ..., \mathbf{F}_n$ at the point O and a bundle of moments of couples $\mathbf{Q}_1, \mathbf{Q}_2, ..., \mathbf{Q}_n$, which can be added since they originate at the same point O. We obtain a resultant sliding vector

$$\mathbf{F} = \mathbf{F}_1 + \mathbf{F}_2 + ... + \mathbf{F}_n$$

and the moment of the resultant couple

$$\mathbf{Q} = \mathbf{Q}_1 + \mathbf{Q}_2 + ... + \mathbf{Q}_n.$$

Consequently, *an arbitrary system of sliding vectors can be reduced to an equivalent system consisting of a resultant sliding vector applied at O and a resultant couple with moment equal to the sum of the moments of all the given sliding vectors about the origin.*

17. The condition of equivalence. Since elementary operations are invertible, two systems S and S' of sliding vectors are equivalent if, upon a reduction to the origin O, their resultant vectors $\mathbf{F}$ and $\mathbf{F}'$ and the moments of the resultant couples $\mathbf{Q}$ and $\mathbf{Q}'$ are equal:

$$\mathbf{F} = \mathbf{F}', \quad \mathbf{Q} = \mathbf{Q}'.$$

If the resultant vector and the moment of the resultant couple are zero, i.e.

$$\mathbf{F} = 0, \quad \mathbf{Q} = 0,$$

then the system is equivalent to zero.

18. Variation of the reduction point. Assume that when reducing a system of sliding vectors to the origin O, we obtain a resultant sliding vector $\mathbf{F}$ (with projections X, Y, Z on the coordinate axes) and a moment of the resultant couple $\mathbf{Q}$ (L, M, N) (Fig. 16). To give the system a new origin O', we apply two sliding vectors $\mathbf{F}$ and $-\mathbf{F}$ at O'. The vector $\mathbf{F}$ applied at O and the vector $-\mathbf{F}$ applied at O' form a couple with moment

$$-(\overrightarrow{OO'} \times \mathbf{F}).$$

Thus, when it is reduced to a new point O', the system is reduced to a sliding vector $\mathbf{F}$ applied at O' and to a resultant couple with moment

$$\mathbf{Q'} = \mathbf{Q} - \overrightarrow{OO'} \times \mathbf{F}. \tag{1.7}$$

This is the sum of the moment $\mathbf{Q}$ and the moment $-(OO' \times \mathbf{F})$ of the translation of the vector $\mathbf{F}$ to the point O'. The moments are free vectors and we can translate them to O'.

If we designate the coordinates of the point O' as x, y, z, then the projections L', M', N' of the resultant moment $\mathbf{Q'}$ can be expressed, according to (1.7), by the formulas

$$L' = L - (yZ - zY), \quad M' = M - (zX - xZ), \quad N' = N - (xY - yX). \tag{1.8}$$

19. The wrench. When the point of reduction of the system is changed, the magnitude and the direction are invariant and thus the projections of the resultant vector $\mathbf{F}$ are invariant as well as the scalar product

$$\mathbf{F} \cdot \mathbf{Q} = \mathbf{F'} \cdot \mathbf{Q'}.$$

In other words, when the point of reduction is changed, the projection of the moment of the resultant couple on the direction of the resultant vector does not change. The moment of the resultant couple varies when the reduction point is changed if $\mathbf{F}$ is nonzero. If the resultant vector $\mathbf{F}$ is zero, then, when the system is reduced to different points of space, the moment of the resultant couple is the same $\mathbf{Q}$; the system is reduced to one couple with moment $\mathbf{Q}$.

If $\mathbf{F}$ is nonzero, the locus of points O' at which the resultant vector $\mathbf{F}$ and the moment of the resultant couple $\mathbf{Q'}$ are collinear can be determined from the condition of parallelism of the vectors $\mathbf{Q'}$ and $\mathbf{F}$, by virtue of (1.8), from the equations

$$\frac{L - (yZ - zY)}{X} = \frac{M - (zX - xZ)}{Y} = \frac{N - (xY - yX)}{Z}. \tag{1.9}$$

The combination of the resultant vector $\mathbf{F}$ and the moment of the resultant couple $\mathbf{Q}$, directed along the line of action of $\mathbf{F}$, for the points O' of this line, is a *wrench* and the line itself is the *axis of the wrench*, or the *central axis*.

20. Geometric construction. Let $\mathbf{F}$ and $\mathbf{Q}$ be a resultant vector and the moment of a resultant couple when the system is reduced to the origin O (Fig. 17). To construct a wrench geometrically, we resolve the moment

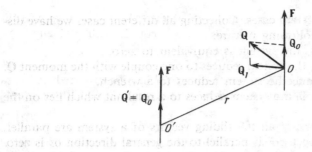

Figure 17

Q into Q_0 along the line of action of the vector F and Q_1 in the direction orthogonal to F, i.e. $Q = Q_0 + Q_1$. On the perpendicular to the plane of the vectors F and Q_1 we find a point O' of the axis of the wrench, i.e. a point such that

$$Q' = Q - \overrightarrow{OO'} \times F = Q_0.$$

From this, after the vector multiplication of this relation by F and using the condition $F \cdot \overrightarrow{OO'} = 0$ (i.e. the presupposed orthogonality of the directions of F and $\overrightarrow{OO'}$) and the condition $F \times Q_0 = 0$ (i.e. the parallelism of the directions of F and Q_0), we obtain

$$\overrightarrow{OO'} = \frac{F \times Q}{F^2}.$$

This formula completely defines the position of the point O' on the axis of the wrench. The line which is parallel to F and passes through O' is the axis of the wrench. The wrench consists of a sliding vector F and a free moment of the resultant couple Q' which are applied at the point O'.

For the points of the axis of the wrench the moment of the resultant couple is the least in magnitude:

$$|Q_0| < |Q|.$$

The wrench does not possess a moment of the resultant couple $Q_0 = 0$ if for nonzero F the invariant is eliminated, i.e. $F \cdot Q = 0$. In this case the system reduces to one resultant vector F which lies on the axis of the wrench. Then we say that the system admits of a resultant (a principal vector). The equation of the axis of the wrench (1.9), or of the line of action of the resultant, degenerates into a system of simultaneous equations $L' = 0, M' = 0, N' = 0$, or

$$L - (yZ - zY) = 0, \quad M - (zX - xZ) = 0, \quad N - (xY - yX) = 0. \quad (1.10)$$

21. Review of different cases. Collecting all different cases we have discussed, we get the following picture:

1°. $\mathbf{F} = 0$, $\mathbf{Q} = 0$, the system is equivalent to zero.

2°. $\mathbf{F} = 0$, $\mathbf{Q} \neq 0$, the system reduces to one couple with the moment $\mathbf{Q}$.

3°. $\mathbf{F} \neq 0$, $\mathbf{F} \cdot \mathbf{Q} \neq 0$, the system reduces to a wrench.

4°. $\mathbf{F} \neq 0$, $\mathbf{F} \cdot \mathbf{Q} = 0$, the system reduces to a resultant which lies on the axis of a wrench.

22. Parallel vectors. If all the sliding vectors of a system are parallel, then their resultant vector $\mathbf{F}$ is parallel to the general direction or is zero. The moments of different vectors about the origin O are perpendicular to the general direction of the parallel vectors and, therefore, the resultant moment $\mathbf{Q}$ of the system is also perpendicular to this direction. Consequently, if $\mathbf{F}$ is nonzero, then the vectors $\mathbf{F}$ and $\mathbf{Q}$ are perpendicular: $\mathbf{F} \cdot \mathbf{Q} = 0$; the system admits of a resultant (principal vector) applied at some point of the axis of the wrench. If the resultant vector $\mathbf{F}$ were zero, the system would reduce to one couple or would be equivalent to zero.

Assume that α, β, γ are the direction cosines of the direction of the given system of parallel sliding vectors $\mathbf{F}_\nu$ whose projections on the coordinate axes are X_ν, Y_ν, Z_ν (Fig. 18). We designate the algebraic magnitude of the vector $\mathbf{F}_\nu$ as P_ν, assuming it to be positive if the vector $\mathbf{F}_\nu$ is in the direction of α, β, γ and negative otherwise. Then

$$X_\nu = P_\nu\alpha, \quad Y_\nu = P_\nu\beta, \quad Z_\nu = P_\nu\gamma.$$

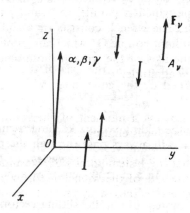

Figure 18

If the coordinates of the point of action of the sliding vector $\mathbf{F}_\nu$ are designated as x_ν, y_ν, z_ν, then the moment of the vector $\mathbf{F}_\nu$ about the origin is

$$\mathbf{Q}_\nu = \overrightarrow{OA}_\nu \times \mathbf{F}_\nu = \begin{vmatrix} \mathbf{i} & \mathbf{j} & \mathbf{k} \\ x_\nu & y_\nu & z_\nu \\ P_\nu\alpha & P_\nu\beta & P_\nu\gamma \end{vmatrix}.$$

Hence the projections L_ν, M_ν, N_ν of the moment $\mathbf{Q}_\nu$ on the coordinate axes are

$$L_\nu = P_\nu(y_\nu\gamma - z_\nu\beta), \; M_\nu = P_\nu(z_\nu\alpha - x_\nu\gamma), \; N_\nu = P_\nu(x_\nu\beta - y_\nu\alpha).$$

The system of parallel sliding vectors at the origin O reduces to the resultant vector $\mathbf{F} = \sum \mathbf{F}_\nu$ with projections

$$X = \sum P_\nu\alpha, \; Y = \sum P_\nu\beta, \; Z = \sum P_\nu\gamma,$$

and to the moment of the resultant couple $\mathbf{Q} = \sum \mathbf{Q}_\nu$ with projections

$$L = \sum L_\nu, \; M = \sum M_\nu, \; N = \sum N_\nu.$$

If the resultant vector is nonzero, i.e. $R = \sum P_\nu \neq 0$, then the obvious relation $XL + YM + ZN = 0$ means that the system reduces to one resultant (principal) sliding vector $\mathbf{F}$ which lies on the axis of the wrench. In this case, the equations of the axis of the wrench are defined by formulas (1.10):

$$\sum P_\nu(y_\nu\gamma - z_\nu\beta) - (yZ - zY) = 0.$$

We can obtain the other two similar equations by means of the cyclic permutation of the letters. Thus we arrive at the equation of the central axis

$$\frac{\sum P_\nu x_\nu - Rx}{\alpha} = \frac{\sum P_\nu y_\nu - Ry}{\beta} = \frac{\sum P_\nu z_\nu - Rz}{\gamma}. \tag{1.11}$$

Whatever the direction of α, β, γ, the axis of the wrench passes through the point x, y, z defined by the equations

$$Rx = \sum P_\nu x_\nu, \; Ry = \sum P_\nu y_\nu, \; Rz = \sum P_\nu z_\nu. \tag{1.12}$$

Thus, if we rotate the sliding vectors keeping them parallel, the line of action of the resultant will always pass through the point x, y, z. That is why this point is called the *centre of the system of parallel vectors*. It is customary to consider it in a narrow sense, as a point of action of the resultant.

If $R = 0$, then the system of parallel sliding vectors is equivalent to one couple with moment $\mathbf{Q}$.

1.5 Bound Vectors

23. The coordinates of a bound vector. A vector $\mathbf{F}$ bound at a point of action A is defined by the projections X, Y, Z on the coordinate axes and the coordinates x, y, z of its origin. These six quantities are independent and sufficient for a complete definition of the bound vector.

We add vectors bound to the same point by the polygon rule. In mechanics problems, free, sliding, and bound vectors have definite dimensions which must be attentively followed.

24. The derivative of a vector. Assume that $\mathbf{F}$ is a vector which emanates from the origin of a rectangular coordinate system xyz and whose projections $X(t)$, $Y(t)$, $Z(t)$ on these axes are continuous and differentiable functions of t. If we assign an increment Δt to t, then the vector $\mathbf{F}$ will get, in the xyz coordinate system, an increment

$$\Delta\mathbf{F} = \mathbf{F}(t + \Delta t) - \mathbf{F}(t).$$

We designate the projections of this increment on the coordinate axes as ΔX, ΔY, ΔZ.

If Δt tends to zero, then the ratio $\Delta\mathbf{F}/\Delta t$ will tend, in the coordinate system, to a vector with projections dX/dt, dY/dt, dZ/dt.

This limiting vector is known as the *derivative of the vector* with respect to t relative to the coordinate system and is written as $d\mathbf{F}/dt$.

The ordinary rules for *differentiating a sum* and a *product* can be applied to the differentiation of vector sums and vector products, viz.

$$\frac{d(\mathbf{F}_1 + \mathbf{F}_2 + \ldots)}{dt} = \frac{d\mathbf{F}_1}{dt} + \frac{d\mathbf{F}_2}{dt} + \ldots,$$

$$\frac{d(\mathbf{F}_1 \times \mathbf{F}_2)}{dt} = \frac{d\mathbf{F}_1}{dt} \times \mathbf{F}_2 + \mathbf{F}_1 \times \frac{d\mathbf{F}_2}{dt},$$

$$\frac{d(\mathbf{F}_1 \cdot \mathbf{F}_2)}{dt} = \frac{d\mathbf{F}_1}{dt} \cdot \mathbf{F}_2 + \mathbf{F}_1 \cdot \frac{d\mathbf{F}_2}{dt}.$$

When defining the derivative of a vector, we only mentioned the projections, and, therefore, the derivative of a vector is constructed as a free vector whose origin is a fixed point in the coordinate system being considered. According to this definition, the derivative of a vector is bound up with the coordinate system being considered and, consequently, the reference system for the derivative must always be specified (especially when several reference systems are used), e.g. $(d\mathbf{F}/dt)_{xyz}$.

CHAPTER 2

KINEMATICS

25. Space and time. We do not have in nature either a fixed space or its metric, neither do we have a uniform motion with which we might calculate equal time intervals. These important circumstances give rise to problems when determining space and time which we shall not touch upon until we accumulate sufficient knowledge in mechanics.

Using reasonable approximations, classical mechanics considers the space relative to which the motion of mechanical systems is reckoned to be fixed and Euclidean.

A mean solar day is accepted as a unit of absolute time.

2.1 Kinematics of a Particle

26. The trajectory of a particle. We shall consider the motion of a particle in some definite system of rectangular and rectilinear coordinates xyz, which we assume to be fixed.

The locus of a moving particle is its *trajectory*. The analytic definition of the motion of a particle is complete if its coordinates x, y, z are defined as continuous functions of time t:

$$x = \varphi_1(t), \quad y = \varphi_2(t), \quad z = \varphi_3(t).$$

These equations define the position of a moving particle at every moment t and are equations of the trajectory in parametric form. If we choose a point M_0 on the trajectory and measure the length of the arc s along the trajectory from it to the moving particle M, then the motion of the particle M can be defined as the way s varies with time t, i.e., $s = s(t)$.

In what follows, we shall assume the functions φ_α and s to be continuous and to have continuous derivatives of the first two orders.

27. Displacement. Velocity. Let M and M' be the positions of a moving particle at the times t and $t + \Delta t$ respectively (Fig. 19). The vector $\overrightarrow{MM'}$ is the *displacement* of the particle in the time interval Δt. This vector originating at the point M is a chord subtending the positions of the moving particle at the moments t and $t + \Delta t$. We divide the displacement $\overrightarrow{MM'}$ by Δt. The

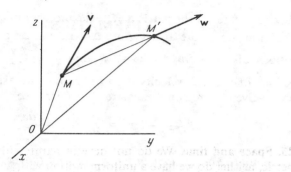

Figure 19

vector $\mathbf{w} = \dfrac{\overrightarrow{MM'}}{\Delta t}$ is the *average velocity* of the particle M in the time interval Δt. The average velocity is a vector which is applied at the point M and has the same direction as the displacement $\overrightarrow{MM'}$.

When Δt tends to zero, the limit of the average velocity is the *instantaneous velocity* (or simply velocity) of the particle M at the moment t, which is written as

$$\mathbf{v} = \lim_{\Delta t \to 0} \frac{\overrightarrow{MM'}}{\Delta t}.$$

In the limit, the direction of the chord MM' is the same as the direction of the tangent to the trajectory, and, therefore, the velocity $\mathbf{v}$ of the particle M is the vector applied at the point M and passing along a tangent to the trajectory in the direction of the motion.

We can define the position of the particle M by the vector $\overrightarrow{OM}$ which originates at O (Fig. 19). The displacement $\overrightarrow{MM'}$ in the time interval Δt is equal to the increment $\Delta \overrightarrow{OM}$ of the vector $\overrightarrow{OM}$

$$\overrightarrow{MM'} = \overrightarrow{OM'} - \overrightarrow{OM} = \Delta\overrightarrow{OM},$$

whence we have

$$\mathbf{v} = \lim_{\Delta t \to 0} \frac{\Delta\overrightarrow{OM}}{\Delta t} = \frac{d\overrightarrow{OM}}{dt}.$$

Consequently, *the velocity of a moving particle is equal to the derivative of the radius vector of the moving particle with respect to time and is a vector applied at the moving particle.*

28. The projection of velocity on the coordinate axes. Assume that x, y, z are the coordinates of the particle M, and $x + \Delta x$, $y + \Delta y$ and $z + \Delta z$

are the coordinates of the particle M'. The projections of the displacement $\overrightarrow{MM'}$ on the coordinate axes xyz are Δx, Δy, and Δz respectively, and the projections of the average velocity $\mathbf{w}$ are $\dfrac{\Delta x}{\Delta t}$, $\dfrac{\Delta y}{\Delta t}$, $\dfrac{\Delta z}{\Delta t}$. Hence the projections of the true velocity $\mathbf{v}$ on the coordinate axes are the limits of the above-given expressions as Δt tends to zero, i.e.

$$v_x = \frac{dx}{dt}, \quad v_y = \frac{dy}{dt}, \quad v_z = \frac{dz}{dt}.$$

Theorem. *The projections of the velocity on the rectangular coordinate axes are equal to the first derivatives with respect to time of the corresponding coordinates of the moving particle.*

Since the coordinate axes are assumed to be orthogonal, the magnitude of the velocity v can be found in terms of the projections from the formula

$$v = \sqrt{\left(\frac{dx}{dt}\right)^2 + \left(\frac{dy}{dt}\right)^2 + \left(\frac{dz}{dt}\right)^2}.$$

We use s to denote the length of the arc of the trajectory (with the appropriate sign) from a fixed point M_0 on the trajectory. Then

$$ds = \pm \sqrt{(dx)^2 + (dy)^2 + (dz)^2}.$$

Consequently, the magnitude of the velocity can be found with the use of the formula

$$v = \frac{ds}{dt}.$$

If v is positive, then the velocity is in the direction of increasing s.

29. Uniform curvilinear motion. The motion is *uniform* if the velocity is constant. In this case

$$\frac{ds}{dt} = \text{const} = a.$$

Let s_0 be the value of s at the moment $t = 0$. Integrating the equation given above, we obtain

$$s = s_0 + at.$$

Consequently, when the motion is uniform the length of the path traversed is proportional to time and the velocity a is equal to the length of the path traversed in unit time.

Motion which is not uniform is called *variable* or *nonuniform*.

30. Theorem on the projection of velocity. If the motion is in a straight line, we can take the trajectory to be the x-axis. Then $s = x$ and the equation of motion has the form

$$x = f(t).$$

Hence the velocity of a particle moving along the x-axis can be represented by the formula

$$v = \frac{dx}{dt} = f'(t).$$

When the particle M moves in space, dx/dt is the projection of its velocity on the x-axis and at the same time it is equal to the velocity of the orthogonal projection M_1 of the particle M on the x-axis since x is the abscissa of the particle M_1. Consequently, *if we project a moving particle and its velocity on a fixed axis, then the projection of the velocity will be equal to the velocity of the projection.*

31. Acceleration. A vector $\overrightarrow{OP}$ equal in magnitude and direction to the velocity **v** of a moving particle M is applied at the origin O of a fixed system of coordinates (Fig. 20). When the particle M moves along its trajectory, a point P describes a curve known as the *hodograph* of the moving particle M. By construction the coordinates of the point P are equal to the derivatives of the coordinates of the particle M: $\dfrac{dx}{dt}$, $\dfrac{dy}{dt}$, $\dfrac{dz}{dt}$.

The velocity of the point P of the hodograph is equal to the derivative of the radius vector $\overrightarrow{OP}$ with respect to time:

$$\frac{d\overrightarrow{OP}}{dt} = \frac{d\mathbf{v}}{dt}$$

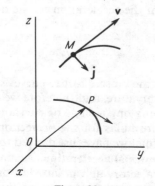

Figure 20

and, consequently, has $\dfrac{d^2x}{dt^2}$, $\dfrac{d^2y}{dt^2}$, $\dfrac{d^2z}{dt^2}$, respectively, as the projections on the coordinate axes.

The vector $\mathbf{j}$, which is equal in magnitude and direction to the velocity of the point of the hodograph but is applied at M, is the *acceleration* of the particle M:

$$\mathbf{j} = \frac{d\mathbf{v}}{dt} .$$

According to this definition, *the projections of the acceleration* $\mathbf{j}$ *of a moving particle on the coordinate axes at a time* t *are equal to the second derivatives of the coordinates of the particle with respect to time*:

$$j_x = \frac{d^2x}{dt^2} , \; j_y = \frac{d^2y}{dt^2} , \; j_z = \frac{d^2z}{dt^2} .$$

32. Motion in a circle. Let us consider an instructive kind of motion, namely that of a particle along a circle of radius ϱ with a fixed centre O, i.e.

$$x = \varrho \cos \theta(t), \; y = \varrho \sin \theta(t).$$

Then the projections of the velocity of the particle are

$$v_x = x' = - \varrho \theta' \sin\theta, \; v_y = y' = \varrho \theta' \cos\theta,$$

where the primes designate the derivatives with respect to time and $v = \dfrac{ds}{dt} = \dfrac{d}{dt} \varrho\theta = \varrho\theta'$. The projections of the acceleration on the x, y axes are

$$j_x = x'' = - \varrho\theta'' \sin\theta - \varrho\theta'^2 \cos\theta,$$
$$j_y = y'' = \varrho\theta'' \cos\theta - \varrho\theta'^2 \sin\theta.$$

Hence the projections of the acceleration on the tangent j_τ and on the normal j_n to the circle are

$$j_\tau = \varrho\theta'' = \frac{dv}{dt} , \; j_n = \varrho\theta'^2 = \frac{v^2}{\varrho} .$$

The circle which is the closest to the curve is the osculating circle of the curve, or circle of curvature, and therefore we can approximate motion along a curve by motion along a circle of curvature and, consequently, we would expect that the formulas for the projections of the acceleration on the tangent and the normal we have just derived to have a general meaning.

33. Tangential and normal acceleration. Assume that τ_0 is a unit vector of the tangent to the trajectory at a point M in the direction of increasing arcs s, $\mathbf{n}_0$ is a unit vector of the principal normal in the direction of the concavity of the trajectory and, consequently, towards its centre of curvature, $\mathbf{b}_0$ is a unit vector of the binormal, and ϱ is the radius of curvature.

We know Frenet's formula

$$\frac{d\tau_0}{ds} = \frac{\mathbf{n}_0}{\varrho}$$

and have

$$\mathbf{v} = v\tau_0, \quad \mathbf{j} = \frac{dv}{dt} = \frac{dv}{dt}\,\tau_0 + \frac{v^2}{\varrho}\,\mathbf{n}_0.$$

Hence the projections of the acceleration on the axes of a moving trihedral composed of a tangent, a normal, and a binormal to the trajectory are evident:

$$j_\tau = \frac{dv}{dt}, \quad j_n = \frac{v^2}{\varrho}, \quad j_b = 0.$$

2.2 Relative Motion of a Particle

34. Absolute, relative, and transportation motions. In addition to fixed axes, we may consider a moving rigid body and a system of rectangular coordinates associated with it.

The motion of a particle relative to a moving system of axes is known as *relative motion*.

The motion of a particle relative to fixed axes is *absolute motion*.

The *transportation motion* of a particle in time $(t, t + \Delta t)$ is its motion relative to fixed axes which the particle would have if at time t and in an interval $(t, t + \Delta t)$ the particle were rigidly connected with a moving system of axes and, consequently, it would move with the system.

A trajectory, velocity, and acceleration are said to be absolute, relative or transportation depending on whether they refer to an absolute, relative or transportation motion.

35. Theorem. *The absolute velocity of a moving particle is equal to the geometric sum of the relative and the transportation velocity.*

Consider a rigid body S and a particle M in it[*]. Let the body and the particle be moving. Assume that M_1 is the position of the particle M at the moment $t + \Delta t$ and S_1 is the position of the rigid body at the same moment, and let M_0 be the position the particle M would have by the mo-

---

[*] The particle M is not assumed to be fixed in the body.—*Ed.*

ment $t + \Delta t$ if it did not move relative to the rigid body. The vectors $\overrightarrow{MM_1}$, $\overrightarrow{MM_0}$, $\overrightarrow{M_0M_1}$ are, respectively, the displacements of the particle M in time Δt in its absolute, transportation, and relative motions (Fig. 21). We must assume that the body S is rigid in order to have an undeformed trajectory M_0M_1 of the relative motion in the position S_1. We have

$$\overrightarrow{MM_1} = \overrightarrow{MM_0} + \overrightarrow{M_0M_1}.$$

Dividing the left-hand and right-hand sides of this relation by Δt, we obtain

$$\mathbf{w}_a = \mathbf{w}_t + \mathbf{w}_r,$$

where $\mathbf{w}_a$, $\mathbf{w}_t$, $\mathbf{w}_r$ are vectors equal, respectively, in value and direction to the vector of the average absolute velocity, the vector of the average transportation velocity, and the vector of the average relative velocity.

When Δt tends to zero, the last relation yields an equality

$$\mathbf{v}_a = \mathbf{v}_t + \mathbf{v}_r. \tag{2.1}$$

The absolute velocity $\mathbf{v}_a$ is equal to the geometric sum of the transportation $\mathbf{v}_t$ and the relative $\mathbf{v}_r$ velocity.

The theorem on the addition of velocities is an important theorem of mechanics. We have to solve a large number of problems to realize that a relative motion is considered with respect to a rigid body (or to a system of moving axes) and that the motion of the rigid body generates a transportation of a particle. A number of interesting problems dealing with compound motions of a particle arise because the absolute motion of a particle can be represented as several compound motions in which the transportation or relative velocities are not completely defined.

We shall consider several problems on the application of this theorem when we discuss instantaneous motions of a rigid body.

Example 1. Two intersecting straight lines are in a translational motion with velocities $\mathbf{v}_1$ and $\mathbf{v}_2$ in a fixed plane. We have to construct geometrically the absolute velocity of the point M when they meet (Fig. 22).

We assume the motion of the point M to be compound. In a compound motion when the motion of the point M along the first straight line is assumed to be relative, the transportation velocity $\mathbf{v}_1$ of M is defined in magnitude and direction (it equals the translational velocity of the first straight line), whereas the relative velocity of M is only defined in direction. A similar circumstance obtains in a compound motion when the motion of the point M along the second straight line is assumed to be relative, namely, the transportation velocity $\mathbf{v}_2$ of the point M is defined in magnitude and direction whereas the relative velocity of the point M is only defined in direction.

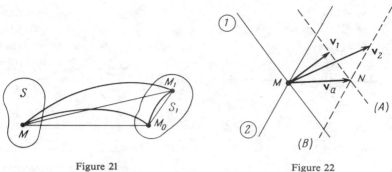

Figure 21   Figure 22

To construct the absolute velocity of the point M from the formula $\mathbf{v}_a = \mathbf{v}_t + \mathbf{v}_r$, we must apply the vector $\mathbf{v}_t = \mathbf{v}_1$ at the point M and the vector $\mathbf{v}_r$ at its terminus. We only know the direction of $\mathbf{v}_r$ and therefore we obtain a straight line A along which the terminus of the absolute velocity $\mathbf{v}_a$ must lie. The terminus of the absolute velocity $\mathbf{v}_a$ of the point M must also lie on the straight line B, which is parallel to the second straight line and passes through the terminus of the vector $\mathbf{v}_2$ applied at the point M. To satisfy these two requirements, the terminus of the vector $\mathbf{v}_a$ must be the intersection N of the straight lines A and B and, consequently, $\mathbf{v}_a = \overrightarrow{MN}$.

Example 2. To construct a tangent to a curve, we may regard the curve as the trajectory of a point whose absolute motion is compound. Sometimes one such representation is sufficient, but often we have to consider several simultaneous representations of a motion in the form of compound motions.

The first case can be illustrated by the construction of a tangent to an Archimedean spiral or to the conchoid of Nicomedes, and the second case, by the construction of a tangent to an ellipse, a hyperbola, a parabola or a lemniscate.

Example 3. *Constructing a tangent to a conchoid.* A ray rotates about a fixed point O (Fig. 23). A segment AB of a ray with a constant length moves so that one endpoint A slides along a straight line aa', the other endpoint B then describes a curve called the conchoid of Nicomedes. We may consider the motion of the point B to be compound and consisting of a relative motion along the ray OB and the rotation of the ray about the point O. Let $\mathbf{v}_t$ be the transportation velocity of the point B. The relative velocity of B is $\mathbf{v}_r$, i.e. the relative velocity of the point A, which is easy to construct. Hence $\mathbf{v}_a = \mathbf{v}_t + \mathbf{v}_r$ defines the direction of a tangent to the conchoid of Nicomedes.

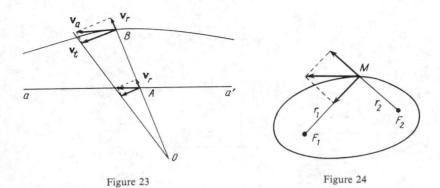

Figure 23   Figure 24

Example 4. *Constructing a tangent to an ellipse* (Fig. 24). We associate a moving coordinate system with r_1. Then the relative velocity of the point M is parallel to r_1 and the transportation velocity is perpendicular to r_1. We associate a second system with r_2. The direction of the relative velocity is the same as that of the transportation velocity.

Since $r_1 + r_2 = 2a$, $\dfrac{dr_1}{dt} = -\dfrac{dr_2}{dt}$, we can construct a tangent.

36. Assume that a point is moving relative to the first reference system S_1, which moves relative to a second system S_2 which, in turn, moves relative to a third system S_3, and so on, up to some last, fixed, system S.

The velocity of the point relative to the system S_2 can be determined by adding velocities: it is the sum of the relative velocity v' relative to S_1 and the transportation velocity v_1 which the point would have relative to S_2 if it were fixed in S_1. The velocity of the point relative to S_3 is the geometric sum of the relative velocity with respect to S_2, i.e. the velocity $v' + v_1$, and the transportation velocity v_2 which the point would have relative to S_3 if it were fixed in S_2. Consequently, the required velocity is equal to

$$v' + v_1 + v_2.$$

Continuing this process, we find that the absolute velocity of the moving point is the sum of its relative velocity with respect to the first moving reference system and all the successive transportation velocities generated by the motion of the first system relative to the second, of the second system relative to the third, and so on.

When we say that a body or a particle is undergoing several simultaneous motions, it means we can add the velocities in a compound motion.

Example. We can define the position of a point in space using spherical coordinates, i.e. the radius r, the angle θ which the radius makes with the

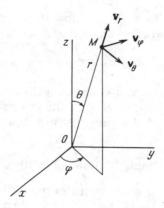

Figure 25

z-axis, and the angle φ between the x-axis and the projection of the radius on the xy-plane (Fig. 25).

The spherical coordinates r, θ, φ define three motions: when only r varies, the point M moves along the radius, when only the angle θ varies, the point M describes a circle in the plane which passes through the z-axis and the point M, with centre at O, and when only the angle φ varies, the point M describes a circle resulting from the rotation of M about the z-axis. We can project the velocity of the moving point M onto the tangents to the trajectories of the three indicated mutually orthogonal motions and obtain projections v_r, v_θ, v_φ respectively. We can determine these projections using the theorem on the addition of velocities in a compound motion. To do this, we can assume the straight line OM to be the system S_1, the plane passing through the z-axis and the radius $\mathbf{r}$ to be the system S_2 and fixed space to be the system S.

The velocity of the point M relative to the system S_1

$$v' = v_r = \frac{dr}{dt}$$

and is directed along the radius vector. The transportation velocity $\mathbf{v}_1$ which the point M would have relative to S_2 if it were fixed on the radius goes along the tangent to a circle of radius r and centre at O, which lies in S_2, in the direction of increasing θ and its magnitude is

$$v_1 = v_\theta = r \frac{d\theta}{dt} .$$

The transportation velocity $\mathbf{v}_2$ which the point M would have relative to the fixed system S were it fixed in S_2 goes along a tangent to a circle of radius $r \sin \theta$ resulting from the rotation of the point M about the z-axis in the direction of increasing angle φ and its magnitude is

$$v_2 = v_\varphi = r \sin \theta \, \frac{d\varphi}{dt} \, .$$

Since the velocities $\mathbf{v}'$, $\mathbf{v}_1$, and $\mathbf{v}_2$ are mutually orthogonal, they are equal to the orthogonal projections of $\mathbf{v}$ on the respective directions of v_r, v_θ, v_φ and this is written in the above-given formulas.

2.3 Simple Motions of a Rigid Body

37. A *rigid body* is a set of particles the distances between which do not change in motion. A rigid body can contain all the particles of a geometric figure (a curve, a surface or a volume) on the assumption that the distances between the particles of the figure do not change in motion.

If we fix two points of a rigid body, it will only be able to rotate about the straight line which passes through these points. This straight line is the rotation axis of the rigid body. If we fix a third point which is not on the straight line, then the body will be fixed. Thus the position of a rigid body is defined by the position of three points which do not lie on the same straight line.

38. If a rigid body moves from one position to another, it is *displaced*, and every one of its particles is displaced. If all the particles of a body are equally displaced, then the particles will remain at the same distances from one another and, consequently, the rigid body may be displaced. A displacement of a rigid body in which all its particles are equally displaced is called a translation. A free vector equal in magnitude and direction to the displacement of the particles of a rigid body is called its *translation vector*.

Two successive translations of a rigid body yield another translation.

We say that a rigid body is undergoing translation if its motion between two points in time is a translation. When a rigid body is undergoing a translation, any two of its particles A and B will, in time Δt, be equally displaced by an amount equal to the displacement of the body $\Delta \mathbf{u}$. If we divide this displacement by Δt, we get the *average velocity of the translation of the rigid body*

$$\mathbf{w} = \frac{\Delta \mathbf{u}}{\Delta t} \, .$$

The average velocities of the particles of a rigid body in translation are evidently equal and parallel to the average velocity of the body. At the limit, all the particles of a rigid body, at a moment t, have velocities equal in magnitude and direction to the *velocity*

$$\mathbf{v} = \lim_{\Delta t \to 0} \frac{\Delta \mathbf{u}}{\Delta t}$$

of the rigid body moving in translation, and $\mathbf{v}$ is a free vector by definition.

The position of a rigid body can be defined by the position of three rectangular axes rigidly associated with the body. In a translation, these axes are displaced parallel to themselves. Conversely, if rectangular axes (or even only two intersecting straight lines) associated with the body are displaced parallel to themselves, then the rigid body is undergoing a translation.

39. Rotation about a fixed axis. If two points of a rigid body are fixed during motion, the body will be rotating about the axis which passes through the fixed points. This motion is a *rotation*.

When a rigid body rotates about a fixed axis AR (Fig. 26), its every point, e.g. M, describes a circle in a plane which is perpendicular to the axis, with a velocity which is perpendicular to the plane MAR and proportional to the distance between the point and the axis. The velocity of the point M, which is at a distance h from the axis, is $h\omega$ where ω is the rate of change of the rotation angle θ of the rigid body about the AR axis:

$$\omega = \frac{d\theta}{dt}.$$

The angle θ is the angle through which the plane MAR, connected with the body, rotates from the initial position. The quantity ω is also known as the magnitude of the angular velocity of the body.

To define the rotation of the rigid body, we must know three quantities: *the position of the rotation axis, the magnitude of the angular velocity, and the direction of rotation.* These three quantities can be represented by

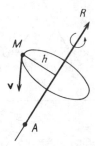

Figure 26

one vector. To do this, we lay off, from an arbitrary point A on the axis, a vector $\overrightarrow{AR}$ equal to ω in magnitude and directed so that the rotation of the body about AR is positive (counterclockwise) for an observer whose feet are at A and whose head is at R. The vector $\overrightarrow{AR}$, which defines the rotation of the rigid body, is known as the angular velocity vector of the rigid body and denoted by $\omega = \overrightarrow{AR}$.

According to this definition, the angular velocity vector is a sliding vector since its origin can be chosen arbitrarily on the rotation axis.

The velocity of an arbitrary point M of a rigid body rotating about the axis is equal to the moment of the vector of the angular velocity ω relative to the point M.

Indeed, the velocity is equal in magnitude to $h\omega$, is perpendicular to the plane containing M and ω, and is directed to the side from which the rotation about AR is seen to be counterclockwise:

$$\mathbf{v} = \overrightarrow{MA} \times \omega = \omega \times \overrightarrow{AM}.$$

40. Let p, q, r be the projections of the angular velocity ω on the x, y, z axes of a rectangular system of coordinates.

If the vector ω is applied at a point A with coordinates x_0, y_0, z_0, then the linear velocity $\mathbf{v}$ of the point M with coordinates x, y, z is equal to the vector product of the vectors ω and $\overrightarrow{AM}$ with projections p, q, r and $x - x_0$, $y - y_0$, $z - z_0$ respectively (Fig. 27):

$$\mathbf{v} = \omega \times \overrightarrow{AM} = \begin{vmatrix} \mathbf{i} & \mathbf{j} & \mathbf{k} \\ p & q & r \\ x - x_0 & y - y_0 & z - z_0 \end{vmatrix}.$$

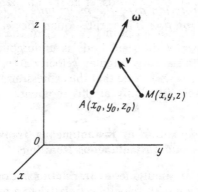

Figure 27

The projections of the vector **v** on the coordinate axes are therefore equal to the determinants obtained from the matrix

$$\left\|\begin{array}{ccc} p & q & r \\ x - x_0 & y - y_0 & z - z_0 \end{array}\right\|,$$

i.e.

$$\begin{aligned} v_x &= q(z - z_0) - r(y - y_0), \\ v_y &= r(x - x_0) - p(z - z_0), \\ v_z &= p(y - y_0) - q(x - x_0). \end{aligned} \qquad (2.2)$$

These are *Euler's formulas*.

When the point A is at the origin of the coordinate system, i.e. at the point O, the angular velocity vector ω passes through the point O, the coordinates x_0, y_0, z_0 are zero and the above formulas assume the form

$$\mathbf{v} = \omega \times \overrightarrow{OM}, \quad v_x = qz - ry, \quad v_y = rx - pz, \quad v_z = py - qx.$$

41. An instantaneous translation and an instantaneous rotation of a rigid body. It may happen so that at some moment t the velocities of all the particles of a rigid body are geometrically equal to one another. In this case we say that at the moment t the rigid body is undergoing an *instantaneous translation*. We must not forget that the expression "instantaneous translation" only describes the state of the velocities of the body's particles at the moment t.

It may so happen that at some moment the velocities of all particles of a rigid body are the same as they would be if the body were rotating as defined by an angular velocity vector ω. Then we say that at this moment the body is undergoing *instantaneous rotation with an instantaneous angular velocity vector* ω. In other words, if at the moment t the velocities of the body's particles are defined by formulas (2.2) with the quantities x_0, y_0, z_0 and p, q, r equal for all points (x, y, z) of the rigid body, then we say that at the moment t the rigid body is undergoing an instantaneous rotation with an instantaneous angular velocity $\omega(p, q, r)$ which passes through the point (x_0, y_0, z_0). Note that the expression "instantaneous rotation" indicates the velocities only at this moment.

2.4 Composition of Instantaneous Translations and Instantaneous Rotations

42. Composition of simultaneous translations. Consider a rigid body undergoing a translation with velocity $\mathbf{v}'$ relative to a moving reference system S_1 and assume that this reference system is undergoing a translation with velocity v_1 relative to a second system S_2, which, in turn, is being translated with velocity v_2 relative to a system S_3, and so on. According to the theorem on the composition of velocities in a compound motion, the absolute velocity v of the particles of the rigid body is equal to the sum $v' + v_1 + v_2 + \ldots$ of the velocities of the indicated motions and, consequently, is the same for all the particles of the rigid body.

Hence, *if a rigid body is undergoing several simultaneous translations, then its absolute motion is also a translation. The velocity of this resultant absolute translation is equal, at every moment, to the geometric sum of the component translations.* We also consider instantaneous translations, and then the theorem on the composition of velocities can be applied in this case as well: several instantaneous simultaneous translations can be reduced to a single resultant instantaneous translation.

43. Composition of simultaneous rotations. Assume that a rigid body is undergoing two simultaneous rotations. We shall consider the states of the velocities at the moment t for these rotations. We assume that at the moment t the rigid body is undergoing an instantaneous rotation with an instantaneous angular velocity ω' relative to a moving system S_1, and S_1 is undergoing a rotation with an instantaneous angular velocity ω_1 relative to a fixed coordinate system S. In this case we say that at the moment t the rigid body is undergoing two simultaneous rotations with angular velocities ω' and ω_1 respectively.

We assume that the sliding vectors ω' and ω_1 meet at the point O. According to the theorem on composition of velocities in compound motion, the absolute linear velocity v of a particle M of the rigid body is

$$v = \omega' \times \overrightarrow{OM} + \omega_1 \times \overrightarrow{OM} = \Omega \times \overrightarrow{OM},$$

where

$$\Omega = \omega' + \omega_1.$$

In other words, *if the instantaneous angular velocities ω' and ω_1 intersect, then the resultant motion of the rigid body is an instantaneous rotation with an instantaneous angular velocity Ω equal to the geometric sum of the instantaneous angular velocities ω' and ω_1.* This is the gist of the theorem on the composition of instantaneous rotations about intersecting axes.

If the instantaneous angular velocities ω' and ω_1 are equal in magnitude, lie on the same straight line, but are opposite in direction, then the resultant instantaneous angular velocity Ω is zero.

Since the intersecting vectors of instantaneous angular velocities ω' and ω_1 can be added according to the parallelogram (or polygon) rule, they are sliding vectors with the principal operation of addition of intersecting sliding vectors. We studied them in detail in Chapter 1.

Consequently, equivalent systems of an arbitrary number of instantaneous angular velocity vectors for the rotation of a rigid body represent the same instantaneous motion of the rigid body.

44. A couple of instantaneous rotations. If two simultaneous rotations with angular velocities ω and $-\omega$ are not on the same straight line and are applied at points A and A' respectively, they form a couple. Then the rigid body undergoes an instantaneous translation with a velocity equal to the moment of the couple since the velocity of the particle M of the body

$$\mathbf{v} = \omega \times \overrightarrow{AM} + (-\omega \times \overrightarrow{A'M}) = \omega \times (\overrightarrow{AM} + \overrightarrow{MA'}) = \omega \times \overrightarrow{AA'}$$

is the same for all particles of the body and is equal to the moment of the couple. It follows that *a couple of instantaneous rotations is equivalent to an instantaneous translation with a velocity equal to the moment of the couple.*

45. Reduction of several simultaneous instantaneous translations and rotations. Assume that a rigid body is simultaneously undergoing several translations with velocities $\mathbf{v}_1$, $\mathbf{v}_2$, ... and rotations with angular velocities ω_1, ω_2, ... applied at points A_1, A_2, ..., respectively. This system of free and sliding vectors of the velocities of the rigid body can be reduced (in accordance with the results in item 16) at the origin O of a fixed system to the vector of the resultant angular velocity

$$\omega = \omega_1 + \omega_2 + ...,$$

which passes through the origin O, and to the vector of the resultant velocity of the translation

$$\mathbf{v} = \mathbf{v}_1 + \mathbf{v}_2 + ... + \overrightarrow{OA_1} \times \omega_1 + \overrightarrow{OA_2} \times \omega_2 + ...$$

The method of reduction consists in adding to the given instantaneous motions the instantaneous rotations defined by the vectors ω_1 and $-\omega_1$, ω_2 and $-\omega_2$, ..., applied at the point O. The vector ω_1 applied at A_1 and the vector $-\omega_1$ applied at O form a couple of instantaneous rotations which is equivalent to an instantaneous translation with velocity equal to the moment of the couple $\overrightarrow{OA_1} \times \omega_1$. In the same way the couple of instantaneous rotations ω_2 with origin at A_2 and $-\omega_2$ with origin at O is equivalent to an instantaneous translation of a rigid body with velocity $\overrightarrow{OA_2} \times \omega_2$, and so on. Then the vectors of the angular velocities ω_1, ω_2, ... originating at O are added at this point as well as the free vectors of velocities of the instantaneous translations $\mathbf{v}_1$, $\mathbf{v}_2$, ..., $\overrightarrow{OA_1} \times \omega_1$, $\overrightarrow{OA_2} \times \omega_2$, ...

A rigid body's instantaneous motion can be reduced to a new point O' by means of the following operations which transform the system of instantaneous motions to an equivalent system: at the point O' we place two rotation vectors with velocities ω and $-\omega$, which are equal, lie on the same straight line, and are opposite in direction. The vector ω applied at O and the vector $-\omega$ applied at O' form a couple of rotations which is equivalent to the instantaneous translation of the rigid body with velocity equal to the moment of the couple $\omega \times \overrightarrow{OO'}$. Then the system reduces to one instantaneous rotation with angular velocity ω which passes through O' and to an instantaneous translation with velocity

$$\mathbf{v}' = \mathbf{v} + \omega \times \overrightarrow{OO'}. \tag{2.3}$$

The translation velocity $\mathbf{v}'$ depends on the choice of the point O', and the angular velocity of rotation ω is the same in magnitude and direction for any point of reduction.

We can reduce the motion of a rigid body to a single helical motion or to a screw if ω is different from zero. Indeed, if we take point O' (Fig. 28) defined by the relation (see item 20)

$$\overrightarrow{OO'} = \frac{\omega \times \mathbf{v}}{\omega^2},$$

as a reduction point, then, according to (2.3), the motions of the rigid body will reduce, at the point O', to a helical motion which consists of rotation with angular velocity ω passing through the point O' and translation with velocity

$$\mathbf{v}' = \mathbf{v} + \omega \times \overrightarrow{OO'} = \frac{(\omega \cdot \mathbf{v})}{\omega} \left(\frac{\omega}{\omega} \right) = v \cos \alpha \left(\frac{\omega}{\omega} \right),$$

directed along the vector with angular velocity ω. The parameter of the screw is the quantity $\beta = v'/\omega$ and the lead of the screw is $2\pi\beta$.

Example. Consider three bevel pinions (Fig. 29). Pinions 1 and 2 have a radius R and rotate about the same vertical with angular velocities ω_1 and ω_2 respectively. Pinion 3 of radius r can rotate about a horizontal axis $\overrightarrow{OA}$ which, in turn, can rotate about a vertical axis. We have to find the instantaneous angular velocity of pinion 3.

We can assume the rotation of pinion 3 to be a compound motion and consider its motion relative to pinions 1 and 2. In the first case, the transportation is the rotation with angular velocity ω_1 and the relative motion is the rolling of pinion 3 over pinion 1, i.e. rotation about the OB axis. Therefore the terminus of the vector of the angular velocity ω of rotation of pinion 3 must lie on a straight line which is parallel to OB and drawn

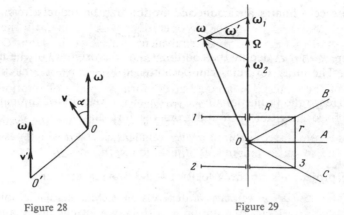

Figure 28   Figure 29

through the terminus of the vector ω_1. In the second case, the transportation is the rotation with angular velocity ω_2 and the relative motion is the rolling of pinion *3* over pinion *2*, i.e. rotation about the *OC* axis. Therefore the terminus of the vector of the angular velocity ω of rotation of pinion *3* must lie on a straight line which is parallel to *OC* and drawn through the terminus of the vector ω_2. This completely defines the vector ω.

From this we can directly determine the magnitudes of its vertical Ω and horizontal ω' projections:

$$\Omega = \frac{\omega_1 + \omega_2}{2}, \quad \omega' = \frac{\omega_1 - \omega_2}{2} \frac{R}{r}.$$

2.5 Distribution of Velocities in a Moving Rigid Body

46. To find the distribution of velocities in a moving rigid body, we consider a fixed system of axes $x_1 y_1 z_1$ and the system of axes xyz which is rigidly associated with the moving rigid body. The cosines of the angles between these axes can be tabulated:

	x	y	z
x_1	α	α_1	α_2
y_1	β	β_1	β_2
z_1	γ	γ_1	γ_2

We designate the coordinates of a particle M of the rigid body as x_1, y_1, z_1 and x, y, z, respectively, with respect to the fixed and moving axes.

These coordinates are connected by the translation relations

$$x_1 = x_0 + x\alpha + y\alpha_1 + z\alpha_2, \quad ...,$$

where x_0, y_0, z_0 denote the coordinates of the origin O of the moving system. The other formulas can be obtained by means of a circular permutation.

Hence the projections v_{x_1}, v_{y_1}, v_{z_1} of the velocity of the particle M on the fixed axes can be found from the formulas

$$v_{x_1} = \frac{dx_0}{dt} + x\,\frac{d\alpha}{dt} + y\,\frac{d\alpha_1}{dt} + z\,\frac{d\alpha_2}{dt}, \quad ....$$

The projections of the velocity on the moving axes are

$$v_x = v_{x_1}\alpha + v_{y_1}\beta + v_{z_1}\gamma, \quad ....$$

Differentiating the obvious formulas $\alpha^2 + \beta^2 + \gamma^2 = 1$, ..., $\alpha_1\alpha_2 + \beta_1\beta_2 + \gamma_1\gamma_2 = 0$, ... with respect to time, we get $\alpha\,\dfrac{d\alpha}{dt} + \beta\,\dfrac{d\beta}{dt} + \gamma\,\dfrac{d\gamma}{dt} = 0$, ... and

$$\alpha_2\,\frac{d\alpha_1}{dt} + \beta_2\,\frac{d\beta_1}{dt} + \gamma_2\,\frac{d\gamma_1}{dt} = -\left(\alpha_1\,\frac{d\alpha_2}{dt} + \beta_1\,\frac{d\beta_2}{dt} + \gamma_1\,\frac{d\gamma_2}{dt}\right) = p,$$

$$\alpha\,\frac{d\alpha_2}{dt} + \beta\,\frac{d\beta_2}{dt} + \gamma\,\frac{d\gamma_2}{dt} = -\left(\alpha_2\,\frac{d\alpha}{dt} + \beta_2\,\frac{d\beta}{dt} + \gamma_2\,\frac{d\gamma}{dt}\right) = q,$$

$$\alpha_1\,\frac{d\alpha}{dt} + \beta_1\,\frac{d\beta}{dt} + \gamma_1\,\frac{d\gamma}{dt} = -\left(\alpha\,\frac{d\alpha_1}{dt} + \beta\,\frac{d\beta_1}{dt} + \gamma\,\frac{d\gamma_1}{dt}\right) = r.$$

In these designations we obtain

$$v_x = v_x^0 + qz - ry, \quad v_y = v_y^0 + rx - pz, \quad v_z = v_z^0 + py - qx,$$

where v_x^0, v_y^0, v_z^0 are the projections of the velocity of the point O on the moving axes of coordinates. In vector notation, these formulas yield

$$\mathbf{v} = \mathbf{v}^0 + \omega \times \overrightarrow{OM}, \tag{2.4}$$

where ω denotes a vector with projections p, q, r. Comparing these formulas with those of Euler, we consequently find that (**Euler's theorem**) *the motion of a rigid body at time t is equivalent to an instantaneous translation with velocity* $\mathbf{v}^0$ *and instantaneous rotation with an instantaneous angular velocity* ω *about the axis which passes through the point O.*

The axis of an instantaneous screw can be found as the locus of the points whose velocities $\mathbf{v}$ are parallel to the vector ω of the instantaneous angular velocity. The equation of the axis of the instantaneous screw in the moving axes has the form

$$\frac{v_x^0 + qz - ry}{p} = \frac{v_y^0 + rx - pz}{q} = \frac{v_z^0 + py - qx}{r}.$$

The locus of the axis of an instantaneous screw in a moving system of coordinates is known as a *loose axoide*.

If we designate the projections of an instantaneous angular velocity ω on the fixed axes as p_1, q_1, r_1, the projections of v^0 on the fixed axes as $v_{x_1}^0$, $v_{y_1}^0$, $v_{z_1}^0$, and the coordinates of the origin O of the moving axes in the system of fixed coordinate axes as x_0, y_0, z_0, then the equation of the axis of the instantaneous screw is

$$\frac{v_{x_1}^0 + q_1(z_1 - z_0) - r_1(y_1 - y_0)}{p_1} = \frac{v_{y_1}^0 + r_1(x_1 - x_0) - p_1(z_1 - z_0)}{q_1}$$
$$= \frac{v_{z_1}^0 + p_1(y_1 - y_0) - q_1(x_1 - x_0)}{r_1}.$$

The locus of the axis of an instantaneous screw in a fixed coordinate system is a *fixed axoide*.

The axis of an instantaneous screw becomes indeterminate if the instantaneous angular velocity ω is zero. Then the motion of the rigid body reduces to one instantaneous translation or an instantaneous moment of rest.

Example. Euler's theorem can be used to find the derivatives with respect to t of the unit vectors i, j, k of the system of the moving coordinate axes relative to a system of fixed axes. According to the definition of the derivative of a vector with respect to the system $x_1 y_1 z_1$ (see item 24) the derivative di/dt is the velocity with respect to $O_1 x_1 y_1 z_1$ of the terminus of the vector which is equal and parallel to i and whose origin is at the origin O_1 of the fixed axes.

We can obtain the absolute velocity of the terminus of the vector i, transferred to the point O_1 from formula (2.4) if we make the points O and O_1 coincide ($v^0 = 0$):

$$\frac{di}{dt} = \omega \times i,$$

and, similarly,

$$\frac{dj}{dt} = \omega \times j, \quad \frac{dk}{dt} = \omega \times k.$$

47. At the moment t the fixed axoide Σ and the loose axoide Σ' have a common generatrix, the axis AB of an instantaneous screw for the moment t being considered (Fig. 30).

Let us prove that the axoides touch each other. Indeed, let us assume that a point P moving along the loose axoide is such that at every moment t it is on the axis of the instantaneous screw existing at time t. Let the curve S', which lies on the loose axoide, be the relative trajectory of the point P, and the curve S, which lies on the fixed axoide, be the trajectory of the absolute motion of the point P. According to the rule of addition of velocities, the absolute velocity $\mathbf{v}_a$ of the point P is related to the transportation $\mathbf{v}_t$ and the relative $\mathbf{v}_r$ velocity of the point P by

$$\mathbf{v}_a = \mathbf{v}_t + \mathbf{v}_r.$$

The absolute velocity $\mathbf{v}_a$, which is a tangent to S, lies in the tangent plane to the fixed axoide Σ. The relative velocity $\mathbf{v}_r$ is a tangent to S' and lies in a plane tangential to Σ'. The transportation velocity $\mathbf{v}_t$ of the point P, which lies on the axis of the instantaneous screw, is equal to the translation velocity of the instantaneous helical motion and is directed along the common, for the moment t, generatrix of the axoides Σ and Σ'. Since the last equality holds true for any of the indicated motions of the point P, we must infer that the tangent planes to the axoides Σ and Σ', which both pass through the axis AB of the instantaneous screw, coincide. This means that at the moment t the axoides Σ and Σ' themselves touch the axis AB of the instantaneous screw.

An instantaneous motion of a rigid body is produced by an instantaneous rotation of the loose axoide about the axis of the instantaneous screw with angular velocity ω and a simultaneous translation of the loose axoide in the direction of the instantaneous screw.

48. Continuous motion of a body parallel to a fixed plane. If all the points of a rigid body are displaced parallel to a fixed plane P, then we say that the rigid body undergoes a plane-parallel motion. In this case a section S of the body by the plane P is a plane figure (associated with the body) which moves in its own plane (Fig. 31).

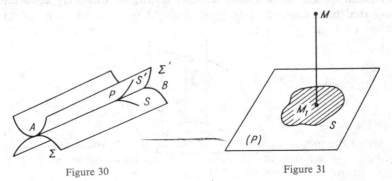

Figure 30   Figure 31

The motion of the section S defines the motion of the whole body. A straight line, which projects a point M of the body onto the plane of the section S, is associated with the rigid body, is perpendicular to the fixed plane P, moves parallel to itself, and all of its points describe similar trajectories at equal velocities.

If the instantaneous motion of the rigid body is not a translation, then the instantaneous screw reduces to an instantaneous rotation with an angular velocity orthogonal to the fixed plane P. In this case, the axoides, both fixed Σ and loose Σ', are cylindrical surfaces with generatrices orthogonal to the fixed plane (Fig. 32).

The trace of the intersection of the axis of the instantaneous rotation and the fixed plane P is the *instantaneous centre of rotation* of the plane figure. The intersections of the axoides Σ and Σ' and the fixed plane define the curves c_f and c_m which are the loci of the instantaneous centres of rotation in the fixed plane P and in the section S of the rigid body respectively. The curves c_f and c_m are the *space centrode* and *body centrode* respectively. In the section S the instantaneous centre of rotation describes a relative trajectory c_m (a body centrode) and in the fixed plane (P) it describes a space centrode c_f. Then the transportation velocity $\mathbf{v}_t$ of the instantaneous centre of rotation C is zero, being the velocity of the point of the body at the instantaneous centre C. Hence

$$\mathbf{v}_a = \mathbf{v}_r,$$

or

$$\frac{ds_f}{dt} = \frac{ds_m}{dt},$$

where s_f and s_m denote the lengths of the arcs of the centrodes c_f and c_m respectively. Consequently, the centrodes c_f and c_m touch each other at the instantaneous centre of rotation C, and the plane-parallel motion of the rigid body reduces to a rolling without sliding of the body centrode c_m along the space centrode c_f.

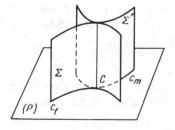

Figure 32

In a plane-parallel motion, the velocity of a point M of the body is orthogonal to a straight line CM, where C is an instantaneous centre of rotation. Therefore, if at the moment t we know the directions of the velocities of two points M_1 and M_2 of the body and if these velocities are not parallel to each other and not orthogonal to the segment M_1M_2, then the instantaneous centre of rotation can be constructed as the intersection point of the perpendiculars drawn from the points M_1 and M_2 to their velocities.

2.6 The Coriolis Theorem. Distribution of Accelerations in a Moving Rigid Body

49. Consider a fixed $x_1y_1z_1$ and a moving xyz coordinate system. Assume that a particle M has coordinates x_1, y_1, z_1 with respect to the fixed system and coordinates x, y, z with respect to the moving system. As before, when we proved Euler's theorem we tabulate the cosines of the angles between the coordinate axes as

	x	y	z
x_1	α	α_1	α_2
y_1	β	β_1	β_2
z_1	γ	γ_1	γ_2

We use x_0, y_0, z_0 to denote the coordinates of the point O and get

$$x_1 = x_0 + x\alpha + y\alpha_1 + z\alpha_2 .$$

It follows that the projections of the absolute acceleration $\mathbf{j}_a$ of the particle M on the fixed axes are expressed by the formulas

$$\frac{d^2x_1}{dt^2} = \frac{d^2x_0}{dt^2} + x\frac{d^2\alpha}{dt^2} + y\frac{d^2\alpha_1}{dt^2} + z\frac{d^2\alpha_2}{dt^2} + \alpha\frac{d^2x}{dt^2} + \alpha_1\frac{d^2y}{dt^2}$$
$$+ \alpha_2\frac{d^2z}{dt^2} + 2\left(\frac{dx}{dt}\frac{d\alpha}{dt} + \frac{dy}{dt}\frac{d\alpha_1}{dt} + \frac{dz}{dt}\frac{d\alpha_2}{dt}\right), \ \dots$$

We obtain the projections of the transportation acceleration $\mathbf{j}_t$ on the fixed axes when we fix the values of the coordinates x, y, z and, consequently,

$$j_{tx_1} = \frac{d^2x_0}{dt^2} + x\frac{d^2\alpha}{dt^2} + y\frac{d^2\alpha_1}{dt^2} + z\frac{d^2\alpha_2}{dt^2}, \ \dots$$

We get the projections of the relative acceleration $\mathbf{j}_r$ on the fixed axes when we fix x_0, y_0, z_0 and the values of all cosines between the axes:

$$j_{rx_1} = \alpha\frac{d^2x}{dt^2} + \alpha_1\frac{d^2y}{dt^2} + \alpha_2\frac{d^2z}{dt^2}, \ \dots$$

The projections of the complementary acceleration $\mathbf{j}_c$ on the fixed axes are

$$j_{cx_1} = 2 \left(\frac{dx}{dt} \frac{d\alpha}{dt} + \frac{dy}{dt} \frac{d\alpha_1}{dt} + \frac{dz}{dt} \frac{d\alpha_2}{dt} \right), \dots$$

As a result we obtain the **Coriolis theorem**

$$\mathbf{j}_a = \mathbf{j}_t + \mathbf{j}_r + \mathbf{j}_c,$$

i.e., *the absolute acceleration of a particle is equal to the geometric sum of the relative, transportation, and complementary (or Coriolis) accelerations.*

Given the projections p, q, r of the instantaneous angular velocity ω of the rotation of a moving system on the moving coordinate axes (item 46), the projections of the Coriolis acceleration on the moving coordinate axes $j_{cx} = j_{cx_1}\alpha + j_{cy_1}\beta + j_{cz_1}\gamma, \dots$ are

$$j_{cx} = 2 \left(q \frac{dz}{dt} - r \frac{dy}{dt} \right), \ j_{cy} = 2 \left(r \frac{dx}{dt} - p \frac{dz}{dt} \right),$$

$$j_{cz} = 2 \left(p \frac{dy}{dt} - q \frac{dx}{dt} \right).$$

But $\dfrac{dx}{dt}$, $\dfrac{dy}{dt}$, $\dfrac{dz}{dt}$ are the projections of the relative velocity $\mathbf{v}_r$ of the particle M on the moving axes. Therefore

$$\mathbf{j}_c = 2(\omega \times \mathbf{v}_r).$$

The Coriolis acceleration is double the vector product of the angular velocity of the rotation of the moving coordinate system by the relative velocity of the moving particle. Consequently, the Coriolis acceleration is orthogonal both to ω and to $\mathbf{v}_r$ and vanishes when at least one of the following three quantities is eliminated: the sine of the angle between ω and $\mathbf{v}_r$, the quantity ω or the quantity $\mathbf{v}_r$.

50. Since we have derived formulas for the derivatives of the unit vectors, we can give another proof of the Coriolis theorem.

We shall define the position of the particle M in the coordinate systems, which we considered in the preceding item by the radius vector $\mathbf{R}$ in the system of fixed axes $x_1 y_1 z_1$, the radius vector $\mathbf{r}$ in the system of moving axes xyz, and the radius vector ρ of the origin O of the moving axes xyz (Fig. 33):

$$\mathbf{R} = \rho + \mathbf{r}.$$

Hence

$$\mathbf{v}_a = \frac{d\mathbf{R}}{dt} = \frac{d\rho}{dt} + \frac{d\mathbf{r}}{dt},$$

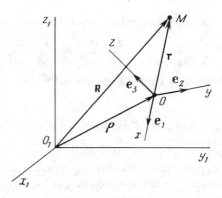

Figure 33

but

$$\frac{d\mathbf{r}}{dt} = \frac{dx}{dt}\mathbf{i} + \frac{dy}{dt}\mathbf{j} + \frac{dz}{dt}\mathbf{k} + x\frac{d\mathbf{i}}{dt} + y\frac{d\mathbf{j}}{dt} + z\frac{d\mathbf{k}}{dt}.$$

From the formulas for the derivatives of unit vectors, we have

$$\frac{d\mathbf{r}}{dt} = \mathbf{v}_r + \boldsymbol{\omega} \times \mathbf{r},$$

and, consequently,

$$\mathbf{v}_a = \frac{d\boldsymbol{\rho}}{dt} + \mathbf{v}_r + \boldsymbol{\omega} \times \mathbf{r}.$$

Differentiating the expression we have obtained for $\mathbf{v}_a$ once again with respect to t in the fixed system $x_1 y_1 z_1$, we get

$$\mathbf{j}_a = \frac{d\mathbf{v}_a}{dt} = \frac{d^2\boldsymbol{\rho}}{dt^2} + \frac{d\mathbf{v}_r}{dt} + \frac{d\boldsymbol{\omega}}{dt} \times \mathbf{r} + \boldsymbol{\omega} \times \mathbf{v}_r + \boldsymbol{\omega} \times \mathbf{r},$$

but

$$\frac{d\mathbf{v}_r}{dt} = \mathbf{j}_r + \boldsymbol{\omega} \times \mathbf{v}_r,$$

and, consequently,

$$\mathbf{j}_a = \frac{d^2\boldsymbol{\rho}}{dt^2} + \boldsymbol{\omega}' \times \mathbf{r} + \boldsymbol{\omega} \times (\boldsymbol{\omega} \times \mathbf{r}) + \mathbf{j}_r + 2(\boldsymbol{\omega} \times \mathbf{v}_r).$$

(Here $\boldsymbol{\omega}' = d\boldsymbol{\omega}/dt$.) By fixing the position of the particle M in the moving system, we find that the transportation acceleration $\mathbf{j}_t$ is

$$\mathbf{j}_t = \frac{d^2\rho}{dt^2} + \omega' \times \mathbf{r} + \omega \times (\omega \times \mathbf{r}),$$

where $d^2\rho/dt^2$ is the absolute acceleration of the origin O of the moving axes, $\omega' \times \mathbf{r}$ is the rotational acceleration defined by the angular acceleration

$$\omega' = \frac{d\omega}{dt} = \frac{dp}{dt}\,\mathbf{i} + \frac{dq}{dt}\,\mathbf{j} + \frac{dr}{dt}\,\mathbf{k},$$

and the term $\omega \times (\omega \times \mathbf{r})$ is the centripetal acceleration.

Let us draw the vectors of the angular velocity ω and of the angular acceleration ω' through the origin O of the moving axes (Fig. 34). The rotational acceleration $\omega' \times \mathbf{r}$ is numerically equal to $\omega'h'$, where h' is the distance between the particle M and the vector of the angular acceleration ω'. Since the centripetal acceleration $\omega \times (\omega \times \mathbf{r})$ is orthogonal to the vectors ω and $\omega \times \mathbf{r}$, it is directed along the perpendicular h dropped from the point M onto the vector of the angular acceleration ω and is numerically equal to $\omega^2 h$.

The complementary (Coriolis) acceleration is defined as

$$\mathbf{j}_c = 2(\omega \times \mathbf{v}_r).$$

Comparing the formula for the Coriolis acceleration $\mathbf{j}_c = 2(\omega \times \mathbf{v}_r)$ with Euler's formula $\mathbf{v} = \Omega \times \overrightarrow{OM}$ for the velocity of a particle M of a rigid body which rotates with angular velocity Ω whose vector passes through the point O, we can formulate a rule that can be of use when we determine the direction of the Coriolis acceleration in specific cases. In magnitude and direction the Coriolis acceleration $\mathbf{j}_c$ is equal to double the velocity of the terminus of the relative velocity vector $\mathbf{v}_r$ if the latter is rotated with angular velocity ω passing through the origin of the relative velocity vector $\mathbf{v}_r$.

51. By way of example, we shall consider the projections of the acceleration on the radius vector and the transversal direction.

We assume the motion of the particle M to be compound consisting of a transportation motion together with the ray OM, which rotates about a fixed pole O with angular velocity $d\theta/dt$, and a relative motion of the particle M along the ray OM (Fig. 35). Assume that the relative acceleration $\mathbf{j}_r = d^2r/dt^2$ goes along the radius in the direction of increasing r. In a transportation along a circle of radius r with centre at O the normal component of the acceleration

$$j_{tn} = r\left(\frac{d\theta}{dt}\right)^2$$

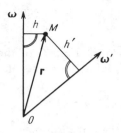

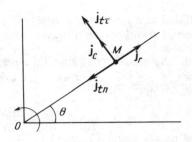

Figure 34   Figure 35

is directed along the radius to the centre O and the tangential component of the transportation acceleration

$$j_{t\tau} = r \frac{d^2\theta}{dt^2}$$

is orthogonal to r in the direction of increasing θ, the Coriolis acceleration is in the same direction,

$$j_c = 2 \frac{d\theta}{dt} \frac{dr}{dt}$$

since the relative velocity $v_r = dr/dt$ is directed along the ray in the direction of increasing r.

Hence, the projections of the acceleration in polar coordinates in the plane in the radial direction and the transversal direction are

$$\frac{d^2 r}{dt^2} - r \left(\frac{d\theta}{dt}\right)^2, \ r \frac{d^2\theta}{dt^2} + 2 \frac{d\theta}{dt} \frac{dr}{dt}$$

respectively.

Let us consider one more example, that of the acceleration of a particle M moving along a radius r of a fixed sphere (Fig. 36).

We assume the motion of the particle M to be compound motion and will consider the relative motion of the particle with respect to the plane of the meridian NM which rotates in space about the axis ON with angular velocity $d\varphi/dt$. The relative velocity $v_r = r \, d\theta/dt$ is tangential to the meridian at the point M and goes in the direction of increasing θ. The tangential and normal relative accelerations are

$$j_{r\tau} = \frac{dv_r}{dt} = r \frac{d^2\theta}{dt^2}, \ j_{rn} = \frac{v_r^2}{r} = r \left(\frac{d\theta}{dt}\right)^2.$$

The trajectory of the transportation is the parallel of the radius $r \sin \theta$. The transportation velocity $v_t = r \sin \theta \dfrac{d\varphi}{dt}$ is tangential to the parallel and goes in the direction of increasing φ. The tangential and the normal transportation accelerations are

$$j_{tr} = r \sin \theta \frac{d^2 \varphi}{dt^2}, j_{tn} = r \sin \theta \left(\frac{d\varphi}{dt} \right)^2.$$

The Coriolis acceleration is

$$j_c = 2r \frac{d\theta}{dt} \frac{d\varphi}{dt} \cos \theta$$

and passes along a tangent to the parallel in the direction of increasing φ.

52. Let us now consider an example of the distribution of accelerations in a plane figure which moves in its own plane. An excellent exposition of this subject can be found in the lecture on theoretical mechanics by Vallée Poussin*. Consider a plane figure moving in its plane (Fig. 37). We reduce it to fixed rectangular coordinates Oxy. Assume that x_0, y_0 are the coordinates of an instantaneous centre of rotation C and ω is the magnitude of the instantaneous angular velocity. According to Euler's formulas, the projections of the velocity of particle $M(x, y)$ of the figure are

$$v_x = - \omega (y - y_0), v_y = \omega (x - x_0).$$

The projections of the acceleration of the particle M are equal to their derivatives with respect to time:

$$j_x = \omega y_0' - \omega^2 (x - x_0) - \omega' (y - y_0),$$
$$j_y = -\omega x_0' - \omega^2 (y - y_0) + \omega' (x - x_0). \tag{2.5}$$

We shall seek a point C' with coordinates a, b, whose acceleration is zero. Such a point is known as an *instantaneous acceleration centre*. Equa-

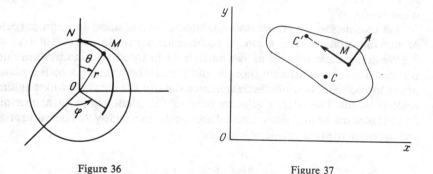

Figure 36   Figure 37

---

* Ch.-J. de La Vallée Poussin, *Leçons de méchanique*, T. 1, 2 (Paris: 1932).

tions (2.5) yield

$$0 = \omega\, y_0' - \omega^2\,(a - x_0) - \omega'\,(b - y_0),$$
$$0 = -\,\omega\, x_0' - \omega^2\,(b - y_0) + \omega'\,(a - x_0).$$

The determinant in the unknowns a, b, equal to $\omega^4 + \omega'^2$, is nonzero if ω and ω' are simultaneously nonzero. In this case, the point $C'\,(a,\,b)$, which is the instantaneous acceleration centre, exists. Subtracting the last relations from formulas (2.5), we have

$$j_x = -\,\omega^2\,(x - a) - \omega'\,(y - b),$$
$$j_y = -\,\omega^2\,(y - b) + \omega'\,(x - a).$$

Consequently, the acceleration of the particle $M\,(x, y)$ consists of two accelerations, the acceleration $C'M\cdot\omega^2$ directed from M to the instantaneous acceleration centre C' and the acceleration $C'M\cdot\omega'$ whose direction is orthogonal to $C'M$ and is defined by the sign of ω', i. e. it is in the direction of rotation if $\omega' > 0$ is counter to the rotation if $\omega' < 0$.

For the particle of a figure with coordinates x_0, y_0, which coincides with the instantaneous rotation centre, the right-hand sides of (2.5) reduce to their first terms $\omega\, y_0'$ and $-\omega\, x_0'$. Consequently, these terms are the projections of the acceleration $\mathbf{j}_c$ of the particle of the figure, which, at this moment, coincides with the instantaneous centre x_0, y_0. If the instantaneous rotation centre were fixed, then the motion of the particle M would be circular and the right-hand sides of the above relations would reduce to the second and the third term. But in this circular motion of the particle M, the normal acceleration $j_n = \omega^2\, r$ is directed along the radius to the instantaneous centre C and the tangential acceleration $j_\tau = r\,\omega'$ is orthogonal to CM and its direction coincides with the rotation defined by the sign of ω'.

We can derive the conclusion geometrically. We imagine a moving system of coordinates with origin at the point of a moving figure which coincides with the instantaneous centre of rotation and with axes which are parallel to the fixed axes. In this moving system there is no Coriolis acceleration for the particles of the figure since the moving system of axes undergoes a translation. The relative motion of the plane figure at a moment t is a rotation about the origin. This yields the relative accelerations

$$j_n = \omega^2\, r, \quad j_\tau = \omega'\, r.$$

It is easy to find the reference frame acceleration j_C. At the moment $t + \Delta t$ the figure rotates with an angular velocity $\omega + \Delta\omega$ about its instantaneous centre C_1 (Fig. 38). Hence the acceleration of the origin of the moving axis is

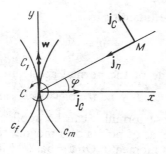

Figure 38

$$j_C = \lim_{\Delta t \to 0} \frac{CC_1(\omega + \Delta\omega) - 0}{\Delta t} = \omega\, w_C, \quad w_C = \lim_{\Delta t \to 0} \frac{CC_1}{\Delta t},$$

in magnitude and is in the direction of the velocity w_C of the origin at the moment t. We have thus arrived at a geometric equality

$$\mathbf{j} = \mathbf{j}_C + \mathbf{j}_n + \mathbf{j}_\tau .$$

53. A circle and a centre of inflection. We place the origin of fixed axes of coordinates at an instantaneous rotation centre C and draw the x-axis in the direction of $\mathbf{j}_C$ (Fig. 38). the projections of j_C on the x, y-axes are

$$j_C = \omega y_0' > 0, \quad -\omega x_0' = 0$$

respectively. Hence $x_0' = 0$ and the velocity $\mathbf{w}$ of the instantaneous centre of rotation C lies on the y-axis. If we direct the positive y-axis along $\mathbf{w}$, then the direction of rotation from the x-axis to the y-axis will coincide with the rotation ω since $\omega y_0' > 0$. In other words, the velocity vector $\mathbf{w}$ is turned relative to the acceleration $\mathbf{j}_C$, through a right angle in the direction of rotation ω.

Assume that r and φ are the polar coordinates of the particle M (assumed to be different from C), i.e. $\mathbf{r}$ is the radius vector $\overrightarrow{CM}$ and φ is the angle of its inclination to the x-axis. The acceleration $j_n = \omega^2 r$ is directed along MC. The positive orientation of the acceleration $j_\tau = r\omega'$ is defined by the direction of the direct rotation about the point C, i.e. is from the x-axis to the y-axis. From this, the total normal acceleration $\mathbf{j}_n$ and the tangential acceleration $\mathbf{j}_\tau$ are

$$\begin{aligned}
j_n &= \omega^2 r - \omega w \cos\varphi, \\
j_\tau &= r\omega' - \omega w \sin\varphi.
\end{aligned} \tag{2.6}$$

We denote the positive quantity by u,

$$u = \frac{w}{\omega},$$

and then the magnitude of the normal acceleration

$$j_n = \omega^2 (r - u \cos \varphi).$$

Let us see how these formulas can be applied.

We first find the locus of points of a moving figure whose normal accelerations are zero at a moment t. On the basis of formulas (2.6), the equation of this curve has the form

$$r = u \cos \varphi.$$

This is an equation of a circle constructed on a segment $CK = u$, as a diameter, lying on the axis Cx (Fig. 39). This circle is known as a *circle of inflection* (de La Hire circle) and the point K is the *centre of inflection*. The points of the figure lying on the circle are at the inflection points of their trajectories at the moment t since their normal accelerations $j_n = v^2/\varrho$ are zero. In addition, the tangent to the trajectory at this point passes through the centre of inflection K since, on the one hand, the tangent is perpendicular to MC and, on the other hand, the angle KMC must rest on the diameter CK if the point M lies on the circle of inflection.

Let us find the locus of points of the figure whose tangential accelerations are zero. On the basis of formulas (2.6), the equation of this locus is

$$r = \frac{\omega w}{\omega'} \sin \varphi.$$

We plot a segment CJ which is of the same magnitude and sign as $\omega w/\omega'$ on the y-axis. The above equation is that of a circle (Bresse circle) constructed on the segment CJ as a diameter.

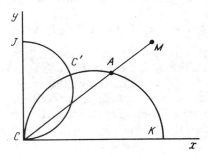

Figure 39

The point K recedes to infinity if the angular velocity ω turns into zero, and the point J recedes to infinity if the angular acceleration ω' is zero and ω is nonzero. Now if both ω and ω' are zero, then the position of J is indeterminate. We exclude this last case and then both circles (which have a point C in common) meet at another point, C', whose total acceleration is zero. Then the point C' is the *centre of accelerations.*

If the velocity of the instantaneous centre of rotation w and the instantaneous angular velocity ω are nonzero, and $\omega' = 0$, then the Bresse circle is the x-axis* and, consequently, the centre of inflections K is the centre of accelerations C'. If $\omega' \neq 0$, and $\omega = 0$, then the centre of accelerations coincides with the instantaneous centre of rotation C.

---

* We mean the degeneration of the Bresse circle into the straight line Cx as $\omega' \rightarrow 0$. — *Ed.*

CHAPTER 3

STATICS

54*. When studying statics and dynamics, we encounter problems which involve force in addition to space and time (see Chapter 2, p. 30).

"Statics is a science concerned with the equilibrium of forces. In general, we understood a force to be any cause which imparts, or tends to impart, motion to bodies to which we assume it to be applied. Therefore we must estimate a force by the quantity of motion which it produces or tends to produce. In the condition of equilibrium a force does not produce an actual action, it only causes a tendency to motion, but it must always be estimated by the effect it would cause if it acted with the absence of any hindrances"**
"Dynamics is a science concerned with accelerating and decelerating forces and on variable motions which they must cause."***

Mechanics is based on Newton's law, which he formulated in 1687 in *Mathematical Principles of Natural Philosophy*.

Newton's laws.

1. Every body preserves in its state of rest or uniform motion in a straight line except in so far as it is compelled to change that state by forces impressed on it.

This law, which we know as Galileo's law of inertia, actually postulates the existence of a reference system in which an isolated particle may be in a state of rest or uniform motion in a straight line when no forces are applied.

2. The rate of change of linear momentum is proportional to the force applied, and takes place in the straight line in which that force acts.

Newton defined momentum as the product of mass by velocity; mass is the quantity of matter in a body. In terms of mathematics, the second law is written as

$$\frac{d}{dt}(m\mathbf{v}) = \mathbf{F},$$

---

* Items 54 and 55 were written by the editors on the basis of Chetaev's lectures. — *Ed.*
** J.L. Lagrange, *Méchanique analytique* (Paris: 1788).
*** Idem.

or, for $m = $ const,

$$m\,\frac{d\mathbf{v}}{dt} = \mathbf{F}.$$

3. The mutual actions of two identical particles are always equal and act in opposite directions.

4. If several forces act on a particle, they can be replaced by a resultant force.

3.1 On Equilibrium

55. Statics studies the conditions of equilibrium of mechanical systems*, i. e. the conditions under which the bodies of a mechanical system acted upon by specified forces under constraints remain in a state of rest relative to a reference system, which is assumed to be fixed.

The laws of statics can be obtained as a special case of dynamics, but we can present statics independently by using several axioms.

1. Several forces applied to a particle can be replaced by a single resultant force, i. e. added according to the parallelogram rule.

2. If a rigid body is in a state of equilibrium, then the addition (or removal) of two equal and opposite forces lying on the same straight line will not disturb the equilibrium.

3. If only two forces act on a body, the body is in equilibrium if and only if the forces are equal in magnitude, lie on the same straight line, and are opposite in direction.

Using some corollaries of these axioms, we shall derive the necessary and sufficient (general) conditions for equilibrium in a rigid body.

Corollaries. 1. It follows from the second axiom that a force acting on a rigid body can be displaced along its line of action. In other words, forces acting on a rigid body can be represented by sliding vectors.

2. It follows from the second axiom that if a body is in equilibrium under the action of a system of forces, it will be in equilibrium under the action of a system equivalent to the first system.

3. Any system of forces can be reduced to equivalent pair of forces.

We reduce a system of forces to the origin of coordinates. By the theory on sliding vectors, the forces reduce to a resultant vector $\mathbf{F} = \Sigma \mathbf{F}_i$ and to the moment of the resultant couple $\mathbf{Q} = \Sigma \mathbf{Q}_i$. Since we can assume that one force of the couple cuts the line of action of $\mathbf{F}$, the system of forces acting on the rigid body can be always reduced to two forces.

---

* See the definition in item 70 of Section 3.5.

4. It follows from property 3 and the third axiom that under the action of a system of forces a rigid body will be in equilibrium if and only if this system of forces is equivalent to zero.

Analytically, if the set of forces $\mathbf{F}_i (X_i, Y_i, Z_i)$, which pass through points $M_i (x_i, y_i, z_i)$, act on a body, then the equations for equilibrium assume the form

$$X = \Sigma X_i = 0, \ Y = \Sigma Y_i = 0, \ Z = \Sigma Z_i = 0 \ (\mathbf{F} = 0),$$

$$\left. \begin{aligned} L &= \Sigma (y_i Z_i - z_i Y_i) = 0, \\ M &= \Sigma (z_i X_i - x_i Z_i) = 0, \\ N &= \Sigma (x_i Y_i - y_i X_i) = 0. \end{aligned} \right\} \ (\mathbf{Q} = 0).$$

If a body is not free, then, according to the axiom, the constraints imposed on the system of particles can be replaced by forces whose action is equivalent to the action of the constraints.

These forces are called *reactions,* or *constraint forces.*

We can divide the forces acting on a rigid body into active forces defined arbitrarily and acting on the body and passive, or constraint, forces arising as a result of the active forces.

Consider some examples.

Example 1. A heavy body is suspended by an elastic inextensible string. The weight is an active force. The tension in the string is a passive force (Fig. 40a).

Example 2. A body lies on a surface. If the surface is smooth, then the reaction, which is a passive force, is directed along a normal to the surface (Fig. 40b).

Example 3. A heavy bar lies on a horizontal table. If the bar is in equilibrium, then the reaction of the table balances the weight, i. e. is directed along a normal to the table and is equal in magnitude to the weight $\mathbf{P}$ (Fig. 40c). We apply a horizontal force $\mathbf{F}$ to the bar. If the plane is not smooth, the body will remain in equilibrium until $\mathbf{F}$ begins to exceed a

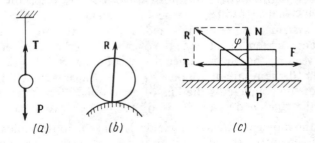

Figure 40

certain value Φ. This means that the reaction $\mathbf{R}$ is opposite to the resultant of the weight $\mathbf{P}$ and the force $\mathbf{F}$. We can represent $\mathbf{R}$ by the sum of the normal reaction $\mathbf{N} = -\mathbf{P}$ and a tangential force of friction $\mathbf{T} = -\mathbf{F}$. If $\mathbf{R}$ makes an angle β with the plane, then

$$\tan \beta = \frac{T}{N} = \frac{F}{P}.$$

If we increase F, then at the moment when $\mathbf{F} = \Phi$, the body begins moving. The value of the friction force, which corresponds to Φ, is known as *static friction* and the corresponding measure φ of the angle β is known as the *angle of friction*,

$$k = \tan \varphi = \frac{\Phi}{P}$$

is the *coefficient of friction.*

Coulomb formulated the laws of friction, which had been found empirically.

56. If only three forces act on a rigid body and the body is in equilibrium, then these three forces must necessarily lie in the same plane, and if two of the three forces meet at a point P, then the third force also passes through that point.

Let us consider, for example, a point O on the line of action of the force $\mathbf{F}_3$ (Fig. 41). The triangles S_1 and S_2 constructed on the forces $\mathbf{F}_1$ and $\mathbf{F}_2$ with vertices at the point O must lie in the same plane, wherever the point O lies on the direction of the force $\mathbf{F}_3$ since the sum of the moments of the forces $\mathbf{F}_1$ and $\mathbf{F}_2$ relative to the point O must be zero according to the condition of equilibrium:

$$\mathbf{m}_1 + \mathbf{m}_2 = 0.$$

It follows that the balanced forces $\mathbf{F}_1$, $\mathbf{F}_2$, $\mathbf{F}_3$ must necessarily be in the same plane.

If the lines of action of $\mathbf{F}_1$ and $\mathbf{F}_2$ meet at the point P, then, by considering the moments of all the forces relative to the point P, i. e. m_P, we have, from the equilibrium condition,

$$m_P(\mathbf{F}_3) = 0,$$

i. e. the force $\mathbf{F}_3$ must pass through the point P.

57. A rigid body with one fixed point. Assume that a fixed point of a rigid body is denoted by O and the fixed axes of Cartesian coordinates xyz are chosen with origin at O (Fig. 42). Assume furthermore that the points $P_\nu(x_\nu, y_\nu, z_\nu)$ of the rigid body are acted upon by active forces $\mathbf{F}_\nu$ with projections X_ν, Y_ν, Z_ν on the coordinate axes and the reaction $\mathbf{R}$ of

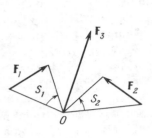

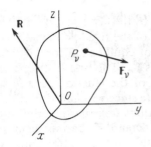

Figure 41   Figure 42

the fixed point has projections X', Y', Z', and the moments of the active forces are L_ν, M_ν, N_ν.

Under the action of the forces $\mathbf{F}_\nu$ and the reaction $\mathbf{R}$, the equilibrium conditions for the body have the form

$$\Sigma X_\nu + X' = 0, \ \Sigma Y_\nu + Y' = 0, \ \Sigma Z_\nu + Z' = 0,$$
$$\Sigma L_\nu = \Sigma (y_\nu Z_\nu - z_\nu Y_\nu) = 0, \quad \Sigma M_\nu = \Sigma (z_\nu X_\nu - x_\nu Z_\nu) = 0,$$
$$\Sigma N_\nu = \Sigma (x_\nu Y_\nu - y_\nu X_\nu) = 0.$$

The first three equations can be used to determine the three unknown projections of the reaction $\mathbf{R}$ at the point O and the last three equations of the moments are the equilibrium conditions in the true sense of the word.

When the last three equations are satisfied, the rigid body with the fixed point O is in equilibrium under the action of the specified forces, irrespective of the projections X', Y', Z' of the reaction which do not appear in the equations.

58. A rigid body with two fixed points. Consider the case when a rigid body has two fixed points O and O' (Fig. 43). Nothing prevents us from taking the coordinate axes with origin at O and assuming the z-axis to be directed along the straight line OO'. The x, y-axes are orthogonal to the z-axis. Assume furthermore that the points $P_\nu(x_\nu, y_\nu, z_\nu)$ of the rigid body are acted upon by the specified active forces $\mathbf{F}_\nu$ with projections X_ν, Y_ν, Z_ν on the coordinate axes. We suppose that under the action of these forces the rigid body is in equilibrium. The points O and O' produce reactions $\mathbf{R}$ and $\mathbf{R}'$ which act on the body in addition to the forces $\mathbf{F}_\nu$.

If the distance between the points is h and the projections of the reactions $\mathbf{R}$ and $\mathbf{R}'$ are X, Y, Z and X', Y', Z' respectively, then the equilibrium equations have the form

$$\Sigma X_\nu + X + X' = 0, \quad \Sigma Y_\nu + Y + Y' = 0, \quad \Sigma Z_\nu + Z + Z' = 0,$$
$$L - hY' = 0, \quad M + hX' = 0, \quad N = 0.$$

Only by the last equation does not include projections of the unknown reactions of the fixed points and, consequently, only this equation is the condition of equilibrium for the body with two fixed points acted upon by the forces F_ν. The first five equations can be used to determine the six projections of the reactions at the supports. This problem is statically indeterminate since we cannot find six unknowns from five equations. From these equations we can evidently find in succession X', Y', X, Y, but the projections Z and Z' cannot be determined separately, we can only find their sum.

59. Let us consider an equilibrium when a rigid body has several points A_s ($s = 1, ..., k$) which lie in a plane. The plane is assumed to be smooth.

We assume the plane of the points A_s to be a coordinate plane xy (Fig. 44). In this coordinate system, we designate the coordinates of the points A_s as a_s, b_s, 0, and the projections of the reactions R_s of the smooth plane at the point A_s are equal to 0, 0, R_s respectively.

We assume that the forces $F_\nu (X_\nu, Y_\nu, Z_\nu)$ ($\nu = 1, ..., n$) act on the points $P_\nu (x_\nu, y_\nu, z_\nu)$ of the rigid body and the body resting on the plane $z = 0$ is in equilibrium due to these forces. In this case the equations of equilibrium of the rigid body have the form

$$\Sigma X_\nu = 0, \quad \Sigma Y_\nu = 0, \quad \Sigma Z_\nu + \Sigma R_s = 0,$$
$$L + \Sigma b_s R_s = 0, \quad M - \Sigma a_s R_s = 0, \quad N = 0.$$

Here L, M, N denote the projections of the net moment of the system of the given forces F_ν relative to the origin. The projections of the moment of the reactions R_s have been calculated with the use of the familiar rule from the matrix

$$\begin{Vmatrix} a_s & b_s & 0 \\ 0 & 0 & R_s \end{Vmatrix}.$$

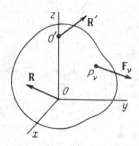

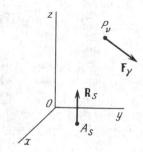

Figure 43   Figure 44

The problem will be statically determinate if the number of points A_s is three or less since we have only three equilibrium equations which involve R_s and from which we can find the reactions R_s.

For the case $k = 3$, we determine the solvability condition for the equations:

$$R_1 + R_2 + R_3 + \Sigma Z_\nu = 0,$$
$$a_1 R_1 + a_2 R_2 + a_3 R_3 - M = 0,$$
$$b_1 R_1 + b_2 R_2 + b_3 R_3 + L = 0.$$

This system of linear equations has a definite solution for R_1, R_2, R_3 if the determinant composes of the coefficients of the unknowns is nonzero:

$$\begin{vmatrix} 1 & 1 & 1 \\ a_1 & a_2 & a_3 \\ b_1 & b_2 & b_3 \end{vmatrix} \neq 0.$$

According to the formula from analytic geometry, the determinant on the left-hand side of the inequality is equal, up to a sign, to double the area of the triangle formed by the points A_1, A_2, A_3. Consequently, the condition for the static determinacy of the problem is that the three points A_1, A_2, A_3 must not lie on the same straight line, i. e. would form a triangle of nonzero area.

3.2 Elements of Graphic Statics

The equations of equilibrium of a rigid body are

$$\Sigma F_\nu = 0, \quad \Sigma m_\nu = 0.$$

When an equilibrium is established, the polygon of forces and the polygon of moments are closed. This geometric property is well studied for a plane case, in the field of graphic statics.

60. The node. Let us consider a light string with a node A which is subjected to an external force (Fig. 45). The left and right ends of the string are fixed. If the node and the string are in equilibrium, then the tensions T and T' act on the node from the left and right ends of the string. We construct a force diagram. To do this, we construct somewhere in the plane a vector $\mathbf{F}$ which is equal and parallel to the force $\mathbf{F}$ acting on the node. Through the terminus of this vector we draw a line parallel to the right end of the string and through the vector's origin we draw a line parallel to the left end of the string. The lines meet at a point O called a *pole*. When the node A is in equilibrium the triangle of forces $\mathbf{F}$, $\mathbf{T}$, $\mathbf{T}'$ must be closed by virtue of the equilibrium equations, and therefore in the force diagram, the sides of the constructed triangle yield $\mathbf{F}$, $\mathbf{T}'$, $\mathbf{T}$ in magnitude

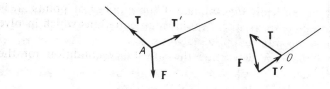

Figure 45

and direction. The sense of traversal in the triangle is defined by the force **F**.

61. A funicular polygon. Let, for the sake of simplicity, three forces $\mathbf{F}_1$, $\mathbf{F}_2$, $\mathbf{F}_3$ be given in a plane (Fig. 46). The problem is to construct a funicular polygon whose nodes are acted upon by the specified forces. The polygon must be such that when its ends are fixed, it is in equilibrium due to the action of the forces.

With this aim in view, we construct, somewhere in the plane, a vector $\mathbf{F}_1$, apply a vector $\mathbf{F}_2$ to its terminus, and apply a vector $\mathbf{F}_3$ to the terminus of $\mathbf{F}_2$. We take a point O as the pole of the force diagram and connect O with the ends of the constructed vectors. We thus obtain rays which we label $1'$, $2'$, $3'$, $4'$ in the order along the line $\mathbf{F}_1$, $\mathbf{F}_2$, $\mathbf{F}_3$. From a point A_1 lying on the force $\mathbf{F}_1$ we draw rays 1 and 2 parallel to the like rays $1'$, $2'$ of the force diagram. The ray 2 will cut the line of action of the force $\mathbf{F}_2$ and define a point A_2. From this point we draw a straight line 3 parallel to the ray $3'$ of the diagram to the point A_3 of intersection with the line of action of the force $\mathbf{F}_3$. From the point A_3 we draw a straight line 4 parallel to the ray $4'$ of the diagram.

The polygon $A_1 A_2 A_3$ is a funicular polygon which is in equilibrium, when the ends 1 and 4 are fixed, under the action of the forces $\mathbf{F}_1$, $\mathbf{F}_2$, $\mathbf{F}_3$ applied at the nodes A_1, A_2, A_3 respectively. Indeed, the node A_1 in the force diagram is associated with a triangle with sides $1'$ and $2'$ subtended by the vector $\mathbf{F}_1$. Traversing this triangle in the direction of the force $\mathbf{F}_1$, we obtain, from the magnitude and direction of the sides of the triangle of forces, the magnitudes and directions of the tensions $\mathbf{T}_1$ and $\mathbf{T}_2$ which

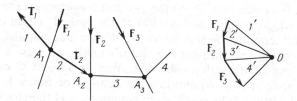

Figure 46

act on the node A_1 in direction of the sides *1* and *2* of the funicular polygon. The node A_2 in the diagram is associated with a triangle in which the rays drawn to the pole O are subtended by the vector $\mathbf{F}_2$, and so on. The triangles of forces in the diagram, which correspond to the nodes of the funicular polygon, are not depicted separately but together, one triangle adjoining another.

The funicular polygon is a special solution, i. e. we could have obtained a funicular polygon of some other kind if we chose another pole O for the force diagram and another initial node A_1 on the direction of the force $\mathbf{F}_1$.

62. Composition of forces. A funicular polygon is not interesting by itself but by the possibility of solving some important problems in statics, e.g. a problem on composition of forces, or on the moment existing with respect to a given point.

Assume that we have to add two forces $\mathbf{F}_1$ and $\mathbf{F}_2$ given in a plane (Fig. 47). To do this, we first construct a force diagram, in other words, a polygon of vectors of forces $\mathbf{F}_1$ and $\mathbf{F}_2$, choose a pole O of the diagram and connect the ends of the vectors $\mathbf{F}_1$ and $\mathbf{F}_2$ with the pole O by segments *1 ′*, *2 ′*, *3 ′* respectively. The vector $\mathbf{F}$, which connects the origin of the vector $\mathbf{F}_1$ with the terminus of the vector $\mathbf{F}_2$, is equal in magnitude and direction to the sum of the vectors $\mathbf{F}_1$ and $\mathbf{F}_2$:

$$\mathbf{F}_1 + \mathbf{F}_2 = \mathbf{F}.$$

It remains for us to find the point of application of the sum of the forces. We construct, as just explained, a funicular polygon $A_1 A_2$ with sides *1*, *2*, *3*. We extend the extreme sides *1* and *3* of the funicualr polygon until they meet at S. We shall prove that the resultant of the forces $\mathbf{F}_1$ and $\mathbf{F}_2$ passes through the point S. We know that the resultant force is equal and parallel to the vector $\mathbf{F}$ of the diagram. The funicular polygon we have

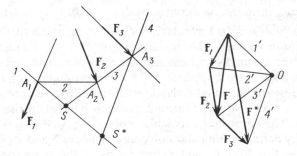

Figure 47

constructed is in equilibrium due to the forces F_1, F_2 acting at the nodes A_1 and A_2, respectively, when the ends are fixed. The tensions in the strings *1, 2, 3* are defined by the rays *1'*, *2'*, *3'* of the diagram. The directions of the tensions acting on the node, say, A_1, can be found by traversing the triangle of forces subtended by the vector F_1. We resolve the force F_1 into forces directed along the sides *1* and *2* of the funicular polygon $F_1 = F_{11} + F_{12}$. The component F_{11} (according to the triangle of forces subtended by the vector F_1) along the side *1* of the funicular polygon is equal and parallel to the side *1'* of the force diagram and is directed from the origin of the vector F_1 to the pole. The component F_{12} of the force F_1 along the side *2* is equal and parallel to the side *2'* of the diagram and is directed from the pole O to the terminus of the vector F_1. We resolve the force F_2 into components directed along the sides *2, 3* of the funicular polygon:

$$F_2 = F_{22} + F_{23}.$$

The component F_{22} is equal and parallel to the side *2'* of the force diagram and is directed from the origin of the vector F_2 to the pole O, and the component F_{23} is equal and parallel to the side *3'* of the diagram and directed from the pole to the terminus of the vector F_2. The forces F_{12} and F_{22} lie on the side *2* of the funicular polygon, are equal in magnitude and opposite in direction. Therefore they cancel out. The forces F_{11} and F_{23} meet at the point S. We can displace them, assume them to be applied at the point of intersection S and add them according to the parallelogram rule, directly using the force diagram:

$$F_{11} + F_{23} = F.$$

Thus the resultant of the forces F_1 and F_2 passes through the point S where the extreme sides of the funicular polygon intersect, its magnitude and direction can be determined from the force diagram with the use of the vector F which closes the triangle of forces F_1 and F_2.

We add one more force F_3 and complete the construction of the previous funicular polygon with a funicular polygon for a system of three forces. With this aim in view, we draw a vector F_3 in the force diagram from the terminus of the vector F_2 and connect the terminus of the vector F_3 with the previous pole O by a segment *4'*. From the point A_3 of intersection of the side *3* of the funicular polygon with the line of action of the force F_3 we draw a side *4* of the funicular polygon parallel to the side *4'* of the diagram. The constructed funicular polygon A_1, A_2, A_3 with sides *1, 2, 3, 4* directly defines a funicular polygon with nodes S and A_3 constructed for the forces $F = F_1 + F_2$ and F_3 by means of the same force diagram *1'*, *3'*, *4'* constructed on the vectors F and F_3 with the same pole O.

The sum of the forces F_1, F_2, F_3 is equal to the sum of two forces, namely, the force $F = F_1 + F_2$, applied at the point S, and the force F_3. Accordingly, the sum of the forces F and F_3 is equal to the force

$$F^* = F + F_3 = F_1 + F_2 + F_3,$$

which is equal and parallel to the vector F^* of the force diagram which connects the origin of the vector F_1 with the terminus of the vector F_3 in the polygon and which is applied at the point S^* of intersection of the extreme sides *1* and *4* of the funicular polygon.

If we have n forces F_1, ..., F_n, then, adding more forces, one by one, and repeating the previous construction, we infer that the sum of the forces F_1, ..., F_n is equal and parallel to the vector

$$F = F_1 + ... + F_n,$$

which is constructed by the polygon rule and applied at the point of intersection of the extreme sides of the funicular polygon constructed for the given system of forces F_1, ..., F_n.

For the system of forces F_1, ..., F_n, which have a resultant force F, the extreme sides of all funicular polygons constructed on these forces meet at points which lie on the same straight line, on the line of action of the resultant F.

We use a funicular polygon to construct the sum of forces when, for instance, we have to find the centre of gravity of a compound body being given the centres of gravity of its component parts.

63. The moment. Assume that we have to find the moment of the force F about a point C (Fig. 48). We construct a vector F, which is equal and parallel to the force F and of the same direction. We choose a pole O and connect the ends of the vector F with the pole O by segments *1′* and *2′*. We draw a straight line *1* parallel to the side *1′* to the point A, where it cuts the force F, and a straight line *2* from the point A parallel to the side *2′*. Through the point C we draw a straight line parallel to the force F. On this line the sides *1* and *2* of the funicular polygon intercept a segment y. From the similarity of two triangles, (A, y) and (O, F), we have

$$\frac{F}{H} = \frac{y}{h}.$$

Hence the magnitude of the moment of force F about the point C is

$$M = Fh = Hy.$$

The moment is positive if the segment y is directed upwards, i. e. from the intersection with the side *1* to the intersection with the side *2*.

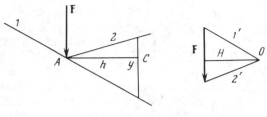

Figure 48

Since the expression for the moment includes the length of the perpendicular drawn from the pole O of the force diagram to the vector $\mathbf{F}$, it is easier to solve the problem involving moments of forces for a system of parallel forces which have a common perpendicular H. In engineering, this may occur when gravity is an external force. For instance, suppose that we have two parallel forces $\mathbf{F}_1$ and $\mathbf{F}_2$ and have to find the moment of these forces about the point C (Fig. 49). We first construct a funicular polygon. We draw a line through the point C parallel to the acting forces and use y_{ik} to designate the segment of the vertical between the sides i and k of the funicular polygon directed from the side i to the side k. From what we proved above, the moment of the force $\mathbf{F}_1$ about C is Hy_{12} and the moment of the force $\mathbf{F}_2$ about C is Hy_{23}. The resultant moment is equal to the product of H by the segment y_{13} between the extreme sides of the funicular polygon:

$$M = H(y_{12} + y_{23}) = Hy_{13}.$$

64. If a system of parallel forces $\mathbf{F}_1$, ..., $\mathbf{F}_n$ is equivalent to zero, then the corresponding force diagram and the funicular polygon are closed. It is obvious that the polygon of forces of the diagram is closed since the

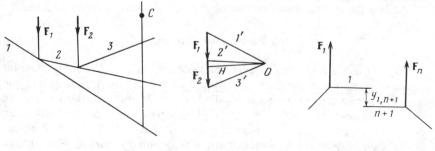

Figure 49   Figure 50

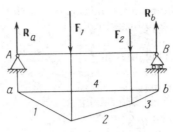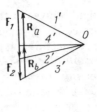

Figure 51

resultant vector $\mathbf{F} = \mathbf{F}_1 + \ldots + \mathbf{F}_n = 0$. We can prove that the funicular polygon for the balanced parallel forces $\mathbf{F}_1, \ldots, \mathbf{F}_n$ is closed as follows (Fig. 50). The resultant moment M of forces $\mathbf{F}_1, \ldots, \mathbf{F}_n$ about the point C is $M = Hy_{1, n+1}$. For balanced forces the moment M is zero, whence it follows that $y_{1, n+1} = 0$, in other words, the sides 1 and $n + 1$ of the funicular polygon coincide, and so on.

Problem. Assume that two vertical forces $\mathbf{F}_1$ and $\mathbf{F}_2$ act on a horizontal beam AB (Fig. 51). We have to find the reactions at the supports $\mathbf{R}_a$, $\mathbf{R}_b$. We construct a funicular polygon. On the verticals passing through the supports A and B we mark nodes a and b of the funicular polygon. Since in the equilibrium position of the beam the acting forces $\mathbf{F}_1$ and $\mathbf{F}_2$ and the unknown reactions at the supports $\mathbf{R}_a$, $\mathbf{R}_b$ must be balanced, the funicular polygon constructed on these forces must be closed and, consequently, its side 4 must pass through the marked nodes a and b. The side 4 in the force diagram must be associated with side $4'$ which is parallel to it. The side $4'$ defines the reactions $\mathbf{R}_a$ and $\mathbf{R}_b$, the reaction $\mathbf{R}_b$ is equal and parallel to the side of the diagram which lies between $3'$ and $4'$ (the sides 3 and 4 of the funicular polygon meet at the node b), the reaction $\mathbf{R}_a$ is equal and parallel to the side of the diagram which is subtended by the segments $4'$ and $1'$. Let H be the altitude of the force diagram (it is not shown in the figure). We draw a vertical through the section of the beam. The sides 4, 2 of the funicular polygon intercept a segment y_{24}. We have to find Hy_{24}. The sides 4, 1, 2 of the funicular polygon are conjugates of the forces $\mathbf{R}_a$, $\mathbf{F}_1$, the sides 4 and 2 being the extreme sides of the funicular polygon constructed on the forces $\mathbf{R}_a$, $\mathbf{F}_1$, and therefore Hy_{24} is equal to the moment of these forces about the section C, thus being callel the bending moment.

What is the greatest bending moment?

3.3 Balance of Trusses

65. A *truss* is a system of bars connected by hinges. We shall consider the case when there are only two hinges to a bar. The points where the bars are connected by hinges are called *joints*. If the external forces, the resistance forces included, are only applied at the joints of the truss, then we speak of all the bars as *members* of the truss. If the external forces are applied both at the hinges and at points along the bars, then we speak of the *beams* of the truss.

Since a rigid body is in equilibrium under the action of two forces when the forces are equal in magnitude, opposite in direction, and act along a straight line connecting the points of action, a *member of a truss can only be under tension or compression*. This does not apply to beams. We shall only consider *flat trusses*. A truss is *statically determinate* (a truss without extra members) if none of its bars may be removed without violating its rigidity (the property of its geometrical invariability). If there are n hinges in a statically determinate truss, then it consists of $2n - 3$ bars. Indeed, one bar has two hinges, the addition of a joint yields two bars, and then the truss has $1 + 2(n - 2) = 2n - 3$ bars (Fig. 52).

There are several methods of determining the internal forces in the members of a truss.

66. Ritter's method (method of sections) is the method of three moments and has its simplest application to a truss which can be cut by a contour intersecting only three members.

Assume that the truss shown in Fig. 53 is in equilibrium under the action of the indicated forces applied at the joints. We cut the truss, as indicated, along three members x, y, z into two parts (we assume the members x, y, z to be stretched; on this assumption, we assign the plus sign to the forces X, Y, Z acting along the member). The members y and z meet at a point R_1, the members z and x meet at a point R_2 on their extension, the members x and y meet at a point R_3.

In the section in question the point R_1, R_2, R_3 are Ritter's points about which we consider the moments. When the right-hand part of the truss is removed, the left-hand part is in equilibrium under the action of the forces $\mathbf{N}, \mathbf{P}_1, \mathbf{P}_2$ which act on the joints of the left-hand part of the truss and the internal forces $\mathbf{X}, \mathbf{Y}, \mathbf{Z}$ which act along the members x, y, z. The equation of the moments of these forces about the point R_1 is

$$m_{R_1}(\mathbf{X}) + m_{R_1}(\mathbf{N}) + m_{R_1}(\mathbf{P}_1) = 0,$$

but

$$m_{R_1}(\mathbf{X}) = -X\delta,$$

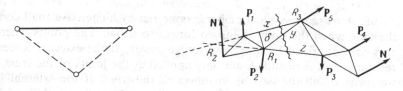

Figure 52   Figure 53

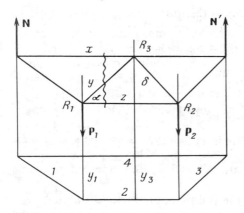

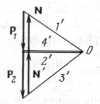

Figure 54

where δ is the length of the perpendicular dropped from the point R_1 onto the member x. Hence

$$X = \frac{m_{R_1}(\mathbf{N}) + m_{R_1}(\mathbf{P}_1)}{\delta}.$$

The equations of the moments for the points R_2 and R_3 define the internal forces $\mathbf{Y}$ and $\mathbf{Z}$ respectively, which act along the members y, z.

67. Culmann suggested a graphical construction of the solution. For a truss shown in Fig. 54, which is under the action of parallel forces $\mathbf{P}_\alpha$, we construct a funicular polygon and find the reactions $\mathbf{N}$ and $\mathbf{N}'$. We consider a section along three members x, y, z. Assume that we are interested in the internal force $\mathbf{Z}$ in the member z. Accordingly, we consider a point R_3. A vertical straight line parallel to the forces acting on the joints of the truss intercepts a segment y between the sides 2 and 4 of the funicular polygon. The equation of the moments about the points R_3 is

$$Z\delta - Hy = 0,$$

where δ is the distance between the top and the bottom chord of the truss and H is the altitude of the force diagram. Hence $Z = Hy / \delta$ is positive and, consequently, the member z is in tension. The equation of the moments about the point R_1 is

$$X\delta + Hy = 0,$$

or $X = -Hy / \delta.$ Consequently, the member x is in compression.

The point R_2 (see Fig. 53) recedes to infinity since the members x, z are parallel, and therefore, we can regard the following equation relating to the projections on the vertical axis as the third equation of equilibrium, which defines the internal force Y acting on the member y:

$$Y \sin \alpha + (N - P_1) = 0.$$

From this we have $Y = (P_1 - N) / \sin \alpha$.

3.4 Centre of Mass

68. The centre of mass of a system is a point $G(\xi, \eta, \zeta)$ with coordinates

$$\xi = \frac{\sum m_k x_k}{M}, \quad \eta = \frac{\sum m_k y_k}{M}, \quad \zeta = \frac{\sum m_k z_k}{M}, \quad M = \sum m_k,$$

where the summation is over all the points of the system.

We divide the system into two parts: I is m_i and II is m_j. Then $M_1 = \sum m_i$, $M_2 = \sum m_j$, $M = M_1 + M_2$. If $G_1(\xi_1, \eta_1, \zeta_1)$ and $G_2(\xi_2, \eta_2, \zeta_2)$ are the centres of mass of parts I and II respectively, then

$$\sum m_i x_i = \xi_1 M_1, \quad \sum m_j x_j = \xi_2 M_2,$$
$$\sum m_i y_i = \eta_1 M_1, \quad \sum m_j y_j = \eta_2 M_2,$$
$$\sum m_i z_i = \zeta_1 M_1, \quad \sum m_j z_j = \zeta_2 M_2,$$

$$\xi = \frac{\sum m_i x_i + \sum m_j x_j}{M_1 + M_2} = \frac{\xi_1 M_1 + \xi_2 M_2}{M_1 + M_2}, \quad \eta = \frac{\eta_1 M_1 + \eta_2 M_2}{M_1 + M_2},$$
$$\zeta = \frac{\zeta_1 M_1 + \zeta_2 M_2}{M_1 + M_2}.$$

The centre of mass of a body is the centre of mass of the centres of mass of its parts.

Centre of mass of a triangle. The centres of mass of homogeneous strips (Fig. 55), which are parallel to the base of the triangle, lie on a median. Their net centre of mass lies on a median, and the centre of mass of the triangle lies on the intersection O of the medians;

$$\triangle AOB \sim \triangle aOb, \ \frac{AO}{Oa} = \frac{AB}{ba}; \ \triangle ABC \sim \triangle baC, \ \frac{AB}{ba} = \frac{BC}{aC} = \frac{2}{1};$$

$$\frac{AO}{Oa} = \frac{2}{1}, \ AO = 2 \cdot Oa,$$

$$Aa = AO + Oa = 2 \cdot Oa + Oa = 3 \cdot Oa, \ Oa = \frac{1}{3} Aa.$$

($\sim$ means "is similar to").

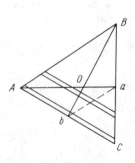
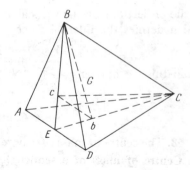

Figure 55

Figure 56

Centre of mass of a pyramid (Fig. 56).

$$\triangle\, Gcb \sim \triangle\, GCB, \quad \frac{BG}{Gb} = \frac{BC}{cb} = \frac{BE}{cE} = \frac{3}{1}, \quad Bb = 4Gb.$$

69. Pappus' (Guldin's) theorems. Solid bodies (Fig. 57).

$$\xi = \frac{\int x\, dm}{\int dm}, \quad \eta = \frac{\int y\, dm}{\int dm}, \quad \zeta = \frac{\int z\, dm}{\int dm}.$$

1. $dA = ds \cdot 2\pi y, \quad A = 2\pi \int y\, ds = 2\pi \eta l.$

The lateral surface of a solid of revolution is equal to the product of the length of the arc of the circle described by the centre of mass, $2\pi n$, by the length of the curve l.

Centre of mass of the arc of a circle (Fig. 58).

$$x = R\cos\alpha, \quad \xi = \frac{\displaystyle\int_{-\alpha}^{\alpha} R\cos\alpha R\, d\alpha}{R\cdot 2\alpha} = \frac{R^2 \sin\alpha\Big|_{-\alpha}^{\alpha}}{2R\alpha} = R\,\frac{\sin\alpha}{\alpha}.$$

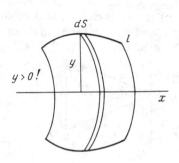
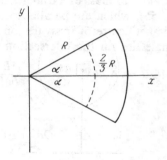

Figure 57

Figure 58

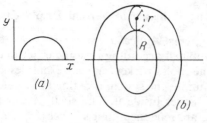

Figure 59

Centre of mass of a semicircle (Fig. 59a).

$$4\pi R^2 = 2\pi\eta\cdot\pi R, \ \eta = \frac{2}{\pi} R.$$

Lateral surface of a torus (Fig. 59b).

$$S_{\text{lat}} = 2\pi\eta\cdot 2\pi r = 2\pi R\cdot 2\pi r = 4\pi^2 Rr.$$

2. $dU = d\sigma\cdot 2\pi y$, $U = 2\pi\int y\,d\sigma = 2\pi\eta S.$

The volume of a solid of revolution is equal to the product of the length of the arc of the circle $2\pi\eta$ described by the centre of mass by the area S (Fig. 60).

Centre of mass of a semicircular plate.

$$\frac{4}{3}\pi R^3 = 2\pi\eta \cdot \frac{\pi R^2}{2}, \ \eta = \frac{4}{3\pi} R.$$

Centre of mass of a circular sector (Fig. 58).

$$\xi = \frac{2}{3} R \frac{\sin\alpha}{\alpha}.$$

Centre of mass of a circular segment (Fig. 61).

$$\xi = \frac{2}{3} R \frac{\sin^3\alpha}{\alpha - \sin\alpha\cos\alpha}.$$

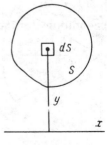

Figure 60

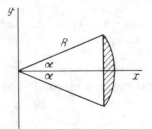

Figure 61

3.5. The Principle of Virtual Displacements

70. Mechanical systems. *A mechanical system* is a set of particles isolated for study and combined according to some criterion. The solar system, mechanisms, machine-tools, rockets are all examples of mechanical systems. In the latter example, the system is defined by a reference limit surface which contains the masses inside it. The shell of a rocket and the plane of its exhaust nozzle are reference limit surfaces. When a rocket is flying, gas is exhaused through the nozzle; in a way, the system loses part of its mass. This is an example of a system with variable mass.

71. Constraints. The particles of a system may have special *constraints,* i. e. restrictions which prevent it from moving freely.

Example 1. A free particle can move along the x, y, and z axes, there are no restrictions to its motion.

Example 2. A particle must remain on a sphere

$$x^2 + y^2 + z^2 = R^2;$$

its motion is constrained, for it may only move over the surface of the sphere and may not move along its radius.

Example 3. A particle must remain inside a sphere:

$$x^2 + y^2 + z^2 \leqslant l^2.$$

This is an example of inoperative unilateral constraints. In problems on motion and equilibrium, when a body is subjected to unilateral constraints, we always seek displacements which would satisfy the equalities at the moment being considered since if the conditions turn to inequalities, the constraints which restrict the motion of the particles become inoperative.

Example 4. A *rigid body* consists of n particles ($i = 1, ..., n$) the distances between which do not vary:

$$(x_1 - x_2)^2 + (y_1 - y_2)^2 + (z_1 - z_2)^2 = r_{12}^2 ,$$
$$(x_2 - x_3)^2 + (y_2 - y_3)^2 + (z_2 - z_3)^2 = r_{23}^2 ,$$
$$(x_3 - x_1)^2 + (y_3 - y_1)^2 + (z_3 - z_1)^2 = r_{31}^2 .$$

For the other $n - 3$ particles we have $3 (n - 3)$ conditions for the constancy of the lengths:

$$r_{i1} , \ r_{i2} , \ r_{i3} \quad (i = 4, ..., n).$$

The total number of conditions is

$$3 (n - 3) + 3 = 3n - 6,$$

i. e. the body has six virtual displacements.

Example 5. *The general case:* the constraints imposed on the motion of the particles of a system are defined by the independent equations

$$f_j(x_1, y_1, z_1, ..., x_n, y_n, z_n) = 0 \quad (j = 1, ..., s).$$

If $s = 3n$, then these equations define $3n$ coordinates and the system may have one or several definite positions and will not be displaced. For a system to be able to move, it is necessary that $k = 3n - s > 0$, where k is the number of degrees of freedom.

72. Lagrange's method of defining constraints:

$$x_i = x_i(q_1, ..., q_k), \quad y_i = y_i(q_1, ..., q_k),$$
$$z_i = z_i(q_1, ..., q_k) \quad (i = 1, ..., n),$$

where $q_1, ..., q_k$ are independent parameters, or Lagrange's coordinates. Their elimination must always yield the constraints equations

$$f_j = 0 \quad (j = 1, ..., 3n - k = s).$$

73. Virtual displacements. Virtual work of a force. Let M be a particle subjected to the action of a force **F**. We assume that the particle is imparted an infinitesimal displacement $\overrightarrow{MM'}$ compatible with the constraints. This displacement is known as a *virtual displacement* as distinct from a *true displacement* which the particle actually undergoes due to the action of the applied forces.

The *elementary work* done by the force **F** during a virtual displacement $\overrightarrow{MM'}$ is the quantity

$$\mathbf{F} \cdot \overrightarrow{MM'} = F \cdot MM' \cos(\mathbf{F} \cdot \overrightarrow{MM'}).$$

The virtual displacement $\overrightarrow{MM'}$ is δs and the projections of the vector δs are δx, δy, δz, where x, y, z are the coordinates of the particle M. The infinitesimals denoted by δ will be called variations to distinguish them from differentials. If we designate the projections of the force **F** as X, Y, Z, then the analytic expression for the work done by the force **F** during a virtual displacement δs is

$$X\delta x + Y\delta y + Z\delta z.$$

We assume that the virtual displacement δs takes time δt, and the vector

$$\mathbf{v} = \frac{\delta s}{\delta t}$$

is the virtual velocity of the particle M. The definition of the elementary work done by the force **F** during the virtual displacement $\overrightarrow{MM'}$ must include the force **F** applied to the particle M and the virtual displacement $\overrightarrow{MM'}$. In this case we do not assume anything concerning the effect of the force on the particle after the particle has been displaced.

The notion of the work done by a force without the definition of a displacement, even if we consider a true displacement, is ambiguous in many cases.

Examples. A heavy body is subjected to a true translation $\overrightarrow{MM'}$ leaving a table (Fig. 62). What is the true work of the force of reaction **R** if the reaction disappears as soon as the body M leaves the table?

The body shown in Fig. 63a slides with velocity **v** on a fixed point M. What is the true work of the reaction **R** if the point of application of the reaction remains at the same point M of a fixed space and the particle of the body (which was at M) is displaced, but after the displacement the reaction force is no longer applied to it? (If the body is upturned (Fig. 63b) everything is clear.)

A cylinder rolls without sliding over a rough surface (Fig. 64). What is the true work of the reaction **R** if the true velocity of the point of tangency M of the cylinder is zero?

If the virtual work of the force X, Y, Z for δx, δy, δz is the variation of a function $U(x, y, z)$, i. e.

$$X\delta x + Y\delta y + Z\delta z = \delta U(x, y, z),$$

then $U(x, y, z)$ is a *force function*. In this case

$$X = \frac{\partial U}{\partial x}, \quad Y = \frac{\partial U}{\partial y}, \quad Z = \frac{\partial U}{\partial z}.$$

If the force function U exists, then

$$\frac{\partial Y}{\partial x} = \frac{\partial X}{\partial y}, \quad \frac{\partial Z}{\partial y} = \frac{\partial Y}{\partial z}, \quad \frac{\partial X}{\partial z} = \frac{\partial Z}{\partial x},$$

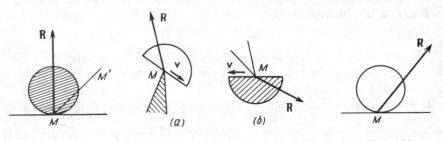

Figure 62   Figure 63   Figure 64

and vice versa. The surfaces $U = $ const are known as *level surfaces*. The force **F** acts along the normal to the level surface in the direction of increase of the force function U.

74. Principle of virtual work. We consider a system of particles x_ν, y_ν, z_ν ($\nu = 1, ..., n$) subject to the action of the forces X_ν, Y_ν, Z_ν. We assume that the particles are constrained. Let δx_ν, δy_ν, δz_ν be virtual displacements of the νth particle.

The **axiom of constraints** states that the action of the constraints imposed on a mechanical system can be replaced by *reaction forces* $R_{\nu x}$, $R_{\nu y}$, $R_{\nu z}$ ($\nu = 1, ..., n$).

As a consequence, we get the proposition that if we add the reaction forces $\mathbf{R}_\nu$ ($\nu = 1, ..., n$) to the given forces acting on a mechanical system, then we can assume the system to be free of constraints which cause reactions $\mathbf{R}_\nu$.

In what follows we shall deal with *ideal* constraints for which we shall accept the following definition: *the virtual work done by the reaction forces* $\mathbf{R}_\nu$ ($\nu = 1, ..., n$) *of ideal constraints is zero:*

$$\sum_\nu (R_{\nu x} \delta x_\nu + R_{\nu y} \delta y_\nu + R_{\nu z} \delta z_\nu) = 0. \tag{3.1}$$

The definition of ideal bilateral constraints is a generalization of the familiar facts from physics. Constraints of this kind do not dissipate energy during virtual displacements. It is easy to derive from this the main principle of statics for systems with ideal bilateral constraints independent of time. Indeed, we complement the given forces X_ν, Y_ν, Z_ν by all the reaction forces $R_{\nu x}$, $R_{\nu y}$, $R_{\nu z}$, and then, according to the axiom of constraints, we can assume our mechanical system to be a system of free particles which are subjected only to the action of forces $X_\nu + R_{\nu x}$, $Y_\nu + R_{\nu y}$, $Z_\nu + R_{\nu z}$. We have the following equilibrium equations for particles without constraints:

$$X_\nu + R_{\nu x} = 0, \quad Y_\nu + R_{\nu y} = 0, \quad Z_\nu + R_{\nu z} = 0, \quad (\nu = 1, ..., n). \tag{3.2}$$

From these equations we can find the required reaction forces $R_{\nu x}$, $R_{\nu y}$, $R_{\nu z}$ ($\nu = 1, ..., n$) and substitute them into the expression for the work of the ideal constraint forces. Then we get a relation

$$\sum_\nu (X_\nu \delta x_\nu + Y_\nu \delta y_\nu + Z_\nu \delta z_\nu) = 0, \tag{3.3}$$

which is valid *for arbitrary virtual displacements* δx, δy, δz *in the equilibrim position of the mechanical system.*

We have proved the necessity of relation (3.3) for equilibrium. Now we shall prove the sufficiency of (3.3), in other words, we shall prove that a mechanical system at rest which is put in a position satisfying relation (3.3) will not move.

Assume that in a certain position of the mechanical system M_ν relation (3.3) is satisfied. We must show that the system is in equilibrium in this position, i. e. the (principal) resultant of the forces F_ν and R_ν, acting on to the particle M_ν, is zero for every particle ($\nu = 1, ..., n$).

We assume the contrary, i. e. that at least one particle M_ν at rest in this position will move. In this case, the particle M_ν can move in accord with the constraints in the direction of the resultant of the forces F_ν and the constraint forces R_ν if M_ν is at rest. The work of these forces $F_\nu + R_\nu$ done during true elementary displacement of the particle $M_\nu (dx_\nu, dy_\nu, dz_\nu)$ is positive if the particle M_ν actually moves (the same is true of every particle of the system). Consequently, the total work done by the given forces F_ν and the reaction forces R_ν during the true displacement must be positive if the system moves from the state of rest. In fact, the motions of the particles of a mechanical system must in this case be in accord with the constraints imposed on the system. This means that the displacements dx_ν, dy_ν, dz_θ due to these forces must be some of the virtual displacements δx_ν, δy_ν, δz_ν. Then

$$\sum [(X_\nu + R_{\nu x}) dx_\nu + (Y_\nu + R_{\nu y}) dy_\nu + (Z_\nu + R_{\nu z}) dz_\nu] > 0,$$

since for $v_\nu (t) = 0$ we have

$$j_\nu = \lim_{\Delta t \to 0} \frac{v_\nu (t + \Delta t) - v_\nu (t)}{\Delta t} = \frac{F_\nu + R_\nu}{m},$$

and consequently,

$$(F_\nu + R_\nu, v_\nu (t + dt) dt) = \frac{dt^2}{m} (F_\nu + R_\nu)^2.$$

But by definition we have the following relation for ideal constraints

$$\sum (R_{\nu x} dx_\nu + R_{\nu y} dy_\nu + R_{\nu z} dz_\nu) = 0,$$

and, consequently,

$$\sum (X_\nu dx_\nu + Y_\nu dy_\nu + Z_\nu dz_\nu) > 0,$$

and this contradicts the assumption that relation (3.3) holds true for all virtual displacements.

Thus relation (3.3) is both necessary and sufficient for equilibrium. Therefore relation (3.3) is a principle. This principle is known as the **principle of virtual work** and was established by Johann Bernoulli.[*]

---

[*] For a system to be in equilibrium it is necessary and sufficient that the sum of the elementary works of the forces be zero for every virtual displacement of the system. — *Ed.*

The principle of virtual work yields a general method of solving statics problems. It does not include reactions and therefore it allows us to solve statics problems without calculating the unknown reactions of constraints imposed on the mechanical system.

Example 1. A tackle block (Fig. 65), which is an ancient tool for hoisting heavy objects, consists of two blocks in each of which three pulleys can independently rotate about horizontal axes. The upper block of pulleys is fastened to the upper immovable support and a load Q is suspended from the lower block. A rope is passed over the pulleys in succession. We apply a force P to one end of the rope when we want to hoist a load, the other end of the rope being fastened to the upper block. Find the relation between the forces P and Q in equilibrium.

From the point of view of analytical statics, a tackle block is a mechanical system consisting of two points to which the forces P and Q are applied. All the rest concerns the realization of the constraints imposed on the two points. In the equilibrium position the principle of virtual work yields

$$P\delta p + Q\delta q = 0,$$

where δp, δq denote the virtual displacements of the particles to which the forces P and Q are applied, respectively. We assume the direction along the action of the corresponding force to be positive. If a load Q is hoisted a distance $-\delta q$, then the horizontal mark AB, connected with the block of the lower pulleys, is raised to the position $A'B'$ and, hence, six segments of the rope between AB and $A'B'$ must be pulled from the tackle block, i. e.

$$\delta p = -6\delta q.$$

Substituting the value of δp into the formula for the principle of virtual work, we get $P = Q/6$.

Example 2. The principal part of Weston's differential hoist consists of two connected pulleys A whose axle is suspended from an immovable hook (Fig. 66). Their chutes have teeth which grip onto a continuous chain. The chain forms two loops inside one of which is a running pulley B. A load Q to be lifted is suspended from the running pulley and a moving force P is applied to the free loop hanging from the large pulley. The radii of the pulleys A are R and r respectively, and $r < R$. We neglect friction and seek the relationship between the force P and the load Q being held. The system consists of two points to which the forces P and Q are applied. All the rest concerns the realization of the constraints (ideal by hypothesis) imposed on the points. The principle of virtual work yields

$$P\delta p + Q\delta q = 0,$$

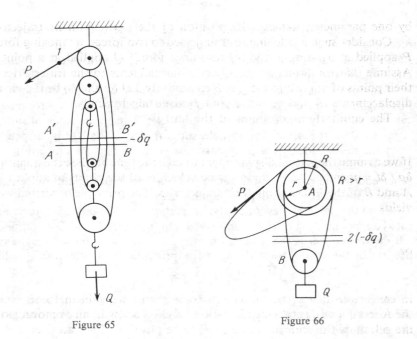

Figure 65   Figure 66

where δp and δq are the algebraic values of the virtual displacements of the points of application of the respective forces P and Q. The displacements are assumed to be positive if they are in the direction of their forces and negative otherwise. When we lift the load a distance $-\delta q$, a length $-2\delta q$ of the chain must be pulled out of the loaded loop. But when the chain is pulled out by δp, the length of the free loop increases by δp and decreases by $r\,\dfrac{\delta p}{R}$ due to the backward motion of the smaller pulley. Hence

$$\delta p - \frac{r}{R}\,\delta p = -2\delta q.$$

Substituting this expression into the formula for the principle of virtual work, we get

$$P = \frac{1}{2}\left(1 - \frac{r}{R}\right)Q.$$

Example 3. Simple machine-tools are systems which have total constraints in the sense that the position of every part of the tool is completely defined by the position of one particle which may only move along a definite curve. Therefore the position of a simple machine-tool can be defined by one parameter, namely, the position of the particle on its trajectory.

Consider such a machine-tool subjected to two forces, a "moving force" P applied at a point A and a "resistance force" Q applied at a point B. Assume that the forces P and Q act along tangents to the trajectories of their points of application A and B respectively. Let δp and δq be the virtual displacements of the points A and B given algebraically.

The equilibrium condition of the tool is

$$P\delta p + Q\delta q = 0.$$

If we assume the virtual displacement to take the time δt, then the quantities $\delta p\,/\,\delta t = u$ and $\delta q\,/\,\delta t = v$ are virtual velocities of the points of application A and B of the forces along their trajectories. The principle of virtual work yields

$$Pu + Qv = 0,$$

or

$$\frac{P}{Q} = -\frac{v}{u},$$

in other words, we lose in velocity as much as we gain in force.

A lever, a wedge, a winch, a tackle block, a screw in an immovable nut are all simple machines.

Example 4. *Torricelli's law:* in a system of heavy bodies, which are in equilibrium, the centre of mass occupies the lowest position possible.

We shall prove that the centre of mass of a system of heavy bodies which is in equilibrium occupies an extremal position and postpone the proof of the second part of Torricelli's law until we give Lagrange's theorem on the stability of equilibrium when the force function is at a maximum. The z-axis is directed vertically upwards. For the equilibrium of a system of heavy bodies with masses m_ν and coordinates x_ν, y_ν, z_ν, the principle of virtual work yields

$$-\sum m_\nu g\delta z_\nu = 0,$$

where g is free fall acceleration. If we consider the z-coordinate of the centre of mass

$$\zeta = \frac{\sum m_\nu z_\nu}{\sum m_\nu},$$

then we can rewrite the principle of virtual work in the form

$$\delta\zeta = 0.$$

This relation proves that the centre of mass of heavy bodies in equilibrium (the constraints being ideal) occupies an extremal position.

Example 5. Consider a problem usually solved by elementary statics. A homogeneous bar AB of length $2l$ and weight P can rotate about its end A (Fig. 67). It rests on a homogeneous bar CD of the same length $2l$ which can rotate about its midpoint E. The points A and E lie on the same vertical a distance $AE = l$ apart. A load $Q = 2P$ is suspended from the end D. Neglecting friction, find the angle φ the bar AB makes with the vertical in the equilibrium position. The system consists of two points to which the forces P and Q are applied. The rest concerns the realization of the ideal constraint imposed on the points.

We introduce the coordinate axes Axy as shown in the figure. The principle of virtual work yields a relation

$$P\delta x_M + Q\delta x_D = 0,$$

but

$$x_M = l\cos\varphi, \quad x_D = l + l\cos(\pi - 2\varphi) = l(1 - \cos 2\varphi).$$

Hence

$$\delta x_M = -l\sin\varphi\,\delta\varphi, \quad \delta x_D = 2l\sin 2\varphi\,\delta\varphi.$$

Substituting these values into the formula for the principle of virtual work, we have

$$\sin\varphi(8\cos\varphi - 1) = 0.$$

The equation has two solutions, namely, $\sin\varphi = 0$, $\cos\varphi = 1/8$.

Example 6. A homogeneous bar of length $2l$ and weight P lies in a hemispherical bowl of radius R. Neglecting friction, find the angle φ the bar makes with the horizontal in equilibrium.

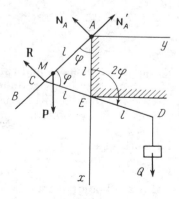

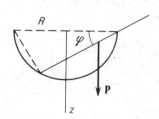

Figure 67   Figure 68

We direct the z-axis vertically downwards (Fig. 68). The coordinate of the centre of gravity

$$z = (2R \cos \varphi - l) \sin \varphi.$$

According to Torricelli's law, we have $\delta z = 0$, whence we obtain the following condition for the angle φ:

$$(2R \cos \varphi - l) \cos \varphi - 2R \sin^2 \varphi = 0.$$

3.6 Constraints and Virtual Displacements

75. Bilateral constraints imposed on the particles x_ν, y_ν, z_ν ($\nu = 1, \ldots, n$) of a mechanical system are expressed by the system of equations

$$f_j(x_\nu, y_\nu, z_\nu) = 0 \quad (j = 1, \ldots, m < 3n),$$

where the f_j are functionally independent.

The virtual infinitesimally close position of a mechanical system defined by the values of the coordinates $x_\nu + \delta x_\nu$, $y_\nu + \delta y_\nu$, $z_\nu + \delta z_\nu$ ($\nu = 1, \ldots, n$) must not disturb the constraints

$$f_j(x_\nu + \delta x_\nu, y_\nu + \delta y_\nu, z_\nu + \delta z_\nu) = 0 \quad (j = 1, \ldots, m)$$

imposed on the system. Expanding the left-hand sides of these equations in Taylor's series and truncating the series at the first-order terms, we have

$$\delta f_j = \sum \left(\frac{\partial f_j}{\partial x_\nu} \delta x_\nu + \frac{\partial f_j}{\partial y_\nu} \delta y_\nu + \frac{\partial f_j}{\partial z_\nu} \delta z_\nu \right) = 0 \quad (j = 1, \ldots, m).$$

The number of independent variables $k = 3n - m$ is known as the *number of degrees of freedom*.

The constrains can be expressed by the system of equations

$$x_\nu = x_\nu(q_1, \ldots, q_k), \ y_\nu = y_\nu(q_1, \ldots, q_k),$$
$$z_\nu = z_\nu(q_1, \ldots, q_k) \quad (\nu = 1, \ldots, n).$$

We assume the variables $q_1, \ldots, q_k$ to be real and independent and they define the position of the system. Variables like q_s are Lagrange's defining (or holonomic) generalized coordinates. Hence we can find the virtual displacements δx_ν, δy_θ, δz_θ for infinitesimal variations of the defining variables δq_s, by varying the relations

$$\delta x_\nu = \sum \frac{\partial x_\nu}{\partial q_s} \delta q_s, \quad \delta y_\nu = \sum \frac{\partial y_\nu}{\partial q_s} \delta q_s,$$

$$\delta z_\nu = \sum \frac{\partial z_\nu}{\partial q_s} \delta q_s \quad (\nu = 1, \ldots, n).$$

Example. The system

$$x = q_1 + l \cos q_2, \quad y = l \sin q_2, \quad z = 0$$

is a system with two degrees of freedom (Fig. 69).

Constraints can be defined directly, in terms of kinematics, and in the form of relations which must be satisfied by displacements which are virtual for the constraints. For instance, according to Euler's theorem, virtual velocities for a rigid body must satisfy the relations

$$\delta x_\nu = [u + q z_\nu - r y_\nu]\, \delta t, \quad \delta y_\nu = [v + r x_\nu - p z_\nu]\, \delta t,$$

$$\delta z_\nu = [w + p y_\nu - q x_\nu]\, \delta t \quad (\nu = 1, ..., n),$$

where u, v, w are the virtual velocities of the origin and p, q, r are the virtual angular velocities of the rotation about the coordinate axes. Or, in vector terms,

$$\delta \mathbf{r}_\nu = (\mathbf{v} + \boldsymbol{\Omega} \times \mathbf{r}_\nu)\, \delta t,$$

where the $\mathbf{r}_\nu$ are the radii vectors of the particles of the rigid body, $\mathbf{v}$ is the virtual velocity vector of the particle which is at the origin at that moment, and $\boldsymbol{\Omega}$ is the vector of the virtual instantaneous angular velocity of the body.

76. Equations of equilibrium of a rigid body without constraints. Assume that forces $\mathbf{F}_\nu$ act on the particles M_ν of a rigid body, which are defined by the radii vectors $\mathbf{r}_\nu$ ($\nu = 1, ..., n$). According to the principle of virtual work, in the equilibrium position of a rigid body we must have

$$\sum (\mathbf{F}_\nu \cdot \delta \mathbf{r}_\nu) = 0.$$

If we substitute the values

$$\delta \mathbf{r}_\nu = (\mathbf{v} + \boldsymbol{\Omega} \times \mathbf{r}_\nu)\, \delta t,$$

we obtain

$$\delta t \sum \mathbf{F}_\nu \cdot \left(\mathbf{v} + \boldsymbol{\Omega} \times \mathbf{r}_\nu \right) = 0,$$

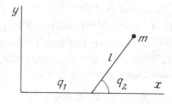

Figure 69

or

$$\mathbf{v} \cdot \sum \mathbf{F}_\nu + \boldsymbol{\Omega} \cdot \sum \mathbf{r}_\nu \times \mathbf{F}_\nu = 0.$$

The translation velocity $\mathbf{v}$ and the velocity of the virtual rotation $\boldsymbol{\Omega}$ are arbitrary. The expression, which is linear with respect to the arbitrary $\mathbf{v}$ and $\boldsymbol{\Omega}$, can only be zero if

$$\sum \mathbf{F}_\nu = 0, \quad \sum \mathbf{r}_\nu \times \mathbf{F}_\nu = 0.$$

These equilibrium equations for a rigid body are the geometrical equilibrium conditions of statics.

77. Lagrange's equations. If the constraints are defined by Lagrange's variables $q_1, ..., q_k$ in the form

$$x_\nu = x_\nu(q_1, ..., q_k), \quad y_\nu = y_\nu(q_1, ..., q_k),$$
$$z_\nu = z_\nu(q_1, ..., q_k) \quad (\nu = 1, ..., n),$$

then we can obtain the equilibrium equations for these systems from the principle of virtual work

$$\sum (X_\nu \delta x_\nu + Y_\nu \delta y_\nu + Z_\nu \delta z_\nu) = 0$$

by substituting into it the explicit values for the virtual displacements

$$\delta x_\nu = \sum \frac{\partial x_\nu}{\partial q_s} \delta q_s, \quad \delta y_\nu = \sum \frac{\partial y_\nu}{\partial q_s} \delta q_s,$$
$$\delta z_\nu = \sum \frac{\partial z_\nu}{\partial q_s} \delta q_s \quad (\nu = 1, ..., n).$$

We have

$$\sum \delta q_s \sum \left(X_\nu \frac{\partial x_\nu}{\partial q_s} + Y_\nu \frac{\partial y_\nu}{\partial q_s} + Z_\nu \frac{\partial z_\nu}{\partial q_s} \right) = 0.$$

Hence, if we introduce generalized forces

$$Q_s = \sum \left(X_\nu \frac{\partial x_\nu}{\partial q_s} + Y_\nu \frac{\partial y_\nu}{\partial q_s} + Z_\nu \frac{\partial z_\nu}{\partial q_s} \right),$$

the principle of virtual work reduces to the form

$$\sum Q_s \delta q_s = 0.$$

If the variables q_s are such that δq_s $(s = 1, ..., k)$ may be arbitrary, both positive and negative $(\delta q_s \gtrless 0)$, then the last relation, which is linear and homogeneous relative to the arbitrary δq_s, may only hold true if all the coefficients Q_s of the arbitrary δq_s are zero:

$$Q_s = 0 \quad (s = 1, ..., k).$$

These are Lagrange's equilibrium conditions.

The sense of Q_s. Assume that one variable q_s varies by δq_s. The work done by the forces during this displacement is

$$\delta A_s = Q_s \delta q_s.$$

From this we have

$$Q_s = \frac{\delta A_s}{\delta q_s}.$$

If the forces admit of the force function U,

$$X_\nu = \frac{\partial U}{\partial x_\nu}, \quad Y_\nu = \frac{\partial U}{\partial y_\nu}, \quad Z_\nu = \frac{\partial U}{\partial z_\nu},$$

then the expression for the generalized force has a simple meaning, namely,

$$Q_s = \sum \left(X_\nu \frac{\partial x_\nu}{\partial q_s} + Y_\nu \frac{\partial y_\nu}{\partial q_s} + Z_\nu \frac{\partial z_\nu}{\partial q_s} \right) = \frac{\partial U}{\partial q_s} \quad (s = 1, ..., k).$$

The equilibrium conditions then assume the form

$$\frac{\partial U}{\partial q_s} = 0 \quad (s = 1, ..., k),$$

and, consequently, in equilibrium the force function has an extremum value.

Example 1. A load R is lifted by the vertical winch shown in Fig. 70. Find the force P which must be applied to the handle of the winch to balance the load R if the handle is l long and the radius of the winch drum is r. This system has one degree of freedom. We take the rotation angle φ as the defining coordinate. The elementary work of the forces during the virtual displacement $\delta \varphi$ is

$$\delta A_\varphi = 2Pl \, \delta \varphi + R \delta s = (2Pl - rR) \, \delta \varphi = Q_\varphi \, \delta \varphi.$$

From this $Q_\varphi = 2Pl - rR$. The equilibrium condition $Q_\varphi = 0$ yields

$$P = \frac{rR}{2l}.$$

Assume that the system's rotation $\delta \varphi$ as a rigid body about an axis z is a virtual displacement. For this virtual displacement we have

$$\begin{pmatrix} \delta x_\nu \\ \delta y_\nu \\ \delta z_\nu \end{pmatrix} = \left\| \begin{array}{ccc} 0 & 0 & \delta \varphi \\ z_\nu & y_\nu & z_\nu \end{array} \right\|.$$

The virtual work is

$$\delta A_\varphi = \sum (X_\nu \delta x_\nu + Y_\nu \delta y_\nu + Z_\nu \delta z_\nu) = \delta \varphi \sum (x_\nu Y_\nu - y_\nu X_\nu) = M_z \, \delta \varphi.$$

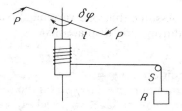

Figure 70

Hence

$$Q_\varphi = M_z.$$

Example 2. What force R should be applied to the rim of the roller with radius r shown in Fig. 71 to balance the couple of forces with moment M applied to the roller?

The system's rotation as a rigid body about an axis is a virtual displacement, and therefore

$$\delta\varphi Q_\varphi = \delta\varphi M - rR\delta\varphi, \quad Q_\varphi = M - Rr.$$

The equilibrium condition $Q_\varphi = 0$ yields $R = M/r$.

Example 3. Two heavy homogeneous bars of length $2a$ and weight P are hinged and pass over a smooth cylinder of radius r as shown in Fig. 72. Find the equilibrium position of the bars.

We have $z = a\cos\theta - \dfrac{r}{\sin\theta}$. The virtual work is

$$\delta A = -2P\delta z = 2P\left(a\sin\theta - \frac{r\cos\theta}{\sin^2\theta}\right)\delta\theta,$$

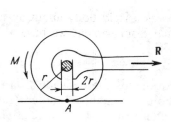

Figure 71

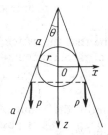

Figure 72

whence we have

$$Q_\theta = 2P \left(a \sin \theta - \frac{r \cos \theta}{\sin^2 \theta} \right), \quad Q_z = -2P.$$

The equilibrium condition is $Q_\theta = 0$, but never $Q_z = 0$. Why? Because z does not satisfy the conditions we imposed on the variables q when deriving Lagrange's equilibrium equations. It is precisely equilibrium that $\delta z = 0$ and for the virtual displacements from the equlibrium position $\delta z < 0$. The coordinate θ satisfies the conditions $\delta\theta \lesseqgtr 0$ for any value of θ, and thus the condition is fulfilled in the equilibrim position too.

78. Equilibrim equations with multipliers. Assume that forces with projections X_ν, Y_ν, Z_ν ($\nu = 1, ..., n$) act on the particles x_ν, y_ν, z_ν of a system and that bilateral constraints

$$f_j(x_\nu, y_\nu, z_\nu) = 0 \quad (j = 1, ..., m)$$

are imposed on the system. The displacements δx_ν, δy_ν, δx_ν are virtual for the constraints and satisfy the equations

$$\sum \left(\frac{\partial f_j}{\partial x_\nu} \delta x_\nu + \frac{\partial f_j}{\partial y_\nu} \delta y_\nu + \frac{\partial f_j}{\partial z_\nu} \delta z_\nu \right) = 0 \quad (j = 1, ..., m). \tag{3.4}$$

We assume the functions f_j to be independent. In other words, we assume that the matrix $\left\| \dfrac{\partial f_j}{\partial x_\nu}, \dfrac{\partial f_j}{\partial y_\nu}, \dfrac{\partial f_j}{\partial z_\nu} \right\|$ has at least one nonzero determinant Δ of order m. Those of the variables x_ν, y_ν, z_ν with respect to which the differentiation is performed in this determinant are *dependent* variables and the other, *independent* variables. We can solve equations (3.4) for the variations of the dependent variables.

In the equilibrium position, the following relation must hold true:

$$\sum (X_\nu \delta x_\nu + Y_\nu \delta y_\nu + Z_\nu \delta z_\nu) = 0.$$

We multiply (3.4) by the undetermined multipliers λ_j and add them to the formulas of the principle of virtual work. We obtain

$$\sum \left[\left(X_\nu + \sum \lambda_j \frac{\partial f_j}{\partial x_\nu} \right) \delta x_\nu + \left(Y_\nu + \sum \lambda_j \frac{\partial f_j}{\partial y_\nu} \right) \delta y_\nu \right.$$
$$\left. + \left(Z_\nu + \sum \lambda_j \frac{\partial f_j}{\partial z_\nu} \right) \delta z_\nu \right] = 0.$$

We choose the λ_j such that the variations of the dependent variables are eliminated. This is possible since the determinant of the coefficients in λ_j is Δ and Δ is nonzero. With this choice of λ_j, the last relation will only contain variations of the independent variables. Since they are arbitrary, it follows from the last relation that the coefficients of the variations of the independent variables are eliminated in the equilibrium position too. In other words, all the coefficients

$$X_\nu + \sum \lambda_j \frac{\partial f_j}{\partial x_\nu} = 0, \quad Y_\nu + \sum \lambda_j \frac{\partial f_j}{\partial y_\nu} = 0,$$

$$Z_\nu + \sum \lambda_j \frac{\partial f_j}{\partial z_\nu} = 0 \quad (\nu = 1, ..., n)$$

are eliminated in the equilibrium position. These equations are complemented by the constraint equations $f_j = 0$ $(j = 1, ..., m)$. There are $3n + m$ equations to define $3n + m$ variables x_ν, y_ν, z_ν $(\nu = 1, ..., n)$, λ_j $(j = 1, ..., m)$.

79. Unilateral constraints. Consider a particle m (x, y, z) which is fastened by an inextensible string l to a point O, which is the origin of coordinates (Fig. 73). The domain of the virtual positions of the particle m is defined by the inequaltiy

$$x^2 + y^2 + z^2 - l^2 \leqslant 0.$$

The string is in tension if

$$x^2 + y^2 + z^2 - l^2 = 0.$$

If the string is not in tension, it does not act on the particle m with any force and can therefore be neglected.

If the string is in tension, it acts on the particle m with the reaction force with projections R_x, R_y, R_z which is directed along the extended string inwards, to the point O, and, consequently, for the virtual displacements of the particle m we have

$$R_x \, \delta x + R_y \, \delta y + R_z \, \delta z \geqslant 0,$$

and the inequality sign arises if the particle m is displaced inside the sphere for a virtual displacement δx, δy, δz.

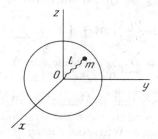

Figure 73

In the general case of a mechanical system with unilaterel constraints we can generalize this elementary example.

Let the particles x_ν, y_ν, z_ν ($\nu = 1, ..., n$) of a mechanical system be subjected to constraints in the form of inequalities

$$f_j(x_1, y_1, z_1, ..., x_n, y_n, z_n) \leqslant 0.$$

We consider the constraints which are in tension for the system's position, i. e. the relations are equalities since constraints which are not in tension, do not restrict the motion of the system. In this case, it is sufficient to consider a system of independent constraints which are in tension. Let $j = 1, ..., m$ for the independent constraints in tension. Then the virtual displacements satisfy the inequalities

$$\delta f_j = \sum \left(\frac{\partial f_j}{\partial x_\nu} \delta x_\nu + \frac{\partial f_j}{\partial y_\nu} \delta y_\nu + \frac{\partial f_j}{\partial z_\nu} \delta z_\nu \right) \leqslant 0.$$

For ideal unilateral constraints we accept, as a definition, an inequality

$$\sum (R_{\nu x} \delta x_\nu + R_{\nu y} \delta y_\nu + R_{\nu z} \delta z_\nu) \geqslant 0,$$

and put the equality sign only for displacements $\delta f_j = 0$ ($j = 1, ..., m$).

In the equilibrium position

$$X_\nu + R_{\nu x} = 0, \quad Y_\nu + R_{\nu y} = 0, \quad Z_\nu + R_{\nu z} = 0 \quad (\nu = 1, ..., n).$$

Using these relations to find the reactions $R_{\nu x}$, $R_{\nu y}$, $R_{\nu z}$ and substituting them into the inequality for ideal unilateral constraints, we get

$$\delta A = \sum (X_\nu \delta x_\nu + Y_\nu \delta y_\nu + Z_\nu \delta z_\nu) \leqslant 0. \tag{3.5}$$

The equality arises for displacements which cause the constraints to be in tension, $\delta f_j = 0$ ($j = 1, ..., m$).

We get an expression

$$\delta A + \sum \lambda_j \delta f_j$$

$$= \sum \left[\left(X_\nu + \sum \lambda_j \frac{\partial f_j}{\partial x_\nu} \right) \delta x_\nu + \left(Y_\nu + \sum \lambda_j \frac{\partial f_j}{\partial y_\nu} \right) \delta y_\nu + \left(Z_\nu + \sum \lambda_j \frac{\partial f_j}{\partial z_\nu} \right) \delta z_\nu \right].$$

The virtual displacements δx_ν, δy_ν, δz_ν include those which cause the constraints to be in tension, $\delta f_j = 0$ ($j = 1, ..., m$). For displacements of this kind it follows from the principle $\delta A = 0$ that in the equilibrium position the equations

$$X_\nu + \sum \lambda_j \frac{\partial f_j}{\partial x_\nu} = 0, \quad Y_\nu + \sum \lambda_j \frac{\partial f_j}{\partial y_\nu} = 0,$$

$$Z_\nu + \sum \lambda_j \frac{\partial f_j}{\partial z_\nu} = 0 \quad (\nu = 1, ..., n) \tag{3.6}$$

must be necessarily satisfied and the constraints must be in tension, i. e. $f_j = 0 \quad (j = 1, ..., m)$.

It follows from (3.5) that for unilateral constraints the reactions are determined as for bilateral constraints.

By virtue of (3.6) we have

$$\delta A + \sum \lambda_j \delta f_j = 0,$$

whence, by virtue of (3.5),

$$\sum \lambda_j \delta f_j = -\delta A \geqslant 0.$$

For displacements which satisfy the inequality $\delta f_j < 0$ (the other arise $\delta f_s = 0, s \neq j$) we have

$$\lambda_j \delta f_j > 0 \quad (j = 1, ..., m),$$

or

$$\lambda_j < 0 \quad (j = 1, ..., m),$$

which are additional conditions for choosing real positions of equilibrium of the system.

Example.
$$f_1 = y - x \leqslant 0, \quad f_2 = x^2 + y^2 - R^2 \leqslant 0,$$
$$X = 0, \quad Y = -mg.$$

1. By hypothesis both constraints are in tension:

$$-\lambda_1 + \lambda_2 \cdot 2x = 0, \quad y - x = 0,$$
$$-mg + \lambda_1 + \lambda_2 \cdot 2y = 0, \quad x^2 + y^2 - R^2 = 0.$$

We have $-mg + 2\lambda_1 = 0$, whence $\lambda_1 = mg/2 > 0$, and this does not satisfy the additional conditions.

2. One constraint is in tension:

$$y - x = 0,$$
$$-\lambda_1 = 0, \quad -mg + \lambda_1 = 0,$$

whence $\lambda_1 = mg > 0$, which must be discarded since it does not satisfy the additional conditions.

3. Only the second constraint is in tension:

$$x^2 + y^2 - R^2 = 0, \quad 2\lambda_2 x = 0, \quad -mg + 2\lambda_2 y = 0,$$

whence $x = 0, y = \pm R, \lambda_2 = \dfrac{mg}{2y} = \dfrac{mg}{\pm 2R}$; since λ_2 must be negative, we get $y = -R, \lambda_2 = -\dfrac{mg}{2R} < 0$.

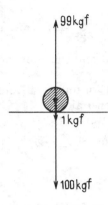

Figure 74

Problem. A stone 1 kgf in weight lies on a smooth horizontal table. Find the reaction of the table (Fig. 74).

We add two forces 99 kgf each acting upwards and downwards. The force of 100 kgf is cancelled out by the unilateral constraint and only the force of 99 kgf directed upwards remains. Where is the error?

The reactions of unilateral constraints must be determined as for bilateral constraints.

80. Finding the reactions. The principle of virtual work makes it possible to find the reactions of constraints. Indeed, we add all the reactions forces $R_{\nu x}$, $R_{\nu y}$, $R_{\nu z}$ to the given forces X_ν, Y_ν, Z_ν and assume that the system is without constraints and is acted upon by the forces $X_\nu + R_{\nu x}$, $Y_\nu + R_{\nu y}$, $Z_\nu + R_{\nu z}$.

Then the principle of virtual work yields

$$\sum \left[(X_\nu + R_{\nu x})\,\delta x_\nu + (Y_\nu + R_{\nu y})\delta y_\nu + (Z_\nu + R_{\nu z})\delta z_\nu \right] = 0.$$

For particles without constraint the displacements δx_ν, δy_ν, δz_ν are arbitrary. Hence we get equations of equilibrium

$$X_\nu + R_{\nu x} = 0, \quad Y_\nu + R_{\nu y} = 0, \quad Z_\nu + R_{\nu z} = 0 \quad (\nu = 1, ..., n),$$

from which we find the reactions $R_{\nu x}$, $R_{\nu y}$, $R_{\nu z}$.

However, the method of determining the separate reactions is more practical. We shall omit the general discussion and illustrate the method with some examples.

Example 1. A rigid body consisting of particles x_ν, y_ν, z_ν has two fixed points O and O' (Fig. 75). The particles x_ν, y_ν, z_ν are acted upon by forces X_ν, Y_ν, Z_ν and the points O and O' are acted upon by the corresponding reactions R_x, R_y, R_z and R_x', R_y', R_z'.

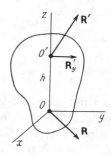

Figure 75   Figure 76

Assume that we want to find the reaction R_y'. The other reactions being retained, the reaction R_y' prevents the rigid body from rotating about the x-axis. We add this reaction R_y' to the given forces:

$$\sum (X_\nu\, \delta x_\nu + Y_\nu\, \delta y_\nu + Z_\nu\, \delta z_\nu) + R_y'\, \delta y_{O'} = 0.$$

The set of virtual displacements of the system is complemented by rotations about the x-axis, which were forbidden before. We shall now have the virtual displacements

$$\delta x_\nu = 0, \quad \delta y_\nu = -z_\nu\, \delta \varphi, \quad \delta z_\nu = y_\nu\, \delta \varphi,$$

where $\delta \varphi$ is the angle of the virtual rotation about the x-axis. Substituting these values into the last relation, we get

$$\delta \varphi\, (M_x - R_y'\, h) = 0, \quad M_x = \sum_\nu (y_\nu Z_\nu - z_\nu Y_\nu),$$

whence, due to the arbitrariness of $\delta \varphi$, we get a formula for R_y':

$$M_x - R_y'\, h = 0.$$

If we try to find R_z', we encounter a peculiarity. By adding the reaction R_z' to the given forces and retaining all the other reactions, we do not remove the constraints imposed on the system since two reactions, R_z and R_z', hinder the translation along the z-axis. The problem is statically indeterminate and our consideration has only disclosed the origin of the static indeterminacy.

Example 2. Find the reactions of two bars considered in Example 5 on p. 88 (see Fig. 67).

1. An isolated rotation about A is

$$-R \cdot 2l \cos \varphi + Pl \sin \varphi = 0.$$

2. A rotation about C is
$$N_A \cdot 2l \cos \varphi - P(2l \cos\varphi - l) \sin \varphi = 0.$$

3. A translation along the bar is
$$N'_A - P \cos \varphi = 0.$$

Example 3. Find the forces in the bars of a truss (Fig. 76). We cut the bar and replace it by a reaction **R**. When the truss is cut, it becomes a mechanism which can rotate. It rotates about a point A. The work of the given forces plus the work of the reaction **R** during this displacement is zero.

From this we can find R.

3.7 Equilibrium of a String

81. Consider an inextensible string (Fig. 77). The elementary work of the force **F**dS is
$$(X\delta x + Y\delta y + Z\delta z)\, ds.$$

The total work is
$$A = \int_0^l (X\delta x + Y\delta y + Z\delta z)\, ds.$$

The principle of virtual work is
$$\int_0^l (X\delta x + Y\delta y + Z\delta z)\, ds = 0.$$

Constraints. The string is inextensible, ds does not vary during the displacement:
$$f \equiv dx^2 + dy^2 + dz^2 - ds^2 = 0,$$
$$\frac{1}{2}\delta f \equiv dx\, d\delta x + dy\, d\delta y + dz\, d\delta z = 0.$$

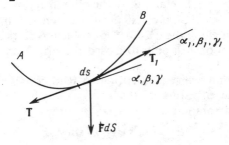

Figure 77

We shall solve the problem using the principle of virtual work and Lagrange multipliers:

$$\int_0^l \left[(X\delta x + Y\delta y + Z\delta z)\, ds + \lambda \left(\frac{dx}{ds}\, d\delta x + \frac{dy}{ds}\, d\delta y + \frac{dz}{ds}\, d\delta z \right) \right] = 0.$$

When the ends A and B are fastened, integration by parts yields

$$\int_0^l \left\{ \left[X\, ds - d\left(\lambda \frac{dx}{ds} \right) \right] \delta x + \left[Y\, ds - d\left(\lambda \frac{dy}{ds} \right) \right] \delta y \right.$$
$$\left. + \left[Z\, ds - d\left(\lambda \frac{dz}{ds} \right) \right] \delta z \right\} = 0.$$

We assume δx to be a dependent variation. We choose $\lambda(s)$ such that

$$Z\, ds - d\left(\lambda \frac{dz}{ds} \right) = 0.$$

Then the principle of virtual work yields another two equations

$$X\, ds - d\left(\lambda \frac{dx}{ds} \right) = 0, \quad Y\, ds - d\left(\lambda \frac{dy}{ds} \right) = 0.$$

We seek the meaning of λ. Three forces, $\mathbf{T}$, $\mathbf{T}_1$ and $\mathbf{F}ds$ act on the element ds of the string (Fig. 77). The equilibrium conditions for ds are

$$-T\alpha + T_1 \alpha_1 + X\, ds = 0,$$

or

$$X\, ds + d(T\alpha) = 0.$$

But $\alpha = dx/ds$, $\beta = dy/ds$, $\gamma = dz/ds$, and, hence,

$$X ds + d\left(T\frac{dx}{ds} \right) = 0, \quad Y ds + d\left(T\frac{dy}{ds} \right) = 0, \quad Z ds + d\left(T\frac{dz}{ds} \right) = 0.$$

Whence we have $T = -\lambda$.

Corollary. Using Frenet's formulas

$$\frac{d\alpha}{ds} = \frac{\alpha'}{\varrho}, \quad \frac{d\beta}{ds} = \frac{\beta'}{\varrho}, \quad \frac{d\gamma}{ds} = \frac{\gamma'}{\varrho},$$

where α', β', γ' are direction cosines of the principal normal and ϱ is the radius of the curvature of the string, we get

$$X = -\alpha \frac{dT}{ds} - \alpha' \frac{T}{\varrho}, \quad Y = -\beta \frac{dT}{ds} - \beta' \frac{T}{\varrho}, \quad Z = -\gamma \frac{dT}{ds} - \gamma' \frac{T}{\varrho},$$

whence follow the natural equilibrium equations for the string

$$F_n = -\frac{T}{\varrho}, \quad F_\tau = -\frac{dT}{ds}, \quad F_b = 0.$$

It is easy to find the tension of the string from the natural equations:

$$dT = -(X\,dx + Y\,dy + Z\,dz).$$

If the forces admit of a force function, then

$$dT = -dU, \quad T = -U + h.$$

Examples. 1. The forces are parallel, say, to the z-axis:

$$T\frac{dx}{ds} = A, \quad T\frac{dy}{ds} = B, \quad \frac{dy}{dx} = \frac{B}{A}, \quad Ay - Bx = C.$$

2. The forces are central, and then the equilibrium is planar:

$$T\left(y\frac{dz}{ds} - z\frac{dy}{ds}\right) = A, \quad T\left(z\frac{dx}{ds} - x\frac{dz}{ds}\right) = B,$$

$$T\left(x\frac{dy}{ds} - y\frac{dx}{ds}\right) = C.$$

The string lies in the plane which passes through the centre of the forces:

$$Ax + By + Cz = 0.$$

82. Catenary curve. Let us consider an extensible homogeneous heavy string suspended at both ends A and B (Fig. 78). Since the forces are parallel, the equilibrium is planar:

$$X = 0, \quad Y\,ds = -p\,ds.$$

Assume that the point A is the origin, the point B has coordinates α and β, and l is the length of the string, $l^2 > \alpha^2 + \beta^2$. The equilibrium equations

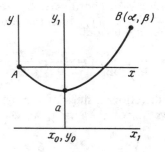

Figure 78

assume the form

$$d\left(T\frac{dx}{ds}\right) = 0, \quad d\left(T\frac{dy}{ds}\right) - p\,ds = 0.$$

Hence

$$T\frac{dx}{ds} = A > 0, \quad T = A\frac{ds}{dx}, \quad d\left(A\frac{dy}{dx}\right) - p\,ds = 0,$$

or

$$d\,(Ay') - p\,ds = 0,$$

where $y' = dy/dx$. For the plane $ds = \sqrt{1 + y'^2}\,dx$, and

$$d\,(pay') - p\sqrt{1 + y'^2}\,dx = 0$$

if $A = pa$. From this we have

$$\frac{dy'}{\sqrt{1 + y'^2}} = \frac{dx}{a}.$$

Integrating, we obtain

$$y' + \sqrt{1 + y'^2} = e^{\frac{x - x_0}{a}}, \quad -y' + \sqrt{1 + y'^2} = e^{-\frac{x - x_0}{a}},$$

or

$$y' = \frac{1}{2}\left(e^{\frac{x - x_0}{a}} - e^{-\frac{x - x_0}{a}}\right).$$

Integrating once again, we get an equation for a catenary curve

$$y - y_0 = \frac{a}{2}\left(e^{\frac{x - x_0}{a}} + e^{-\frac{x - x_0}{a}}\right) - a$$

if y' is zero at the point with coordinates x_0, y_0.

If $x - x_0 = x_1$, $y - y_0 + a = y_1$, then the equation assumes the form

$$y_1 = \frac{a}{2}\left(e^{\frac{x_1}{a}} + e^{-\frac{x_1}{a}}\right).$$

Corollary. Since

$$\sqrt{1 + y'^2} = \frac{1}{2}\left(e^{\frac{x - x_0}{a}} + e^{-\frac{x - x_0}{a}}\right) = \frac{y_1}{a},$$

we have

$$T = A\frac{ds}{dx} = A\sqrt{1 + y'^2} = py_1.$$

Determining the parameters x_0, y_0, a. From the conditions at the end-points A and B we obtain

$$-y_0 = \frac{a}{2}\left(e^{-\frac{x_0}{a}} + e^{\frac{x_0}{a}}\right) - a, \quad \beta - y_0 = \frac{a}{2}\left(e^{\frac{\alpha - x_0}{a}} + e^{-\frac{\alpha - x_0}{a}}\right) - a.$$

Hence

$$\beta = \frac{a}{2}\left(e^{\frac{\alpha - x_0}{a}} + e^{-\frac{\alpha - x_0}{a}} - e^{-\frac{x_0}{a}} - e^{\frac{x_0}{a}}\right).$$

Since $ds = \sqrt{1 + y'^2}dx$, we have

$$l = \int_0^l ds = \int_0^\alpha \frac{1}{2}\left(e^{\frac{x - x_0}{a}} + e^{-\frac{x - x_0}{a}}\right) dx,$$

integrating, we obtain

$$l = \frac{a}{2}\left(e^{\frac{\alpha - x_0}{a}} - e^{-\frac{\alpha - x_0}{a}} - e^{-\frac{x_0}{a}} + e^{\frac{x_0}{a}}\right).$$

We seek the expressions for $l + \beta$ and $l - \beta$:

$$l + \beta = a\left(e^{\frac{\alpha - x_0}{a}} - e^{-\frac{x_0}{a}}\right) = ae^{-\frac{x_0}{a}}\left(e^{\frac{\alpha}{a}} - 1\right),$$

$$l - \beta = a\left(e^{\frac{x_0}{a}} - e^{-\frac{\alpha - x_0}{a}}\right) = ae^{\frac{x_0}{a}}\left(1 - e^{-\frac{\alpha}{a}}\right).$$

Multiplying, we get

$$l^2 - \beta^2 = a^2\left(e^{\frac{\alpha}{a}} - 2 + e^{-\frac{\alpha}{a}}\right) = a^2\left(e^{\frac{\alpha}{2a}} - e^{-\frac{\alpha}{2a}}\right)^2 > 0.$$

We have thus obtained relations from which we can find the parameters of the catenary curve from the length of the string and the points at which it is fastened.

CHAPTER 4

DYNAMICS OF A PARTICLE

4.1 Principal Concepts. General Theorems

83. The *particle* is a principal concept in mechanics. The centre of mass of a sufficiently small body is a particle provided that we are not interested in the motion of the body about its centre of mass.

Newton derived the fundamental equations of motion of a free particle of mass m acted upon by a force. The force is assessed from the motion it causes or tends to cause in the system of coordinates in question, which we assume to be fixed. In the equilibrium position the force is quiescent, it only induces a tendency to move, but it must always be measured by the effect it would produce if there were no hindrance to its action.

Every body continues in its state of rest or uniform motion in a straight line until it is compelled to change that state.

A force **F** is measured by the acceleration **j** it causes:

$$m\mathbf{j} = \mathbf{F}. \tag{4.1}$$

The magnitude of the acceleration is proportional to the force applied and is directed along the line of action of the force. By the definition, **F** is an acceleration force. Dynamics is concerned with acceleration forces and the motion the acceleration forces can cause.

We introduce a rectangular coordinate system xyz. The coordinates of a moving particle are labelled x, y, z and the projections of the force **F** are labelled X, Y, Z respectively. Then Newton's equations in projections assume the form

$$m\frac{d^2x}{dt^2} = X, \quad m\frac{d^2y}{dt^2} = Y, \quad m\frac{d^2z}{dt^2} = Z.$$

For a separate motion of a particle of mass m defined by the equations

$$x = \varphi(t), \quad y = \psi(t), \quad z = \chi(t),$$

we find, by differentiation, the projections X, Y, Z of the force **F** as functions of time t:

$$X = m\frac{d^2x}{dt^2} = m\varphi'', \quad Y = m\frac{d^2y}{dt^2} = m\psi'', \quad Z = m\frac{d^2z}{dt^2} = m\chi'',$$

where primes denote derivatives with respect to time.

Classical mechanics is concerned with a more general case when X, Y, Z are functions of the position of the particle x, y, z, its velocity x', y', z', and time t:

$$X = X(t, x, y, z, x', y', z'),$$
$$Y = Y(t, x, y, z, x', y', z'),$$
$$Z = Z(t, x, y, z, x', y', z').$$

We can write the equations of motion in terms of projections on the coordinate axes:

$$\frac{dx}{dt} = x', \quad \frac{dy}{dt} = y', \quad \frac{dz}{dt} = z',$$
$$\frac{dx'}{dt} = \frac{X}{m}, \quad \frac{dy'}{dt} = \frac{Y}{m}, \quad \frac{dz'}{dt} = \frac{Z}{m}.$$

The general solution of these equations depends on six arbitrary integration constants

$$x = x(t, c_1, \ldots, c_6), \ldots, \quad x' = x'(t, c_1, \ldots, c_6), \ldots.$$

The integration constants can be found from the values of the coordinates x_0, y_0, z_0 and velocities x_0', y_0', z_0' for the initial moment:

$$x_0 = x(t_0, c_1, \ldots, c_6), \ldots, \quad x_0' = x'(t_0, c_1, \ldots, c_6), \ldots.$$

In the natural problems of dynamics, the initial conditions uniquely define the solutions of Cauchy's problem for the equations of motion.

A force is a relative concept associated with the coordinate system which we assume to be fixed and in which we measure the force by the acceleration it produces. We express the force as a function of time, the position of the particle, and its velocity.

84. The *momentum of a particle P* is a vector $m\mathbf{v}$, applied at P, whose direction coincides with that of the velocity and whose length is equal to the product of the mass m by the magnitude of the velocity v. The angular momentum about the origin is the vector

$$\mathbf{K} = \overrightarrow{OP} \times m\mathbf{v}$$

constructed at the origin O.

Newton's law of motion can be written in the form

$$\frac{d}{dt}(m\mathbf{v}) = \mathbf{F}.$$

In other words, the time derivative of the angular momentum relative to a fixed system of coordinates is equal to the force acting on the particle.

Example. If a force $\mathbf{F}$ is parallel to a fixed direction, then the trajectory is planar. Indeed, if the force is parallel to the z-axis, then $X = 0$, $Y = 0$

and, consequently,

$$m \frac{dx}{dt} = A, \quad m \frac{dy}{dt} = B,$$

or

$$A \, dy - B \, dx = 0.$$

After integration we have

$$Ay - Bx = C,$$

where A, B, C are integration constants.

To put it another way, the coordinates of a moving particle satisfy the equation of a plane parallel to the z-axis and passing through the vector of the initial velocity.

85. Theorem on the angular momentum. The area integral. Let us consider a time derivative of the angular momentum:

$$\frac{d\mathbf{K}}{dt} = \frac{d}{dt} \, (\overrightarrow{OP} \times m\mathbf{v}) = \overrightarrow{OP} \times \frac{d(m\mathbf{v})}{dt} = \overrightarrow{OP} \times \mathbf{F}.$$

But $\overrightarrow{OP} \times \mathbf{F} = \mathbf{Q}$ is the moment of the acting force. Therefore the derivative of the angular momentum with respect to time is equal to the moment of the force acting on the particle:

$$\frac{d\mathbf{K}}{dt} = \mathbf{Q}.$$

In other words, the velocity of the terminus of the angular momentum vector is equal to the moment of the acting force.

Example. Assume that a force $\mathbf{F}$ acting on a particle P is such that its line of action passes through a fixed point O. If we assume O to be the origin, then the moment of the force about O will be zero and, consequently, according to the theorem on the angular momentum, we have

$$\overrightarrow{OP} \times m\mathbf{v} = \mathbf{c},$$

where $\mathbf{c}$ is a constant vector. Hence

$$\mathbf{c} \cdot \overrightarrow{OP} = 0$$

and, consequently, the particle P lies in a plane which is orthogonal to the constant vector $\mathbf{c}$ and passes through the origin.

Let us assume that the projection of the moment of the acting force on the z-axis is zero, i.e. $M_z = 0$. We can encounter this case when the force acting on a particle either cuts the z-axis, or is parallel to the z-axis, or is zero. The theorem on the angular momentum about the z-axis yields

$$\frac{d}{dt} \, m\left(x \frac{dy}{dt} - y \frac{dx}{dt} \right) = 0,$$

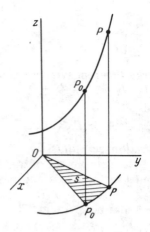

Figure 79

where x, y, z are the coordinates of the moving particle P.

Integrating, we obtain

$$x \frac{dy}{dt} - y \frac{dx}{dt} = c.$$

Let p be the projection of the particle P on the xy-plane and p_0 the projection of the initial particle P_0 (Fig. 79). We consider, the xy-plane, an area s bounded by the projection of the trajectory and two radii vectors Op_0 and Op. Then

$$ds = \frac{1}{2} \begin{vmatrix} x & y \\ x + dx & y + dy \end{vmatrix} = \frac{1}{2} (x\, dy - y\, dx).$$

The preceding integral assumes the form

$$2 \frac{ds}{dt} = c,$$

whence it follows that

$$s = \frac{1}{2} c(t - t_0).$$

To put it otherwise, the area s is swept out with the constant areal speed c.

The constant c of the area integral is then equal to double the area swept out by the radius vector Op in unit time. Conversely, if there is an area integral in the xy-plane with respect to the origin

$$x \frac{dy}{dt} - y \frac{dx}{dt} = c,$$

then

$$x \frac{d^2y}{dt^2} - y \frac{d^2x}{dt^2} = 0, \quad \text{or} \quad xY - yX = M_z = 0.$$

86. Work-kinetic energy theorem (vis viva principle). We multiply the equations of motion by $\frac{dx}{dt}, \frac{dy}{dt}, \frac{dz}{dt}$ respectively and add the results together. We obtain

$$m \left(\frac{d^2x}{dt^2} \frac{dx}{dt} + \frac{d^2y}{dt^2} \frac{dy}{dt} + \frac{d^2z}{dt^2} \frac{dz}{dt} \right) = X \frac{dx}{dt} + Y \frac{dy}{dt} + Z \frac{dz}{dt},$$

or

$$d \frac{mv^2}{2} = X\,dx + Y\,dy + Z\,dz.$$

The expression $\frac{mv^2}{2} = \frac{m}{2}(x'^2 + y'^2 + z'^2) = T$ is known as the *living force* (*vis viva*) or the *kinetic energy* of a particle P, the quantities dx, dy, dz are the projections, on the coordinate axes, of the elementary displacement of the particle P which it actually undergoes when acted upon by the forces X, Y, Z in time dt. The relation we have proved constitutes the work-kinetic energy theorem. The differential of the kinetic energy is equal to the work done by the forces acting during a real elementary displacement.

If the forces have a force function $U(x, y, z)$, then

$$X = \frac{\partial U}{\partial x}, \quad Y = \frac{\partial U}{\partial y}, \quad Z = \frac{\partial U}{\partial z}$$

and, consequently,

$$X\,dx + Y\,dy + Z\,dz = dU.$$

The work-kinetic energy theorem then assumes a very simple form

$$d \frac{mv^2}{2} = dU,$$

whence we get the *integral* (or *law*) *of the kinetic energy*

$$\frac{mv^2}{2} = U + h, \tag{4.2}$$

where the integration constant $h = \frac{mv_0^2}{2} - U_0$ is the *kinetic energy constant* or the *total* (*mechanical*) *energy* H of the particle P which consists of the kinetic energy $mv^2/2$ and the potential energy $-U$:

$$H = \frac{mv^2}{2} + (-U).$$

110 4. Dynamics of a Particle

87. Remark. The first integrals make it possible to simplify the integration of the differential equations of motion.

It should be pointed out that when we replace an equation of motion by the first integral, we can introduce into consideration a secondary solution which results from the mathematical method of finding the first integral. Let us, for instance, consider a simple pendulum, i.e. a particle of mass m moving in a plane and connected to a fixed point O by a massless rigid string (or rod) Om of length l (Fig. 80). Assume that the particle is acted upon by the force of gravity $m\mathbf{g}$ and the reaction force $\mathbf{R}$ of the string directed along the string.

The work-kinetic energy theorem has the form

$$d\left[\frac{1}{2} ml^2 \left(\frac{d\theta}{dt}\right)^2\right] = mgd(l \cos \theta),$$

since the force of reaction $\mathbf{R}$ of the string does not come into play when the displacement of the particle m is true and orthogonal to Om. From the last expression we have a kinetic energy integral

$$\frac{1}{2} l\left(\frac{d\theta}{dt}\right)^2 = g(\cos \theta - \cos \theta_0) \tag{4.3}$$

if at the initial moment $\theta = \theta_0$ and $\theta_0' = 0$.

The position of the particle m is defined by the measure of the angle θ between the segment Om and the vertical, and therefore one differential equation is sufficient to solve any problem concerning the motion of the particle m. We can assume the kinetic energy integral (4.3) to be an equation of this kind, but the integral has a secondary solution which is due to the mathematics by which the kinetic energy integral was obtained and which does not correspond to any mechanical situation, namely,

$$\theta = \theta_0, \quad \frac{d\theta}{dt} = 0.$$

However, a heavy particle cannot be at rest when it is displaced by some angle.

The kinetic energy integral (4.2) defines a domain in the xyz-space in which the particle may move with the constant kinetic energy h,

$$U + h \geqslant 0,$$

since $mv^2/2$ is always nonnegative. In some cases this domain is sufficiently small. For instance, let us consider the equilibrium position of a particle m in which the force function U has an isolated maximum. If this maximum is zero, then in a small neighbourhood of an isolated equilibrium position the function U is negative definite: zero in the equilibrium position and negative for other points. For a small positive h the domain $U + h > 0$ will also be small by continuity and, consequently, in the vicinity of the equilibrium position a particle with small values of the constant kinetic energy $h > 0$ will move in the small neighbourhood $U + h > 0$. An equilibrium possessing this property is said to be *stable*.

88. The motion of a heavy body thrown at an angle to the horizontal. The trajectory of a heavy particle in a vacuum is planar since the forces acting on the particle are parallel. We assume the x-axis to be horizontal and the y-axis to be directed vertically upwards (Fig. 81) and place the origin at the initial position of the particle. At the moment $t = 0$ we have $x_0 = 0$, $y_0 = 0$, $x_0' = v_0 \cos \alpha$, $y_0' = v_0 \sin \alpha$, where v_0 is the initial velocity and α is the angle at which the particle is thrown.

The equations of motion are

$$m \frac{d^2 x}{dt^2} = 0, \quad m \frac{d^2 y}{dt^2} = -mg.$$

When we cancel out m, integrate, and determine the integration constants, we obtain

$$\frac{dx}{dt} = v_0 \cos \alpha, \quad \frac{dy}{dt} = -gt + v_0 \sin \alpha.$$

The integration of these equations yields a law of motion

$$x = t v_0 \cos \alpha, \quad y = -\frac{gt^2}{2} + t v_0 \sin \alpha.$$

Eliminating t, we have the equation of the trajectory

$$y = -\frac{gx^2}{2 v_0^2 \cos^2 \alpha} + x \tan \alpha.$$

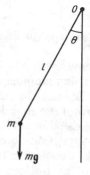

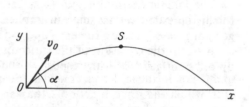

Figure 80   Figure 81

We see that the trajectory of a heavy particle in a vacuum is a parabola with a parameter

$$p = \frac{v_0^2 \cos^2 \alpha}{g}$$

and a vertical axis. The coordinates of the vertex S of the parabola can be obtained from the condition that at S the velocity does not have a projection on the y-axis, i.e. $-gt_S + v_0 \sin \alpha = 0$, and hence the vertex is reached at time $t_S = \frac{v_0 \sin \alpha}{g}$, and, consequently, the coordinates of the vertex are

$$x_S = \frac{v_0^2 \sin 2\alpha}{2g}, \quad y_S = \frac{v_0^2 \sin^2 \alpha}{2g}.$$

The abscissa of the point A is twice the abscissa x_S of the vertex

$$x_A = 2x_S = \frac{v_0^2 \sin 2\alpha}{g},$$

whence we see that x_A is a maximum when $\alpha = 45°$.

The distance OD to the directrix of the parabola

$$OD = y_S + \frac{p}{2} = \frac{v_0^2}{2g}.$$

The vis viva principle

$$v^2 = v_0^2 - 2gy$$

states that a heavy particle in a vacuum must move with the velocity it would have should it have fallen from the directrix without any initial velocity.

Let us find the envelope of the trajectories of a heavy particle thrown from the origin at different angles with an initial velocity v_0. We write the equation of the trajectory as

$$y = -\frac{gx^2}{2v_0^2} (1 + \tan^2 \alpha) + x \tan \alpha.$$

In this equation we assume $\tan \alpha$ to be a variable parameter. To use the general rule of finding the envelope, we must eliminate the parameter, i.e. $\tan \alpha$, from the equation of the trajectory and from the equation obtained by differentiating it with respect to the parameter:

$$-\frac{gx^2}{v_0^2} \tan \alpha + x = 0.$$

The elimination of $\tan \alpha$ yields the equation of the envelope

$$y - \frac{v_0^2}{2g} = -\frac{gx^2}{2v_0^2}.$$

We see that the envelope of the trajectories is a parabola whose axis is the y-axis and whose parameter is v_0^2/g. The origin is the focus of the enveloping parabola, which is known as a safety parabola because a heavy particle thrown from the origin with a fixed initial velocity v_0 cannot fall anywhere outside the parabola. We can obtain these results by considering how to reach the point with coordinates x, y. The equation defining the angle α at which the particle must be thrown for its trajectory to pass through the point (x, y) is

$$y = -\frac{gx^2}{2v_0^2}(1 + \tan^2 \alpha) + x \tan \alpha.$$

This is an algebraic second-degree equation for $\tan \alpha$ and, consequently, we can reach the point by either two techniques when there are two distinct real roots for $\tan \alpha$ or by one technique if the roots are equal, or, else, we cannot reach that point if the roots are complex. The condition under which one technique can be used is

$$x^2 - 4\frac{gx^2}{2v_0^2}\left(y + \frac{gx^2}{2v_0^2}\right) = 0,$$

which is the envelope equation.

We can also consider the motion of a heavy particle in a vacuum in terms of geometry. Let the initial conditions be the same. By virtue of the kinetic energy integral $v^2 = v_0^2 - 2gy$, the velocities of particles in horizontal planes of the level of a force function are equal, the velocity on the directrix $y = OD = \frac{v_0^2}{2g}$ is zero.

We are given the angle at which the particle is thrown and have to find the focus F of the parabola. The parabola is the locus of points which are equidistant from the directrix and the focus. By the hypothesis the point O lies on the parabola and, therefore, the locus of the foci F of the parabolas is a circle of radius OD with centre at the point O (Fig. 82).

A tangent to the parabola bisects the angle which the perpendicular to the directrix makes with a ray drawn through the focus. Hence, if we lay off from v_0, on the other side of the vertical, the angle which $\mathbf{v}_0$ makes with the vertical we get a ray OF on which the focus of the parabola must lie. The intersection of the circle and this ray defines the focus F (Fig. 82).

Consider a problem of a particle falling in a point M. Let MK be the distance of the particle from M to the directrix, which is common for all parabolas (Fig. 83). The focus F of the parabolas which pass through the

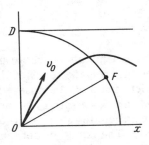

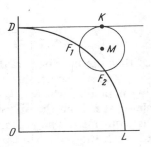

Figure 82   Figure 83

point M must lie on the circle MK of radius MK and centre at M. But the focus must also lie on the circle OD. The intersection of $\bigcirc OD$ and $\bigcirc MK$ defines either two points F_1 and F_2 (Fig. 83) or one point (Fig. 84) when $\bigcirc OD$ and $\bigcirc MK$ are cotangent, or does not define any point when $\bigcirc OD$ and $\bigcirc MK$ do not meet. If the focus F of the parabola is given, then the angle at which the particle is thrown can be geometrically determined, the angle of approach of the target M can be determined as easily.

Let us find the locus of points M which we can get using the single technique. For points of this kind, $\bigcirc MK$ must touch the circle of the foci $\bigcirc OD$. We lay off, on the perpendicular MK, a constant segment ML of length equal to $OD + MK$ (Fig. 84). The distance MO between the point M and the origin is equal to the distance ML between this point and the straight line $D'L$, and, consequently, the points M lie on the parabola with directrix $D'L$ and focus O.

This parabola is the safety parabola.

89. The equation of motion of a particle of mass m, with electric charge ε, in an electromagnetic field with an electric intensity vector $\mathbf{E}$ and magnetic intensity vector $\mathbf{H}$ has the form

$$m\mathbf{j} = \varepsilon\mathbf{E} + \varepsilon(\mathbf{H} \times \mathbf{v}),$$

where $\mathbf{v}$ and $\mathbf{j}$ are the velocity and the acceleration of the moving particle respectively.

We shall consider the case of a constant electromagnetic field, when the vectors $\mathbf{E}$ and $\mathbf{H}$ are constant. We choose the z-axis to be in the direction of the vector $\mathbf{H}$ and the y-axis to be orthogonal to the vectors $\mathbf{H}$ and $\mathbf{E}$. We designate the projections of $\mathbf{E}$ on the x, y, and z axes as P, $Q = 0$, R and the projections of H as $X = 0$, $Y = 0$, Z. Then the equations of motion will have the form

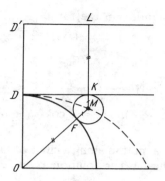

Figure 84

$$m \frac{d^2x}{dt^2} = \varepsilon P - \varepsilon Z \frac{dy}{dt}, \quad m \frac{d^2y}{dt^2} = \varepsilon Z \frac{dx}{dt}, \quad m \frac{d^2z}{dt^2} = \varepsilon R. \qquad (4.4)$$

Integrating each of these equations once and accepting the designations

$$p = \frac{\varepsilon P}{m}, \quad r = \frac{\varepsilon R}{m}, \quad \omega = -\frac{\varepsilon Z}{m},$$

we obtain

$$\frac{dx}{dt} = pt + \omega y + a, \quad \frac{dy}{dt} = -\omega x + b, \quad \frac{dz}{dt} = rt + c, \qquad (4.5)$$

where a, b, c are arbitrary integration constants.

We eliminate dy/dt from the first equation in (4.4)

$$\frac{d^2x}{dt^2} + \omega^2 \left(x - \frac{b}{\omega} - \frac{p}{\omega^2} \right) = 0,$$

and, assuming the expression in brackets to be a new variable, we have

$$x = \frac{b}{\omega} + \frac{p}{\omega^2} + A \sin (\omega t + \alpha),$$

where A and α are integration constants. Directly from the first equation of system (4.5) we have

$$y = -\frac{a}{\omega} - \frac{pt}{\omega} + A \cos (\omega t + \alpha).$$

Integrating the last equation of system (4.5), we obtain

$$z = \frac{rt^2}{2} + ct + c'.$$

We can represent this solution in the form

$$x = x_1 + x_2, \quad y = y_1 + y_2, \quad z = z_1 + z_2,$$

where

$$x_1 = \frac{b}{\omega} + \frac{p}{\omega^2}, \quad y_1 = -\frac{a}{\omega} - \frac{pt}{\omega}, \quad z_1 = \frac{rt^2}{2} + ct + c',$$

$$x_2 = A \sin(\omega t + \alpha), \quad y_2 = A \cos(\omega t + \alpha), \quad z_2 = 0.$$

The solution x_2, y_2, z_2 is a motion along a circle of radius A in the xy-plane with an angular frequency ω. The solution x_1, y_1, z_1 is a motion along a parabola which lies in the plane $x_1 = \frac{b}{\omega} + \frac{p}{\omega^2}$. The axis of the parabola is parallel to the z-axis, for $\varepsilon < 0$ the convexity is in the direction of the z-axis.

90. Central forces. A force is said to be *central* if it passes through a fixed point O which is called the *centre of force*. Under the action of central forces a particle describes a curve which lies in a plane passing through the centre of force O. We assume the plane of the trajectory to be the plane of the coordinate axes x and y with origin at the centre of force O. We assume the central force to be positive if it is repulsive and negative if it is attractive. We use the two first integrals, the area integral and the kinetic energy integral, for the motion due to central forces which depend on the distance r between the moving particle and the centre O since the moment of the central forces about the centre of force is always zero and the central forces which depend on r always have a force function.

The area integral

$$x \frac{dy}{dt} - y \frac{dx}{dt} = C$$

in polar coordinates has the form

$$r^2 \frac{d\theta}{dt} = C. \tag{4.6}$$

The work-kinetic energy theorem has the form

$$d\frac{mv^2}{2} = X\,dx + Y\,dy = F\frac{x}{r}\,dx + F\frac{y}{r}\,dy = F\frac{x\,dx + y\,dy}{r} = F\,dr, \tag{4.7}$$

since the projections of the forces are expressed as

$$X = F\frac{x}{r}, \quad Y = F\frac{y}{r},$$

where $r^2 = x^2 + y^2$. We see that if $F = F(r)$, then the force function is

$$U = \int F(r)\, dr,$$

and, consequently, under this condition there is a kinetic energy integral

$$\frac{mv^2}{2} = U + h. \tag{4.8}$$

The position of a particle in a plane is defined by two coordinates and therefore two integrals, the area integral and the kinetic energy integral, are sufficient for solving problem concerning the motion of a particle acted upon by central forces dependent on r.

It is more convenient to use polar coordinates r and θ in which

$$v^2 = \left(\frac{dr}{dt}\right)^2 + r^2\left(\frac{d\theta}{dt}\right)^2.$$

By virtue of the area integral (4.6) we can exclude dt using the relation

$$\frac{1}{r^4}\left(\frac{dr}{d\theta}\right)^2 = \left(\frac{d\frac{1}{r}}{d\theta}\right)^2$$

and obtain

$$v^2 = C^2\left[\left(\frac{d\frac{1}{r}}{d\theta}\right)^2 + \left(\frac{1}{r}\right)^2\right].$$

If we divide the expression for the vis viva principle (4.7) by $d\theta$, we get

$$\frac{d}{d\theta}\left\{\frac{mC^2}{2}\left[\left(\frac{d\frac{1}{r}}{d\theta}\right)^2 + \left(\frac{1}{r}\right)^2\right]\right\} = F\frac{dr}{d\theta}.$$

Cancelling out $dr/d\theta$ and differentiating, we obtain *Binet's formula*

$$F = -\frac{mC^2}{r^2}\left(\frac{d^2\frac{1}{r}}{d\theta^2} + \frac{1}{r}\right). \tag{4.9}$$

This equation makes it possible to determine r as a function of θ, i.e. an equation of the trajectory, if F depends on r or F depends on r and θ. If F depends only on r, then we can proceed from the kinetic energy integral (4.8), rather than (4.9), to find the trajectory.

4.2 Planetary Motion

91. By thoroughly studying the extensive observations data accumulated by the Danish astronomer Tycho Brahe, Kepler deduced the three fundamental laws of planetary motion.

1. Planets describe conic sections around the sun sweeping equal areas in equal time intervals.

2. Their trajectories are ellipses with the sun at the focus.

3. The squares of the periods of revolution of the planets are proportional to the cubes of the semi-major axes of their orbits.

From these laws Newton deduced his famous law of gravitation. If the trajectories are planar and the planets sweep equal areas in equal time intervals, then the forces are central and the sun is the centre of force.

The force F can be found from Binet's formula. The equation of the ellipse in polar coordinates with the pole at the focus has the form

$$\frac{1}{r} = \frac{1 + e\cos\theta}{p},$$

where p is the parameter and e is the eccentricity of the ellipse. Using Binet's formula, we have

$$F = -\frac{mC^2}{r^2}\left(\frac{d^2\frac{1}{r}}{d\theta^2} + \frac{1}{r}\right) = -\frac{mC^2}{pr^2},$$

where C is the area constant and m is the mass of the planet. If we designate C^2/p as μ, we get

$$F = -\frac{m\mu}{r^2}.$$

We can find from Kepler's third law that μ does not depend on the planet. The area constant C is double the area of the ellipse, i.e. the trajectory $2\pi ab$ divided by the time T of the complete revolution of the planet:

$$C = \frac{2\pi ab}{T},$$

where a is the semi-major axis and b is the semi-minor axis of the planet's orbit. The parameter is $p = b^2/a$ and hence

$$\mu = \frac{C^2}{p} = \frac{4\pi^2 a^3}{T^2}.$$

But according to Kepler's third law a^3/T^2 is the same for all the planets of the solar system, and, consequently, the force attracting the planet is

$$F = -\frac{m\mu}{r^2},$$

is directed towards the sun, is in direct proportion to the mass, and is in inverse proportion to the square of the distance from the sun.

Having deduced the fundamental law of gravitation, Newton found the motion of a particle of mass m attracted by a fixed centre with a force which is in inverse proportion to the square of the distance between the particle and the centre of force.

Let the central force be

$$F = -\frac{m\mu}{r^2}.$$

Then we can write the kinetic energy integral

$$v^2 = \frac{2\mu}{r} + h$$

in the form

$$C^2\left[\left(\frac{d\frac{1}{r}}{d\theta}\right)^2 + \left(\frac{1}{r}\right)^2\right] = \frac{2\mu}{r} + h,$$

where h is a constant. Hence

$$\left(\frac{d\frac{1}{r}}{d\theta}\right)^2 = -\left(\frac{1}{r} - \frac{\mu}{C^2}\right)^2 + \frac{\mu^2}{C^4} + \frac{h}{C^2}.$$

We set

$$\frac{1}{r} - \frac{\mu}{C^2} = u\sqrt{\frac{\mu^2}{C^4} + \frac{h}{C^2}},$$

and write the preceding equation, in a new variable, as

$$\left(\frac{du}{d\theta}\right)^2 = 1 - u^2,$$

or

$$\frac{\pm du}{\sqrt{1 - u^2}} = d\theta.$$

Integrating,

$$\pm \cos^{-1} u = \theta - \theta_0, \quad u = \cos(\theta - \theta_0),$$

we have the equation of the trajectory in the form

$$\frac{1}{r} = \frac{\mu}{C^2} + \sqrt{\frac{\mu^2}{C^4} + \frac{h}{C^2}} \cos(\theta - \theta_0).$$

Comparing this equation with the equation of a conic section in polar coordinates, with the pole at the focus

$$\frac{1}{r} = \frac{1}{p} + \frac{e}{p} \cos(\theta - \theta_0),$$

where e is the eccentricity and p is the parameter of the conic section (focal parameter), we find that the trajectory of the particle is along a conic section with a focus at the centre of force, the orbital parameter being

$$p = \frac{C^2}{\mu}$$

and the eccentricity being

$$e = p\sqrt{\frac{\mu^2}{C^4} + \frac{h}{C^2}} = \sqrt{1 + \frac{hC^2}{\mu^2}}.$$

If h is negative, then the trajectory is an ellipse since $e < 1$. If h is zero, then the trajectory is a parabola since $e = 1$. And if h is positive, then the trajectory is a hyperbola since $e > 1$.

The kinetic energy constant

$$h = v_0^2 - \frac{2\mu}{r_0}$$

depends on the magnitudes of the initial radius vector r_0 and the initial velocity v_0. Consequently, if under certain initial conditions the trajectory of the attracted particle is an ellipse, then the trajectory will remain elliptic when the direction of the initial velocity is changed.

92. Discussion of the problem in terms of geometry. Assume, for definiteness, that the initial distance r_0 and the initial velocity v_0 satisfy the conditions of the elliptic orbit $h < 0$. We know from geometry that

$$e = \frac{c}{a}, \quad p = \frac{b^2}{a} = a(1 - e^2),$$

where a, b are the semi-major and semi-minor axes of the ellipse and c is the focal distance from the centre of the ellipse. Given the parameter $p = C^2/\mu$ and the eccentricity $e = \sqrt{1 + hC^2/\mu^2}$, we obtain

$$a = \frac{p}{1 - e^2} = -\frac{\mu}{h}.$$

In other words, the semi-major axis of the elliptic orbit only depends on the kinetic energy constant. The semi-minor axis b,

$$b^2 = -\frac{C^2}{h},$$

depends on the area constant and the kinetic energy constant.

Let us consider elliptic ($h < 0$) orbits emanating from the same point M_0 with constant velocity v_0. We have seen that all these orbits have the same semi-major axis.

Assume that O is the centre of force, M_0 is the initial position of the point, v_0 is the initial velocity, $h < 0$, and a is the semi-major axis of all the resulting elliptic orbits with the common focus O. We seek the locus of points F, namely the second foci of the orbits.

Since M_0 is a point along the orbit, the definition of the ellipse yields

$$M_0O + M_0F = 2a,$$

or

$$M_0F = 2a - r_0.$$

To put it otherwise, the locus of the second foci F is a circle of radius $2a - r_0$ with centre at M_0 (Fig. 85). We use H to designate the point on the circle the farthest away from the centre O. It lies on a straight line OM_0 at a distance from the centre of force twice the length of the semi-major axis, $OH = 2a$. We can designate the circle of the foci as $\bigcirc M_0H$ since its radius is M_0H and its centre is M_0. It is now easy to find the focus F of the orbit from the given direction of the initial velocity $\mathbf{v}_0$. Since it is directed along a tangent to the orbit at the point M_0, the velocity $\mathbf{v}_0$ must bisect the angle HM_0F.

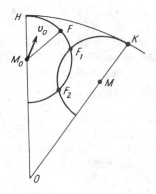

Figure 85

Let us consider how to reach a given point M. We draw $\bigcirc OH$ of radius $OH = 2a$ with centre at O and draw a ray OM. Let K be the point at which the ray meets $\bigcirc OH$. The required elliptic orbit must pass through M, and therefore we must have the following for the other focus F_α of the orbit:

$$MO + MF_\alpha = 2a = MO + MK,$$

or

$$MF_\alpha = MK,$$

i.e. the focus F_α must lie on $\bigcirc MK$ with radius MK and centre at M. The intersection of $\bigcirc MK$ and the circle of the foci $\bigcirc M_0H$ will define the foci F_α of the orbits passing through the point M.

Three cases are possible: (1) $\bigcirc M_0H$ and $\bigcirc MK$ meet at two points, F_1 and F_2, then we can hit the point M in two ways, namely, along the orbit with focus F_1 or along the orbit with focus F_2, (2) $\bigcirc M_0H$ and $\bigcirc MK$ touch each other at the point F, then we can hit the point M in one way, along the orbit with focus F, (3) $\bigcirc M_0H$ and $\bigcirc MK$ do not intersect, then we cannot hit the point M along either orbit.

Let us find the boundary between the points which we can hit in two ways and the points which we cannot hit. The boundary is the locus of points M^* which we can hit in one way.

We draw a ray OK^* from the centre of force to $\bigcirc OH$ (Fig. 86). On the ray OK^* we find the single point M^* at which $\bigcirc M^*K^*$ touches the circle of foci $\bigcirc M_0H$ at the point F^*. The focus F^* of the orbit passing through the point M^* lies on the straight line M_0M^*. We have

$$M^*O + M^*F^* = 2a.$$

Hence

$$M^*O + M^*M_0 = M^*O + M^*F^* + F^*M_0 = HO + HM_0,$$

i.e. the sum of the distances from the point M^* to the centre of force O and to the initial point M_0 is constant and, consequently, the locus of the points M^* is a safety ellipse with foci O and M_0 which passes through the point H and with a semi-major axis equal to $\frac{1}{2}(OH + HM_0)$.

How to reach a point. We have to hit a point M from an initial position M_0 throwing with a given initial velocity v_0 (Fig. 87). The initial velocity v_0 must be such that the point M lies on the safety ellipse, i.e. there should be no excess velocity because there would then be two ways of hitting the point.

It is easy to find the double semi-major axis of the corresponding safety ellipse (see Fig. 87):

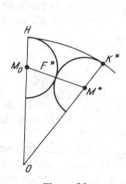

Figure 86

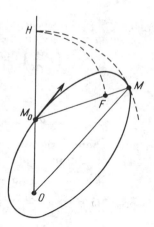

Figure 87

$$2a^* = MO + MM_0.$$

From this we can find the position of the corresponding point H as a point on the straight line OM_0 and on the safety ellipse with

$$OH + HM_0 = MO + MM_0 = 4a - r_0,$$

where a is the semi-major axis of the orbit corresponding to the initial velocity v_0. The parameter a appeared in the last formula and is equal to a quarter of the perimeter R of the triangle OM_0M:

$$a = \frac{MO + MM_0 + OM_0}{4} = \frac{R}{4}.$$

Substituting $a = -\mu/h$ and $h = v_0^2 - \dfrac{2\mu}{r_0}$ into this formula, we obtain

$$v_0 = \sqrt{2\mu\left(\frac{1}{r_0} - \frac{2}{R}\right)}.$$

The focus F of the orbit passing through the point M is the intersection of the orbit M_0H and the straight line M_0M and the angle of throw is defined by the bisector of the angle MM_0H.

93. Determining the orbital period for an ellipse. The closest vertex A of the semi-major axis of the orbit to the sun F is the perihelion. The angle $\theta = AFM$ is the true anomaly (Fig. 88). For an elliptic orbit the kinetic energy integral can be written as

$$\left(\frac{dr}{dt}\right)^2 + \frac{C^2}{r^2} = \frac{2\mu}{r} - \frac{\mu}{a}.$$

We replace C^2 by $\mu p = \mu a(1 - e^2)$ and get

$$\left(r \frac{dr}{dt}\right)^2 = \frac{\mu}{a} \left[a^2 e^2 - (a - r)^2\right],$$

or

$$\sqrt{\frac{\mu}{a}} \, dt = \pm \frac{r \, dr}{\sqrt{a^2 e^2 - (a - r)^2}}.$$

This relation changes sign when we pass through the vertices of the semi-major axis and eliminate $a^2 e^2 - (a - r)^2$. We set

$$a - r = ae \cos u.$$

This is possible since $(a - r)^2$ does not exceed $a^2 e^2$:

$$\frac{1}{a} \sqrt{\frac{\mu}{a}} \, dt = (1 - e \cos u) \, du.$$

Integrating this equation and using the relation $\mu = 4\pi^2 a^3 / T^2$, which we proved above, we get *Kepler's equation*

$$\frac{2\pi}{T} (t - t_0) = u - e \sin u,$$

where t_0 is the time the perihelion is passed, the variable u being zero; T denotes the orbital period of the planet.

The variable u can be found from the equation

$$r = a(1 - e \cos u). \tag{4.10}$$

Indeed, we can regard an ellipse as a deformed circle of radius a whose ordinates have decreased in the ratio b/a. Assume that M is a point of the ellipse, M' is a corresponding point on the circle, O is the centre of the ellipse, and the angle $M'OA = u$ is the *eccentric anomaly* (Fig. 89). Let

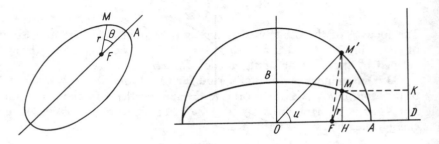

Figure 88   Figure 89

the straight line DK be the directrix of the ellipse. An ellipse is the locus of points M such that the ratio of the focal radius FM to the distance of the points M from the directrix DK is equal to the eccentricity e. Hence

$$r = eMK = e(OD - OH) = e\left(\frac{a}{e} - a\cos u\right) = a(1 - e\cos u).$$

The comparison with formula (4.10) proves that u in Kepler's equation is the eccentric anomaly.

Excluding r from equation (4.10) and the equation of the orbit in polar coordinates

$$\frac{1}{r} = \frac{1 + e\cos\theta}{p},$$

where θ denotes the true anomaly, we get a relationship between the anomalies

$$1 - e\cos u = \frac{1 - e^2}{1 + e\cos\theta},$$

which can be reduced to a symmetric and more convenient form

$$\tan\frac{\theta}{2} = \sqrt{\frac{1+e}{1-e}}\tan\frac{u}{2}.$$

94. The elliptic motion of a particle M in space is defined by six constants. We draw rectangular fixed axes x, y, z through the centre of force O (Fig. 90). The plane of the orbit cuts the xy-plane along the straight line NN', which is called a nodal line. One of the points N of the orbit at which the values of z change from negative to positive when the planet moves is called an *ascending node*. The other point N' is a *descending node*.

The plane of an orbit is defined by the longitude of the ascending node $\Omega = \angle xON$ and by the inclination i of the plane of the orbit to the xy-

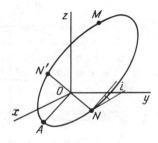

Figure 90

plane; i is measured in terms of the angle between the perpendiculars drawn to ON at the node N, one in the xy-plane in the direction of the positive rotation and the other in the plane of the orbit in the direction of the motion of the particle M along the orbit. Let A be the perihelion. The *longitude* of the perihelion ω is the sum of the angles $\angle xON$ and $\angle NOA$. An ellipse is defined by its semi-major axis a and eccentricity e. To find the time of motion, we must know the time t_0 the planet passes through the perihelion. Thus, to determine an elliptic orbit, it is sufficient to know six constants Ω, i, ω, a, e, t_0 which are known as elements of elliptic motion.

4.3 Motion of a Constrained Particle

95. Let us consider a particle of mass m with coordinates x, y, z, moving in a fixed space with rectangular axes xyz, and subject to a force with projections X, Y, Z and to the constraint $f(x, y, z, t) = 0$. We shall only consider an ideal constraint whose reaction force is normal to the surface $f(x, y, z, t) = 0$ at the moment t, or

$$\frac{R_x}{\partial f/\partial x} = \frac{R_y}{\partial f/\partial y} = \frac{R_z}{\partial f/\partial z} = \lambda. \tag{4.11}$$

If we add the reaction forces due to constraints R_x, R_y, R_z to the given force X, Y, Z, then we can regard the particle m as a free particle but subject to the action of the forces $X + R_x$, $Y + R_y$, $Z + R_z$ and write Newton's equations of motion for a free particle:

$$m\frac{d^2x}{dt^2} = X + R_x, \quad m\frac{d^2y}{dt^2} = Y + R_y, \quad m\frac{d^2z}{dt^2} = Z + R_z.$$

From (4.11) we can find R_x, R_y, R_z and substitute the results into the preceding equation. Then we have

$$m\frac{d^2x}{dt^2} = X + \lambda\frac{\partial f}{\partial x}, \quad m\frac{d^2y}{dt^2} = Y + \lambda\frac{\partial f}{\partial y}, \quad m\frac{d^2z}{dt^2} = Z + \lambda\frac{\partial f}{\partial z}.$$

These equations are known as Lagrange's equations with multipliers. We add the constraint equation $f(x, y, z, t) = 0$ to them and get a system of four equations in four unknowns x, y, z, λ. Differentiating $f(x, y, z, t) = 0$ with respect to t, we find that the projections of the velocity dx/dt, dy/dt, dz/dt satisfy the relation

$$\frac{\partial f}{\partial t} + \frac{\partial f}{\partial x}\frac{dx}{dt} + \frac{\partial f}{\partial y}\frac{dy}{dt} + \frac{\partial f}{\partial z}\frac{dz}{dt} = 0.$$

We multiply Lagrange's equations with the multiplier λ by dx, dy, dz respectively and add them together. After obvious transformations, we obtain the

work-kinetic energy theorem

$$d\,\frac{mv^2}{2} = X\,dx + Y\,dy + Z\,dz - \lambda\,\frac{\partial f}{\partial t}\,dt.$$

The differential of the kinetic energy of the particle is equal to the work done by the given forces during the true displacement $X\,dx + Y\,dy + Z\,dz$ plus the work of the reaction $-\lambda\,\dfrac{\partial f}{\partial t}\,dt$ on the real displacement in time dt.

If the surface on which the particle must remain is fixed, then the function f does not explicitly depend on time, i.e. $\dfrac{\partial f}{\partial t} = 0$. The work-kinetic energy theorem has the form

$$d\,\frac{mv^2}{2} = X\,dx + Y\,dy + Z\,dz$$

since the work of the constraint forces during the true displacement is zero.

96. Motion of a particle over a fixed surface. Motion with no force. Let us consider a case when a surface is fixed and a particle of mass m is subject to the reaction force **R** of the surface and a force **F**. If the surface is perfectly smooth, then its reaction **R** is orthogonal to the surface at the point.

We introduce a moving trihedron consisting of a τ-axis directed along a tangent, an n-axis directed along a normal, and a b-axis directed along a binormal to the trajectory of the particle (Fig. 91). The equations of motion in terms of the projections on the axes of this moving trihedron have the form

$$m\,\frac{dv}{dt} = F_\tau, \quad m\,\frac{v^2}{\varrho} = F_n + R_n, \quad 0 = F_b + R_b, \tag{4.12}$$

where ϱ is the radius of curvature of the trajectory and $F_\tau, \ldots, R_b$ are the projections of the forces on the axes indicated in the indices.

If there are no forces (**F** = 0), then it follows from the last equation in (4.12) that $R_b = 0$, and this means that the projection of the reaction

Figure 91

on the normal to the trajectory is equal to the net reaction, i.e. $R_n = R$, since by definition the reaction of a smooth surface is always orthogonal to the tangent, i.e. $R_\tau = 0$, and this means that the normal n to the trajectory coincides with the normal to the surface. Curves possessing this property are called *geodesic lines*.

Thus, if the surface is fixed, then the motion of the particle under no force is along a geodesic line of the surface. According to the first equation of motion in (4.12) $\left(m \dfrac{dv}{dt} = 0 \right)$, this motion occurs with constant velocity. We can also see this if we consider the kinetic energy integral.

97. Clairaut's theorem. Consider the case when a fixed surface is a surface of revolution about the z-axis and let a particle move over this surface unaffected by any force. We have just established that such a particle will move along a geodesic line with a constant velocity $v = $ const. In addition, the reaction **R** passes through the z-axis or is parallel to the z-axis and, consequently, the moment of the forces acting on the particle about the z-axis is zero, i.e. $M_z = 0$. Therefore there must be an area integral in the xy-plane, i.e. the angular momentum about the z-axis is constant during motion. We resolve the vector of the momentum $m\mathbf{v}$ in a tangential plane into components directed along a tangent to the meridian, $mv \cos i$ and along a tangent to the parallel, $mv \sin i$ (Fig. 92), where i is the angle which **v** makes with the meridian. The angular momentum tangent to the meridian is zero about the z-axis since the tangent to the meridian of the surface of revolution either cuts the z-axis or is parallel to it. The angular momentum tangent to the parallel is equal to $mv \sin i$ multiplied by r, where r is the distance between the particle M and the z-axis, and, consequently,

$$rv \sin i = \text{const.}$$

From this follows Clairaut's theorem

$$r \sin i = \text{const.}$$

To obtain the equations of geodesic lines on a surface of revolution, it is expedient to take cylindrical coordinates like those shown in Fig. 93. The differential of the arc in cylindrical coordinates has the form

$$ds^2 = dr^2 + r^2 d\theta^2 + dz^2.$$

Assume that the equation of the generatrix of the surface of revolution is $z = f(r)$. Then the differential of the arc on the surface can be reduced to the form

$$ds^2 = dr^2(1 + f_r'^2) + r^2 d\theta^2.$$

The kinetic energy principle in the motion of a particle unaffected by any

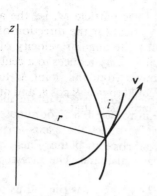

Figure 92

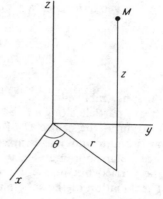

Figure 93

force on the surface of revolution can be written in the form

$$r'^2(1 + f_r'^2) + r^2\theta'^2 = v_0^2$$

and the law of areas in the form

$$r^2\theta' = C,$$

where C is twice the areal velocity of the projection of the particle M on the xy-plane. Hence

$$dt = \frac{r^2 d\theta}{C},$$

and this means that we can exclude time from the vis viva principle:

$$dr^2(1 + f_r'^2) + r^2 d\theta^2 = \frac{v_0^2}{C^2} r^4 d\theta^2,$$

or

$$dr^2(1 + f_r'^2) + \left(\frac{v_0^2}{C^2} r^4 - r^2\right) d\theta^2.$$

It is easy to separate the variables:

$$d\theta = \pm \frac{dr}{r} \sqrt{\frac{1 + f_r'^2}{\frac{v_0^2}{C^2} r^2 - 1}}.$$

Hence the geodesic lines are defined by the relation

$$\theta - \theta_0 = \pm \int_{r_0}^{r} \sqrt{\frac{1 + f_r'^2}{\frac{v_0^2}{C^2} r^2 - 1}} \frac{dr}{r}.$$

98. Spherical pendulum. Consider a problem concerned with the motion of a heavy particle along a fixed sphere. We introduce fixed coordinate axes with origin at the centre O of the sphere. We direct the z-axis vertically upwards and x and y axes arbitrarily in a horizontal plane. Then we introduce polar coordinates r, θ in the horizontal plane. The investigation will be carried out in the coordinates z, r, θ.

The equation of the sphere is $r^2 + z^2 = l^2$, where l is the radius of the sphere. The expression $U = -mgz$, where g is the acceleration due to gravity, is the force function of gravity. The kinetic energy integral

$$\frac{mv^2}{2} = -mgz + h'$$

can be reduced to the relation

$$z'^2 + r'^2 + r^2\theta'^2 = -2gz + h \qquad \left(h = \frac{2h'}{m} \right). \qquad (4.13)$$

In this case we have an area integral in the xy-plane

$$r^2 \frac{d\theta}{dt} = C$$

since the moment of the force of gravity and the reaction force about the z-axis are zero.

To exclude the polar coordinates from consideration, we multiply the equation of the kinetic energy (4.13) by r^2:

$$(rr')^2 + (r^2\theta')^2 + (l^2 - z^2)z'^2 = (-2gz + h)(l^2 - z^2).$$

Given the area integral and the relation $rr' = -zz'$, which results from the differentiation of the constraint equation, we have

$$(lz')^2 = (l^2 - z^2)(h - 2gz) - C^2,$$

or

$$l \frac{dz}{dt} = \pm\sqrt{\varphi(z)}, \qquad (4.14)$$

where

$$\varphi(z) = (l^2 - z^2)(h - 2gz) - C^2.$$

In mechanics the polynomial $\varphi(z)$ has three real roots. Indeed, let z_0 be the initial value of the coordinate z. Since the particle M lies on a sphere of radius l, we have $-l \leqslant z_0 \leqslant l$. The quantity $\varphi(z_0)$ cannot be negative by virtue of (4.14) since in mechanics the initial velocity $(dz/dt)_0$ cannot be imaginary. Let $\varphi(z_0) > 0$. If the area constant C is nonzero, then for $z = -l$ the function $\varphi(z)$ is negative. For sufficiently large z the sign of $\varphi(z) = 2gz^3 + \ldots$, as defined by the leading term, is positive. Hence, according to Lagrange's rule of signs, we infer that the polynomial $\varphi(z)$ has three real roots α, β, γ on the intervals $(-l, z_0)$, $(z_0, +l)$, $(+l, \infty)$ respectively (Fig. 94). In mechanics by virtue of (4.14), z can vary on the interval $(-l, +l)$ where $\varphi(z) \geqslant 0$, and, consequently,

$$\alpha \leqslant z \leqslant \beta.$$

We shall prove that α is always negative. We write out the polynomial $\varphi(z)$ in the form

$$\varphi(z) = 2g\left(z^3 - z^2\frac{h}{2g} - zl^2 + \frac{l^2h - c^2}{2g}\right),$$

from which we immediately have the following relation if the polynomial is represented in terms of the roots of $\varphi(z) = 2g(z - \alpha)(z - \beta)(z - \gamma)$:

$$\alpha\beta + \beta\gamma + \gamma\alpha = -l^2.$$

Hence

$$\gamma = -\frac{l^2 + \alpha\beta}{\alpha + \beta} > 0.$$

Since neither α nor β exceed l, the numerator $l^2 + \alpha\beta$ is always positive and, consequently, it follows from the preceding relation that $\alpha + \beta < 0$, or $\alpha < 0$.

Thus the lower edge of the strip between the parallels $z = \alpha$, $z = \beta$, in which the particle M must move, is below the equator ($\alpha < 0$) (Fig. 95).

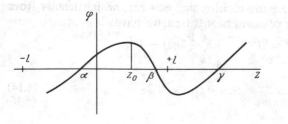

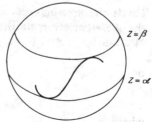

Figure 94   Figure 95

The particle M oscillates in this strip, periodically reaching the upper and the lower parallel, since t is defined by the elliptic integral

$$\int dt = \int \frac{l\,dz}{\pm\sqrt{\varphi(z)}}.$$

If the area constant C is nonzero, then, by virtue of the area integral, the angle θ always varies monotonically on the same side:

$$\frac{d\theta}{dt} = \frac{C}{r^2}.$$

When $C = 0$, we have a simple pendulum which we shall consider later.

4.4 Motion Along a Curve

99. General theorem. Consider a particle M of mass m which must remain on a curve given as the intersection of two surfaces:

$$f_1(t, x, y, z) = 0, \quad f_2(t, x, y, z) = 0. \tag{4.15}$$

The reaction $\mathbf{R}$ of a perfectly smooth line is orthogonal to the curve and, consequently, must lie in the plane which passes through the normals to the surfaces $f_1 = 0$ and $f_2 = 0$. Resolving the reaction force $\mathbf{R}$ into components along the normals to the indicated surfaces and noting that the cosines of the normals are proportional to the partial derivatives

$$\frac{\partial f_i}{\partial x}, \quad \frac{\partial f_i}{\partial y}, \quad \frac{\partial f_i}{\partial z} \quad (i = 1, 2),$$

we have

$$R_x = \lambda_1 \frac{\partial f_1}{\partial x} + \lambda_2 \frac{\partial f_2}{\partial x}, \quad R_y = \lambda_1 \frac{\partial f_1}{\partial y} + \lambda_2 \frac{\partial f_2}{\partial y},$$

$$R_z = \lambda_1 \frac{\partial f_1}{\partial z} + \lambda_2 \frac{\partial f_2}{\partial z},$$

where λ_1 and λ_2 are unknown factors. If we add the reaction force $\mathbf{R}$ to the force $\mathbf{F}$ with projections X, Y, Z, which acts on the particle M, then we can write the equation of motion of the particle as if it is without constraints but acted upon by a force $\mathbf{F} + \mathbf{R}$,

$$m\frac{d^2x}{dt^2} = X + \lambda_1 \frac{\partial f_1}{\partial x} + \lambda_2 \frac{\partial f_2}{\partial x},$$

$$m\frac{d^2y}{dt^2} = Y + \lambda_1 \frac{\partial f_1}{\partial y} + \lambda_2 \frac{\partial f_2}{\partial y}, \tag{4.16}$$

$$m\frac{d^2z}{dt^2} = Z + \lambda_1 \frac{\partial f_1}{\partial z} + \lambda_2 \frac{\partial f_2}{\partial z}.$$

If we substitute the values of the coordinates of the moving particle in the time function t into the constraint equations (4.15), we obtain identities with respect to t which can be differentiated with respect to t any number of times. Differentiating once, we get

$$\frac{\partial f_1}{\partial t} + \frac{\partial f_1}{\partial x}\frac{dx}{dt} + \frac{\partial f_1}{\partial y}\frac{dy}{dt} + \frac{\partial f_1}{\partial z}\frac{dz}{dt} = 0,$$

$$\frac{\partial f_2}{\partial t} + \frac{\partial f_2}{\partial x}\frac{dx}{dt} + \frac{\partial f_2}{\partial y}\frac{dy}{dt} + \frac{\partial f_2}{\partial z}\frac{dz}{dt} = 0.$$

If we multiply the equations of motion by dx, dy, dz respectively and add them together, then after simple transformations which we have already encountered, we obtain

$$d\frac{mv^2}{2} = X\,dx + Y\,dy + Z\,dz + \left(-\lambda_1\frac{\partial f_1}{\partial t} - \lambda_2\frac{\partial f_2}{\partial t}\right)dt.$$

To put it otherwise, the kinetic energy differential is equal to the work of the active forces during the true displacement plus the work of the reaction forces during the true displacement in time dt.

If the surfaces whose intersection defines the curve are fixed and the functions f_1 and f_2 do not explicitly depend on t ($\partial f_i/\partial t = 0$, $i = 1, 2$), then the expression of the work-kinetic energy theorem assumes a simple form

$$d\frac{mv^2}{2} = X\,dx + Y\,dy + Z\,dz.$$

100. Motion of a heavy particle along a curve. The vis viva principle is valid when a heavy particle M moves along a fixed curve. If we direct the z-axis vertically upwards, then the force of gravity will have projections $X = 0$, $Y = 0$, $Z = -mg$, where m is the mass of the particle and g is the acceleration of gravity. The work-kinetic energy theorem yields an equation

$$d\frac{mv^2}{2} = -mg\,dz,$$

or, when m is cancelled out and the equation integrated,

$$\frac{v^2}{2} = -gz + h.$$

Instead of the kinetic energy constant h we introduce a new constant a by a relation $h = ga$. Then

$$v^2 = 2g(a - z),$$

and, consequently, if we introduce a plane π defined by the equation $z - a = 0$, then we can write the preceding equation for the kinetic energy as

$$v^2 = 2gPM,$$

where PM is the distance between the particle M and the plane π (Fig. 96). In other words, the velocity of the particle M is such that it falls vertically without an initial velocity from the point P in the plane π. For a finite closed curve it is always possible to find a large value of the initial velocity v_0 such that the particle will move along the curve for an infinitely long time. Since $a = \dfrac{v_0^2}{2g} + z_0$ is large in this case, the plane π will not intersect the curve and, hence, the velocity will not turn into zero.

We use A' and A to designate the points of intersection of the plane π and the given curve. If the tangent at the point A is not horizontal, then the particle M will reach A. Indeed, from the kinetic energy integral we have

$$\left(\frac{ds}{dt}\right)^2 = 2g(a - z),$$

where s is the length of the arc reckoned along the curve in some direction. Assume that the initial point M_0 is associated with s_0 and $t_0 = 0$. From this we have

$$\frac{ds}{\pm\sqrt{2(a - z)g}} = dt,$$

and after integration we get

$$\sqrt{2g}\, t = \int_{s_0}^{s} \frac{ds}{\sqrt{a - z}}.$$

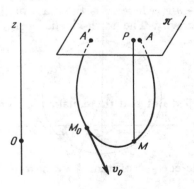

Figure 96

If the arc M_0A is equal to l, then in the vicinity of the point A we have

$$z(s) = z(l) + \frac{1}{\lambda!}(s - l)^\lambda \left(\frac{d^\lambda z}{ds^\lambda}\right)_{s = l} + \cdots,$$

where λ is the order of smallness of the quantity

$$(a - z) = z(l) - z(s) = -\frac{1}{\lambda!}(s - l)^\lambda \left(\frac{d^\lambda z}{ds^\lambda}\right)_{s = l}$$

as $s \to l$. Hence $\lambda/2$ is the order of smallness of the the quantity $\sqrt{a - z}$. We know from integral calculus that if $\lambda/2 \geqslant 1$, then the integral taken in the limits from the point M_0 to A is unbounded, and if $\lambda/2 < 1$, then the integral is bounded. The first case obtains for the point A at which the tangent to the curve is horizontal:

$$\frac{dz}{ds} = 0, \ldots, \quad \frac{\lambda}{2} \geqslant 1.$$

The second case obtains for the point A at which the tangent is not horizontal:

$$\frac{dz}{ds} \neq 0, \quad \lambda = 1, \quad \frac{\lambda}{2} = \frac{1}{2} < 1.$$

101. By way of example, we consider a simple pendulum, or the motion of a heavy particle M along a fixed circle of radius l which lies in a vertical plane. We choose fixed axes with origin at the centre of the circle and direct the z-axis vertically upwards. We base our consideration on the kinetic energy integral

$$v^2 = 2g(a - z).$$

Assume (Fig. 97) that

$$z = -l\cos\theta, \quad a = -l\cos\alpha,$$

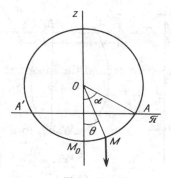

Figure 97

whence

$$a - z = l(\cos\theta - \cos\alpha) = l[(1 - \cos\alpha) - (1 - \cos\theta)]$$

$$= 2l\left[\sin^2\frac{\alpha}{2} - \sin^2\frac{\theta}{2}\right], \quad v = l\frac{d\theta}{dt}.$$

In this case the kinetic energy integral has the form

$$l^2\left(\frac{d\theta}{dt}\right)^2 = 4gl\left(\sin^2\frac{\alpha}{2} - \sin^2\frac{\theta}{2}\right),$$

and hence

$$\sqrt{\frac{g}{l}}\, dt = \frac{d\theta}{2\sqrt{\sin^2\dfrac{\alpha}{2} - \sin^2\dfrac{\theta}{2}}}.$$

We make the substitution

$$\sin\frac{\theta}{2} = u\sin\frac{\alpha}{2}.$$

Then we have

$$\frac{d\theta}{2} = \frac{k\,du}{\cos\dfrac{\theta}{2}} = \frac{k\,du}{\sqrt{1 - u^2k^2}},$$

where $k^2 = \sin^2\dfrac{\alpha}{2} < 1$. Integrating, we obtain

$$\sqrt{\frac{g}{l}}\, t = \int_0^u \frac{du}{\sqrt{(1 - u^2)(1 - k^2u^2)}}.$$

A quarter of the oscillation period T of the pendulum, i.e. the time taken by the pendulum to move from the point M_0 ($\theta = 0$, $u = 0$) to the point A ($\theta = \alpha$, $u = 1$), is

$$\sqrt{\frac{g}{l}}\, T = \int_0^1 \frac{du}{\sqrt{(1 - u^2)(1 - k^2u^2)}}.$$

To calculate this integral, we expand the integrand into a binomial series

$$\frac{1}{\sqrt{1 - k^2u^2}} = 1 + \frac{1}{2}k^2u^2 + \ldots + \frac{1\cdot 3\ldots(2n - 1)}{2\cdot 4\ldots 2n}k^{2n}u^{2n} + \ldots$$

and integrate term-by-term using the Wallis formula

$$\int_0^1 \frac{u^{2n}du}{\sqrt{1 - u^2}} = \frac{\pi}{2}\frac{1\cdot 3\ldots(2n - 1)}{2\cdot 4\ldots 2n}.$$

We obtain

$$T = \frac{\pi}{2} \sqrt{\frac{l}{g}} \left[1 + \frac{1}{2^2} k^2 + \ldots + \left(\frac{1 \cdot 3 \ldots (2n-1)}{2 \cdot 4 \ldots 2n} \right)^2 k^{2n} + \ldots \right].$$

As k and α tend to zero, we get the following expression for infinitesimal oscillations of a simple pendulum:

$$T = \frac{\pi}{2} \sqrt{\frac{l}{g}}.$$

4.5 Lagrange's Equations for a Particle

102. Assume now that the constraint equations (4.15) are identically satisfied with respect to t and q by the relations

$$x = x(q, t), \quad y = y(q, t), \quad z = z(q, t), \tag{4.17}$$

where q is a real variable. Since by virtue of these formulas for Cartesian coordinates, the numerical values of q define the position of the particle at the moment t, it follows that q is the defining (holonomic) coordinate. We can regard (4.17) as a parametric representation of the equations of constraints imposed on the particle. Differentiating, on this assumption, equations (4.15) with respect to the parameter q, we have

$$\frac{\partial f_i}{\partial x} \frac{\partial x}{\partial q} + \frac{\partial f_i}{\partial y} \frac{\partial y}{\partial q} + \frac{\partial f_i}{\partial z} \frac{\partial z}{\partial q} = 0 \quad (i = 1, 2).$$

Multiplying the equations of motion in the form of Lagrange's equations with multipliers (4.16) by $\partial x/\partial q$, $\partial y/\partial q$, $\partial z/\partial q$ respectively and adding the results, we obtain, by virtue of the last relations,

$$m \left(\frac{d^2 x}{dt^2} \frac{\partial x}{\partial q} + \frac{d^2 y}{dt^2} \frac{\partial y}{\partial q} + \frac{d^2 z}{dt^2} \frac{\partial z}{\partial q} \right) = X \frac{\partial x}{\partial q} + Y \frac{\partial y}{\partial q} + Z \frac{\partial z}{\partial q}. \tag{4.18}$$

The expression on the right-hand side is a generalized force

$$Q = X \frac{\partial x}{\partial q} + Y \frac{\partial y}{\partial q} + Z \frac{\partial z}{\partial q}.$$

To transform the left-hand side of (4.18), we consider the total time derivative of the x-coordinate:

$$\frac{dx}{dt} = x' = \frac{\partial x}{\partial q} q' + \frac{\partial x}{\partial t}.$$

Then we have

$$\frac{\partial x'}{\partial q'} = \frac{\partial x}{\partial q}. \tag{4.19}$$

Furthermore,

$$\frac{\partial x'}{\partial q} = \frac{\partial \left(\frac{\partial x}{\partial q}\right)}{\partial q} q' + \frac{\partial \frac{\partial x}{\partial q}}{\partial t} = \frac{d}{dt}\left(\frac{\partial x}{\partial q}\right). \tag{4.20}$$

Relations similar to (4.19) and (4.20) hold true for the y and z coordinates. We can represent the left-hand side of (4.18) in the form

$$\frac{d}{dt} m\left(x'\frac{\partial x}{\partial q} + y'\frac{\partial y}{\partial q} + z'\frac{\partial z}{\partial q}\right)$$

$$- m\left(x'\frac{d\frac{\partial x}{\partial q}}{dt} + y'\frac{d\frac{\partial y}{\partial q}}{dt} + z'\frac{d\frac{\partial z}{\partial q}}{dt}\right).$$

By virtue of (4.19) and (4.20) we have

$$\frac{d}{dt} m\left(x'\frac{\partial x'}{\partial q'} + y'\frac{\partial y'}{\partial q'} + z'\frac{\partial z'}{\partial q'}\right)$$

$$- m\left(x'\frac{\partial x'}{\partial q} + y'\frac{\partial y'}{\partial q} + z'\frac{\partial z'}{\partial q}\right) = \frac{d}{dt}\left(\frac{\partial T}{\partial q'}\right) - \frac{\partial T}{\partial q},$$

where T is the kinetic energy of the particle,

$$T = \frac{m}{2}(x'^2 + y'^2 + z'^2).$$

Substituting this expression into (4.18), we get

$$\frac{d}{dt}\left(\frac{\partial T}{\partial q'}\right) - \frac{\partial T}{\partial q} = Q.$$

These equations were derived by Lagrange.

We are interested in Lagrange's equations because all we need to know to write them in terms of the defining variable q is the expression for the kinetic energy T as a function of the defining coordinate q, its derivative q' and time t, and the generalized force Q. If the forces have a force function U, then $Q = \partial U/\partial q$, and Lagrange's equations assume the form

$$\frac{d}{dt}\left(\frac{\partial T}{\partial q'}\right) - \frac{\partial T}{\partial q} = \frac{\partial U}{\partial q}. \tag{4.21}$$

The kinetic energy T has the form

$$T = T_2 + T_1 + T_0,$$

where

$$T_2 = \frac{m}{2} \left(\left(\frac{\partial x}{\partial q} \right)^2 + \left(\frac{\partial y}{\partial q} \right)^2 + \left(\frac{\partial z}{\partial q} \right)^2 \right) q'^2,$$

$$T_1 = m \left(\frac{\partial x}{\partial q} \frac{\partial x}{\partial t} + \frac{\partial y}{\partial q} \frac{\partial y}{\partial t} + \frac{\partial z}{\partial q} \frac{\partial z}{\partial t} \right) q',$$

$$T_0 = \frac{m}{2} \left(\left(\frac{\partial x}{\partial t} \right)^2 + \left(\frac{\partial y}{\partial t} \right)^2 + \left(\frac{\partial z}{\partial t} \right)^2 \right).$$

Let us try to derive the work-kinetic energy theorem from Lagrange's equation. We multiply (4.21) by q' and then, after a simple transformation, we have

$$\frac{d}{dt} \left(q' \frac{\partial T}{\partial q'} \right) - \frac{\partial T}{\partial q'} q'' - \frac{\partial T}{\partial q} q' = \frac{\partial U}{\partial q} q'.$$

The first term can be transformed in accordance with Euler's theorem on homogeneous functions, and the other terms can be completed to get a total derivative with respect to time:

$$\frac{d}{dt} (2T_2 + T_1) - \frac{dT}{dt} + \frac{\partial T}{\partial t} = \frac{dU}{dt} - \frac{\partial U}{\partial t}$$

or

$$\frac{d}{dt} (T_2 - T_0) + \frac{\partial T}{\partial t} = \frac{dU}{dt} - \frac{\partial U}{\partial t}.$$

This equality expresses the work-kinetic energy theorem.

If T and U do not depend explicitly on time, then

$$\frac{\partial T}{\partial t} = 0, \quad \frac{\partial U}{\partial t} = 0,$$

and the work-kinetic energy theorem assumes the form

$$\frac{d}{dt} (T_2 - T_0) = \frac{dU}{dt}.$$

Integrating, we have a kinetic energy integral

$$T_2 - T_0 = U + h.$$

If the constraints do not depend explicitly on time $\left(\frac{\partial x}{\partial t} = 0, \ \frac{\partial y}{\partial t} = 0, \right.$
$\left. \frac{\partial z}{\partial t} = 0 \right)$, then $T = T_2$ ($T_1 = 0$, $T_0 = 0$) and T and U do not depend on t either $\left(\frac{\partial T}{\partial t} = 0, \ \frac{\partial U}{\partial t} = 0 \right)$. Then the living force integral assumes the

form

$$T = U + h.$$

Example. Consider a ray OM which passes through a fixed point O and rotates about O with a constant angular velocity ω (Fig. 98). Then T can be immediately found with the use of the theorem on the addition of velocities of the particle M in a compound motion

$$T = \frac{mv^2}{2} = \frac{m}{2}\left[\left(\frac{dr}{dt}\right)^2 + \omega^2 r^2\right].$$

If the particle M, which must necessarily remain on the ray OM, is acted upon by a force with a force function $V(r)$, then the expression for the motion includes the kinetic energy integral

$$\frac{m}{2}\left(\frac{dr}{dt}\right)^2 - \frac{m}{2}\omega^2 r^2 = U(r) + h.$$

103. Let us consider the general case. Let the Cartesian coordinates x, y, z of a particle be expressed as some definite functions of the variables q_α ($\alpha = 1, 2, 3$) and time t:

$$x = x(q_1, q_2, q_3, t), \quad y = y(q_1, q_2, q_3, t), \quad z = z(q_1, q_2, q_3, t), \quad (4.22)$$

where the variables q_α are assumed to be independent. By virtue of (4.22), the variables q_α are defining coordinates in the sense that the specification of the variables q_α defines the position of the particle.

Equations (4.22) express the constraints imposed on the particle. If $\alpha = 1$, then the particle is constrained by the condition that it must remain on the curve, if $\alpha = 1, 2$, then the particle must remain on a surface, and if $\alpha = 1, 2, 3$, then relation (4.22) in general (if we do not consider unilateral constraints) expresses the transition from the Cartesian coordinates x, y, z to some curvilinear coordinates q_1, q_2, q_3.

If we assume the constraints imposed on the particle to be ideal, then the reaction of these constraints $\mathbf{R}\,(R_x, R_y, R_z)$ is orthogonal, by definition, to a vector with projections

$$\frac{\partial x}{\partial q_\alpha}, \quad \frac{\partial y}{\partial q_\alpha}, \quad \frac{\partial z}{\partial q_\alpha} \quad (\alpha = 1, 2, 3),$$

or

$$R_x \frac{\partial x}{\partial q_\alpha} + R_y \frac{\partial y}{\partial q_\alpha} + R_z \frac{\partial z}{\partial q_\alpha} = 0 \quad (\alpha = 1, 2, 3). \quad (4.23)$$

Consider the equations of motion

$$m\frac{d^2 x}{dt^2} = X + R_x, \quad m\frac{d^2 y}{dt^2} = Y + R_y, \quad m\frac{d^2 z}{dt^2} = Z + R_z.$$

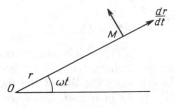

Figure 98

We multiply them by $\dfrac{\partial x}{\partial q_\alpha}$, $\dfrac{\partial y}{\partial q_\alpha}$, $\dfrac{\partial z}{\partial q_\alpha}$ respectively and add the results. By virtue of (4.23), we get

$$m\left(\frac{d^2x}{dt^2}\frac{\partial x}{\partial q_\alpha} + \frac{d^2y}{dt^2}\frac{\partial y}{\partial q_\alpha} + \frac{d^2z}{dt^2}\frac{\partial z}{\partial q_\alpha}\right) = X\frac{\partial x}{\partial q_\alpha} + Y\frac{\partial y}{\partial q_\alpha} + Z\frac{\partial z}{\partial q_\alpha},$$

or, repeating the transformations, we obtain

$$\frac{d}{dt}\left(\frac{\partial T}{\partial q'_\alpha}\right) - \frac{\partial T}{\partial q_\alpha} = Q_\alpha \quad (\alpha = 1,\, 2,\, 3).$$

These are Lagrange's equations. Equations of motion in Lagrange's form are valid for the defining coordinates q_α. To write them, we must know the kinetic energy T and the generalized forces Q_α. Note that it is insufficient to know T and Q_α in terms of any variables to write the equations in Lagrange's form; the coordinates must be defining.

For instance (Neumann's error), when a particle moves in a plane, the kinetic energy T in the variables r (a radius vector) and σ (twice the sector area) has a simple form

$$T = \frac{m}{2}\left(r'^2 + \frac{\sigma'^2}{r^2}\right).$$

If the forces have function $U(r)$, then $Q_r = \dfrac{\partial U}{\partial r}$ and $Q_\sigma = 0$. It is easy to write Lagrange's equations for $q_1 = r$, $q_2 = \sigma$ but they will not be equations of motion since σ is not a defining coordinate.

4.6 Relative Motion

104. Equations of motion. We consider fixed rectangular Cartesian coordinate axes $x_1 y_1 z_1$ in a fixed space and some moving coordinate axes xyz (Fig. 99). The equation of motion of a particle which is without constraints and acted upon by a force $\mathbf{F}$ has the form

$$m\mathbf{j}_a = \mathbf{F}.$$

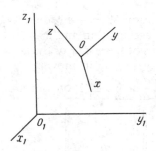

Figure 99

If we take into account the theorem on the composition of motions, we have

$$\mathbf{j}_a = \mathbf{j}_t + \mathbf{j}_r + \mathbf{j}_c,$$

where $\mathbf{j}_a$ is the acceleration relative to the fixed axes $x_1 y_1 z_1$, $\mathbf{j}_r$ is the acceleration relative to the moving coordinate system xyz, $\mathbf{j}_t$ is the transportation acceleration, i.e. the acceleration of the point of the moving system xyz at which a particle with mass m is at the moment t, and $\mathbf{j}_c$ is the Coriolis acceleration. Then

$$m\mathbf{j}_r = \mathbf{F} - m\mathbf{j}_t - m\mathbf{j}_c.$$

The equations of motion relative to moving axes have the same form as those relative to fixed axes, namely,

$$m\mathbf{j}_r = \mathbf{F}_r,$$

where the accelerating force in a moving system

$$\mathbf{F}_r = \mathbf{F} - m\mathbf{j}_t - m\mathbf{j}_c,$$

$-m\mathbf{j}_t$, $-m\mathbf{j}_c$ are Coriolis forces.[*]

If a moving system is in a uniform translation and moves in a straight line relative to a fixed system, then

$$\mathbf{j}_t = 0, \quad \mathbf{j}_c = 0$$

and, consequently, $\mathbf{F}_r = \mathbf{F}$ is Galileo's relativity principle[**].

---

[*] These terms are usually called inertial forces in a relative motion: $-m\mathbf{j}_t$ is the force of moving space, and $-m\mathbf{j}_c$ is the compound centrifugal force. — *Ed.*

[**] Coordinate systems for which Galileo's relativity principle holds true are called Galilean or inertial systems. — *Ed.*

If we use the designation $\mathbf{W} = -m\mathbf{j}_t$ for the Coriolis force associated with the transportation acceleration and express the Coriolis acceleration in terms of the projections of the vector of the instantaneous angular velocity of the moving axes $\omega(p, q, r)$ in moving axes and the projections of the relative velocity $\mathbf{v}_r\left(\dfrac{dx}{dt}, \dfrac{dy}{dt}, \dfrac{dz}{dt}\right)$ on the moving axes

$$\mathbf{j}_c = 2 \begin{Vmatrix} p & q & r \\ \dfrac{dx}{dt} & \dfrac{dy}{dt} & \dfrac{dz}{dt} \end{Vmatrix} = 2(\omega \times \mathbf{v}_r),$$

then we have the following equations of motion in terms of the projections on the moving axes:

$$m\frac{d^2x}{dt^2} = X + W_x - 2m\left(q\frac{dz}{dt} - r\frac{dy}{dt}\right),$$

$$m\frac{d^2y}{dt^2} = Y + W_y - 2m\left(r\frac{dx}{dt} - p\frac{dz}{dt}\right),$$

$$m\frac{d^2z}{dt^2} = Z + W_z - 2m\left(p\frac{dy}{dt} - q\frac{dx}{dt}\right),$$

where W_x, W_y, W_z are the projections of the Coriolis force $\mathbf{W}$.

105. The work-kinetic energy theorem in relative motion. We multiply the equations of motion by dx, dy, and dz respectively and add the results. We get

$$d\frac{mv_r^2}{2} = X\,dx + Y\,dy + Z\,dz + m(W_x dx + W_y dy + W_z dz).$$

The variation of the kinetic energy $d\dfrac{mv_r^2}{2}$ in relative motion is equal to the work done by the forces during a true displacement relative to a moving system of coordinates plus the work of the Coriolis force $\mathbf{W}$ associated with the transportation motion during the same true displacement. The Coriolis force $-m\mathbf{j}_c$ associated with the Coriolis acceleration $\mathbf{j}_c$ does not do any work during the true relative displacement.

106. Variation of the acceleration of gravity with latitude. We consider a particle M which is at rest on the earth. The force of gravity $m\mathbf{g}$ is the geometric sum of two forces, the force of attraction $m\mathbf{a}$ directed away from the particle M to the centre O of the forces of attraction and the force of moving space (Coriolis force) $\mathbf{W} = -m\mathbf{j}_t$ associated with transportation acceleration (Fig. 100). We assume that the earth is a sphere with radius R and centre at O and that the attraction, force is constant for all particles on the earth and passes through O. Let θ be the geocentric latitude of the particle M and φ be the celestial latitude of the particle M, i.e. the angle between the plane of the equator and the vertical (or the direction of the force of gravity) at the point M. We assume the angular velocity of the earth ω to be constant and so the Coriolis force $- m\mathbf{j}_t$ reduces to a centrifugal force, lies on the perpendicular MC drawn from the point M to the axis of rotation of the earth, is numerically equal to $m\omega^2 R \cos\theta$, and directed along CM. From the figure we immediately see that

$$mg \sin\varphi = ma \sin\theta,$$
$$mg \cos\varphi = ma \cos\theta - mR\omega^2 \cos\theta.$$

From this, and given

$$\mu = \frac{R}{a}\omega^2 \cong \frac{1}{289},$$

we have

$$\tan\varphi = \frac{\tan\theta}{1 - \mu},$$

$$g = a\sqrt{1 - 2\mu\cos^2\theta + \mu^2\cos^2\theta} \cong a(1 - \mu\cos^2\theta),$$

i.e. the acceleration due to gravity decreases with a decrease in the geocentric latitude of the location.

107. A heavy particle falling to the earth. We take a point M (Fig. 101) on the earth and direct the z-axis vertically upwards from this point, the y-axis along a tangent to the parallel to the east, and the x-axis along a tangent to the celestial meridian at right angles to the y and z axes to the south. We use φ for the celestial latitude of the location of the point M.

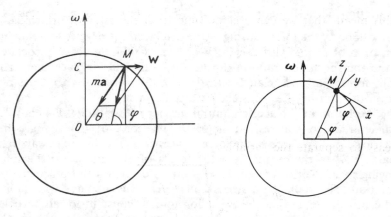

Figure 100   Figure 101

The projections of the angular velocity of the earth ω on the axes are $-\omega\cos\varphi$, 0, $\omega\sin\varphi$ and the projections of the forces of gravity

$$X + W_x = 0, \quad Y + W_y = 0, \quad Z + W_z = -mg.$$

Hence, when the mass m is cancelled out, the equations of the motion of the heavy particle on the earth in the $Mxyz$ axes have the form

$$\frac{d^2x}{dt^2} = 2\omega\sin\varphi\,\frac{dy}{dt},$$
$$\frac{d^2y}{dt^2} = -2\omega\left(\sin\varphi\,\frac{dx}{dt} + \cos\varphi\,\frac{dz}{dt}\right),$$
$$\frac{d^2z}{dt^2} = -g + 2\omega\cos\varphi\,\frac{dy}{dt}.$$

We seek the solution of these equations by successive approximation for the following initial data:

$$t_0 = 0, \quad x_0 = y_0 = z_0 = 0, \quad x_0' = y_0' = z_0' = 0.$$

We assume the first approximation to be a solution of the equations of motion when the initial data are substituted into the right-hand sides:

$$\frac{d^2x}{dt^2} = 0, \quad \frac{d^2y}{dt^2} = 0, \quad \frac{d^2z}{dt^2} = -g.$$

The integration of these equations for the assumed initial data yiedls

$$x = 0, \quad y = 0, \quad z = -\frac{gt^2}{2}.$$

The second approximation is defined by the equations of motion when the first approximation we have found is substituted into their right-hand sides:

$$\frac{d^2x}{dt^2} = 0, \quad \frac{d^2y}{dt^2} = 2\omega gt\cos\varphi, \quad \frac{d^2z}{dt^2} = -g.$$

The integration yields

$$x = 0, \quad y = \frac{1}{3}\omega\cos\varphi\,gt^3, \quad z = -\frac{gt^2}{2},$$

i.e. in the second approximation a heavy particle deviates to the east as it falls onto the rotating earth.

108. The Foucault pendulum. Consider infinitesimal oscillations of the Foucault pendulum or oscillations of the simple pendulum on the rotating earth. We set up local coordinate axes at the point O where the pendulum is suspended. The z-axis is directed vertically upwards, the y-axis is directed to the east along a tangent to the parallel passing through the point O, and the x-axis is orthogonal to the other axes or directed to the south along a tangent to the meridian. Assume that l is the length of the Foucault pendulum, ω is the angular velocity of the earth, φ is the celestial latitude of the point O, N is the reaction of the string of the pendulum, and m is its mass. In these axes the force of gravity acting on the pendulum has projections

$$X + W_x = 0, \quad Y + W_y = 0, \quad Z + W_z = -mg.$$

The coordinates of the pendulum are x, y, z. The projections of the instantaneous angular velocity of the earth ω on the coordinate axes are $-\omega\cos\varphi$, 0, and $\omega\sin\varphi$ respectively. The projections of the Coriolis force associated with the Coriolis acceleration are defined by the matrix

$$-2m \begin{vmatrix} -\omega\cos\varphi & 0 & \omega\sin\varphi \\ \dfrac{dx}{dt} & \dfrac{dy}{dt} & \dfrac{dz}{dt} \end{vmatrix}.$$

Hence we get the following differential equations of motion of the Faucault pendulum:

$$m\frac{d^2x}{dt^2} = -N\frac{x}{l} + 2m\omega\sin\varphi\,\frac{dy}{dt},$$
$$m\frac{d^2y}{dt^2} = -N\frac{y}{l} - 2m\omega\left(\sin\varphi\,\frac{dx}{dt} + \cos\varphi\,\frac{dz}{dt}\right),$$
$$m\frac{d^2z}{dt^2} = -N\frac{z}{l} + 2m\cos\varphi\,\frac{dy}{dt} - mg.$$

If we consider infinitesimal oscillations, we can assume that $z \cong l$. In this approximation the first two equations yield

$$m\frac{d^2x}{dt^2} = -N\frac{x}{l} + 2m\omega\sin\varphi\,\frac{dy}{dt},$$
$$m\frac{d^2y}{dt^2} = -N\frac{y}{l} - 2m\omega\sin\varphi\,\frac{dx}{dt}.$$

Multiplying the first equation by $-y$ and the second by x and adding the results we obtain

$$m\left(x\frac{d^2y}{dt^2} - y\frac{d^2x}{dt^2}\right) = m\frac{d}{dt}\left(x\frac{dy}{dt} - y\frac{dx}{dt}\right)$$
$$= -2m\omega\sin\varphi\left(x\frac{dx}{dt} + y\frac{dy}{dt}\right),$$

or

$$d\left(x\frac{dy}{dt} - y\frac{dx}{dt}\right) = -\omega\sin\varphi\, d(x^2 + y^2).$$

If we introduce $\varrho^2 = x^2 + y^2$ and the polar angle θ of the projection of the point on the xy-plane, we can rewrite the preceding equation as

$$d\left(\varrho^2\frac{d\theta}{dt}\right) = -\omega\sin\varphi\, d\varrho^2.$$

Integrating, we obtain

$$\varrho^2\frac{d\theta}{dt} = -\omega\sin\varphi \cdot \varrho^2 + C.$$

We assume that at the initial moment ϱ is zero. Hence $C = 0$ and we get

$$\frac{d\theta}{dt} = -\omega\sin\varphi,$$

i.e. the plane of oscillation of the pendulum rotates from the east through the south to the west. The complete revolution of the plane takes the time

$$T = \frac{2\pi}{\omega\sin\varphi}.$$

But $2\pi/\omega$ is equal to the period of revolution of the earth about its axis, i.e. 24 hours and, consequently,

$$T = \frac{24}{\sin\varphi}\ \text{hours.}$$

CHAPTER 5

DYNAMICS OF A SYSTEM

5.1 Moments of Inertia

109. Let us consider a plane A, an L-axis, and a point O in space and designate the distance from a particle of mass m of a mechanical system to the plane A, the L-axis, and the point O as δ, Δ, r respectively.

We can compose expressions

$$\Pi = \sum m\delta^2, \quad J = \sum m\Delta^2, \quad \mu = \sum mr^2.$$

The summation is over all the particles of the system. These expressions are the *moments of inertia* with respect to the plane A, the L-axis, and the point O respectively.

The analytic expressions of the moments of inertia about the principal coordinate elements are related as

$$J_z = \Pi_{xz} + \Pi_{zy}, \quad \mu = \Pi_{zx} + \Pi_{zy} + \Pi_{xy}.$$

The moment of inertia about an axis is equal to the sum of the moments of inertia with respect to two orthogonal planes through which the axes pass.

The moment of inertia about a point is equal to the sum of the moments of inertia with respect to the three orthogonal planes which meet at that point.

110. Distributed masses. A sphere. If the distribution of the points is continuous, then the sums pass into the corresponding integrals.

Example. Determine the moment of inertia of a homogeneous sphere of radius R about its centre and diameter. We have (Fig. 102)

$$\mu = \int\limits_0^R \varrho \cdot 4\pi r^2 dr \cdot r^2 = \frac{4}{5}\,\pi\varrho R^5 = \frac{3}{5}MR^2, \quad J_z = \frac{2}{3}\mu,$$

where $M = \dfrac{4}{3}\,\pi R^3 \varrho$ is the mass of the sphere. Since the moments of inertia with respect to the diametral planes are all equal, we have $\Pi = \dfrac{1}{3}\mu$, and $J_z = 2\Pi$.

111. Steiner's theorem (parallel axes rule). Given a mechanical system with coordinates a, b, c, let us draw through its centre of mass axes x', y', z' parallel to the axes of the given system of coordinates (Fig. 103).

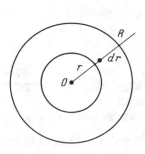
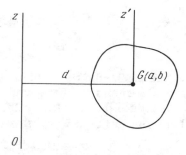

Figure 102   Figure 103

Then $x = x' + a$, $y = y' + b$ and

$$J_z = \sum m[(x' + a)^2 + (y' + b)^2] = \sum m(x'^2 + y'^2) + M(a^2 + b^2),$$

or

$$J_z = J_{z'} + Md^2.$$

The moment of inertia about a z-axis is equal to the moment of inertia about a parallel z'-axis which passes through the centre of mass plus the product of the total mass M by the square of the distance between the two axes.

Corollary. For parallel axes, the axis passing through the centre of mass has the lowest moment of inertia.

Remark. A similar statement can be made for the moments Π.

Example 1. Find the moment of inertia of a homogeneous circular cylinder of density ϱ, radius R, and unit altitude about its axis and generatrix.

We find

$$J_0 = \int\limits_0^R \varrho 2\pi r \, dr \cdot r^2 = \frac{\pi\varrho}{2} R^4 = \frac{MR^2}{2},$$

where $M = \pi R^2 \varrho$ is the mass of the cylinder. According to Steiner's theorem

$$J_A = \frac{MR^2}{2} + MR^2 = \frac{3}{2} MR^2.$$

Example 2. Find the moments of inertia of a homogeneous bar of density ϱ and length l (Fig. 104) about its midpoint G and endpoint O.

We have

$$J_0 = \int\limits_0^l \varrho \, dx \cdot x^2 = \frac{\varrho l^3}{3} = \frac{Ml^2}{3}, \qquad J_0 = \frac{Ml^2}{3} = J_G + M\left(\frac{l}{2}\right)^2.$$

Hence

$$J_G = Ml^2 \left(\frac{1}{3} - \frac{1}{4}\right) = \frac{Ml^2}{12}.$$

Example 3. Find the moments of inertia of homogeneous ellipsoid with the semi-major, semi-mean, and semi-minor axes a, b, c respectively.

We reduce this problem to the problem on the moments of inertia of a homogeneous sphere, which we have solved, since the transformation $x = ax'$, $y = by'$, $z = cz'$ reduces the ellipsoid $\dfrac{x^2}{a^2} + \dfrac{y^2}{b^2} + \dfrac{z^2}{c^2} = 1$ to a sphere of unit radius. We have

$$\Pi_{xy} = \iiint_{\text{ellip}} \varrho z^2 dx\, dy\, dz = \varrho abc^3 \iiint_{\text{sphere}} z'^2 dx'\, dy'\, dz'$$

$$= \frac{4}{15}\, \pi\varrho abc^3 = \left(\frac{4}{3}\, \pi abc\varrho\right) \frac{c^2}{5} = \frac{Mc^2}{5}.$$

112. Ellipsoid of inertia. Consider a system of particles with coordinates x_i, y_i, z_i and masses m_i. We shall omit the index i for simplicity. We consider a ray which passes through the origin O and has the quantities α, β, γ as direction cosines (Fig. 105). The square of the distance from the particle m to the ray is

$$\Delta^2 = (x^2 + y^2 + z^2) - (x\alpha + y\beta + z\gamma)^2,$$

or

$$\Delta^2 = (x^2 + y^2 + z^2)(\alpha^2 + \beta^2 + \gamma^2) - (x\alpha + y\beta + z\gamma)^2.$$

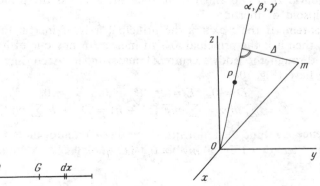

Figure 104   Figure 105

Hence the moment of inertia J of the mechanical system relative to the ray (α, β, γ) is

$$J = \sum m\Delta^2 = A\alpha^2 + B\beta^2 + C\gamma^2 - 2D\beta\gamma - 2E\gamma\alpha - 2F\alpha\beta, \qquad (5.1)$$

where

$$A = \sum m(y^2 + z^2), \; ..., \quad D = \sum myz, \; ...^*$$

On the ray we construct a point P using the formula $OP = 1/\sqrt{J}$. The quantities

$$X = \frac{\alpha}{\sqrt{J}}, \quad Y = \frac{\beta}{\sqrt{J}}, \quad Z = \frac{\gamma}{\sqrt{J}}$$

are its coordinates. Hence, by virtue of (5.1), the equation of the locus of points P is

$$AX^2 + BY^2 + CZ^2 - 2DYZ - 2EZX - 2FXY = 1.$$

If the body is three-dimensional (not straight lines) and finite, then the surface of the locus of points is an *ellipsoid* since it is a second-order surface which does not have points at infinity ($J \neq 0$). An ellipsoid of this kind is known as an *ellipsoid of inertia*, or *momental ellipsoid*, of the system constructed for the point O. The ellipsoid of inertia of the system varies when the point relative to which it is constructed is changed.

An ellipsoid of inertia at the centre of gravity, or central ellipsoid of inertia, is an ellipsoid of inertia constructed relative to the centre of mass of the mechanical system or body.

The principal axes of an ellipsoid of inertia of a body are known as its *principal axes of inertia* relative to the point O.

Remark. If $D = 0$ and $E = 0$, then the z-axis is the principal axis of the ellipsoid of inertia.

Theorem. If the z-axis is the principal axis of inertia for two of its points, then it is the principal axis of inertia for any one of its points and passes through the body's centre of mass.

We have (Fig. 106)

$$D = 0, \quad E = 0, \quad D' = 0, \quad E' = 0,$$
$$D = \sum myz = \sum my'(z' + h) = D' + h \sum my',$$

and hence

$$\sum my' = 0, \quad \text{i.e. } \eta' = 0.$$

---

* A, B, C are the moments of inertia with respect to the coordinate axes and D, E, F are the products of inertia with respect to the same axes. — *Ed.*

Similarly,

$$\sum mzx = \sum m(z' + h)x' = E' + h \sum mx'.$$

From this we have

$$\sum mx' = 0, \quad \text{i.e.} \ \xi' = 0.$$

Here ξ' and η' are the coordinates of the centre of mass, and $\sum mx' = M\xi'$, $\sum my' = M\eta'$. Furthermore, for an arbitrary point of the OO' axis we have

$$D^* = \sum my^*z^* = \sum my(z - h^*) = D - h^* \sum my = 0, \quad E^* = 0.$$

The converse is also true. If the principal axis of an ellipsoid of inertia for the point O' passes through body's centre of mass, then it is the principal axis of inertia for all its points.

Corollary. The principal axes of a central ellipsoid of inertia are principal axes for its every point.

Not every ellipsoid is an ellipsoid of inertia since the inequalities

$$A + B > C, \quad B + C > A, \quad C + A > B$$

must be satisfied for an ellipsoid of inertia.

113. Distribution of the principal axes of inertia in a body. Consider the principal axes xyz of a central ellipsoid of inertia of a mechanical system, the axes originating at the system's centre of mass. Assume that we are interested in the principal axes of inertia of the body relative to the point O', whose coordinates in the system of axes xyz are ξ, η, ζ respectively (Fig. 107). We draw a straight line through the point O' and designate the direction cosines of the line with the coordinate axes as α, β, γ respectively.

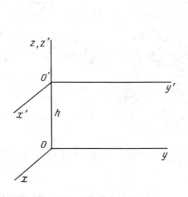

Figure 106

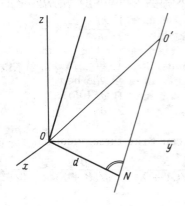

Figure 107

If A, B, C are the moments of inertia of a mechanical system with respect to the xyz axes, then the moment of inertia of the system about the straight line which passes through the centre of mass O and has α, β, γ as its direction cosines is

$$J = A\alpha^2 + B\beta^2 + C\gamma^2.$$

According to Steiner's theorem, the moment of inertia of the system about the straight line which passes through the point O' is

$$J' = J + Md^2,$$

where M is the total mass of the mechanical system and d is the distance between the straight line and the centre of mass O.

From the figure we immediately find that

$$d^2 = OO'^2 - O'N^2 = (\xi^2 + \eta^2 + \zeta^2) - (\xi\alpha + \eta\beta + \zeta\gamma)^2.$$

Hence

$$J' = A\alpha^2 + B\beta^2 + C\gamma^2 - M(\xi\alpha + \eta\beta + \zeta\gamma)^2 + M(\xi^2 + \eta^2 + \zeta^2).$$

To find the principal axes of inertia of a mechnical system about the point O', we must find the extremum of J' provided that $\alpha^2 + \beta^2 + \gamma^2 = 1$, since at least the semi-major axis or the semi-minor axis of the ellipsoid of inertia of the system constructed for the point O' possesses extremal properties.

We use Lagrange's method of the undetermined multiplier λ to find the conditional extremum and consider a function

$$F = J' + \lambda(\alpha^2 + \beta^2 + \gamma^2 - 1).$$

Hence the conditions for the extremum are

$$\frac{1}{2} \frac{\partial F}{\partial \alpha} = (A + \lambda)\alpha - M(\xi\alpha + \eta\beta + \zeta\gamma)\xi = 0,$$

$$\frac{1}{2} \frac{\partial F}{\partial \beta} = (B + \lambda)\beta - M(\xi\alpha + \eta\beta + \zeta\gamma)\eta = 0, \qquad (5.2)$$

$$\frac{1}{2} \frac{\partial F}{\partial \gamma} = (C + \lambda)\gamma - M(\xi\alpha + \eta\beta + \zeta\gamma)\zeta = 0;$$

We can write these equations as a proportion

$$\frac{\alpha}{\xi/(A + \lambda)} = \frac{\beta}{\eta/(B + \lambda)} = \frac{\gamma}{\zeta/(C + \lambda)} = M(\xi\alpha + \eta\beta + \zeta\gamma) = \varkappa. \quad (5.3)$$

Multiplying equations (5.2) by $\dfrac{\xi}{A + \lambda}$, $\dfrac{\eta}{B + \lambda}$ and $\dfrac{\zeta}{C + \lambda}$ respectively and adding the results, we get

$$(\xi\alpha + \eta\beta + \zeta\gamma)\left(1 - \frac{M\xi^2}{A+\lambda} - \frac{M\eta^2}{B+\lambda} - \frac{M\zeta^2}{C+\lambda}\right) = 0. \qquad (5.4)$$

Multiplying equations (5.2) by α, β, γ respectively and adding the results, we obtain

$$A\alpha^2 + B\beta^2 + C\gamma^2 + \lambda - M(\xi\alpha + \eta\beta + \zeta\gamma)^2 = 0.$$

Equation (5.4) is satisfied if one of the factors is eliminated. We assume it to be the second factor, i.e.

$$\frac{M\xi^2}{A+\lambda} + \frac{M\eta^2}{B+\lambda} + \frac{M\zeta^2}{C+\lambda} = 1. \qquad (5.5)$$

If the moments of inertia A, B, C are all different, then equations (5.5), which is of the third degree has three real roots, and if $A > B > C$, then the roots λ_i evidently lie on the intervals

$$-A < \lambda_3 < -B < \lambda_2 < -C < \lambda_1.$$

Substituting the roots λ_i into equation (5.5) and assuming ξ, η, ζ to be running coordinates, we get equations for the second-order central surfaces: an equation of an ellipsoid for $\lambda = \lambda_1$, an equation of a one-sheet hyperboloid for $\lambda = \lambda_2$, and an equation of a two-sheet hyperboloid for $\lambda = \lambda_3$.

If only two of the moments A, B, C are distinct then equation (5.5) is of the second degree and we have only two real roots, and, hence, the investigation is much simpler. When all the moments A, B, C are equal, i.e. when the central ellipsoid of inertia is a sphere, equation (5.5) is linear with respect to λ.

It is easy to consider these cases separately. They can also be obtained from the general case.

Let us return to the case when $A > B > C$. According to (5.3), the root λ_i is associated with the direction cosines α_i, β_i, γ_i. We shall prove that the directions corresponding to the different roots λ_i, λ_k ($i \neq k$) are mutually orthogonal. Indeed,

$$\alpha_i\alpha_k + \beta_i\beta_k + \gamma_i\gamma_k$$
$$= x_i x_k \left(\frac{\xi}{A+\lambda_i}\frac{\xi}{A+\lambda_k} + \frac{\eta}{B+\lambda_i}\frac{\eta}{B+\lambda_k} + \frac{\zeta}{C+\lambda_i}\frac{\zeta}{C+\lambda_k}\right)$$
$$= \frac{x_i x_k}{\lambda_k - \lambda_i}\left[\left(\frac{\xi^2}{A+\lambda_i} + \frac{\eta^2}{B+\lambda_i} + \frac{\zeta^2}{C+\lambda_i}\right)\right.$$
$$\left. - \left(\frac{\xi^2}{A+\lambda_k} + \frac{\eta^2}{B+\lambda_k} + \frac{\zeta^2}{C+\lambda_k}\right)\right] = 0$$

by virtue of (5.5). Consequently, the extremal directions are mutually orthogonal, two of them correspond to the semi-major and semi-minor axes of the ellipsoid of inertia constructed for the point O' and the third extremal direction ζ coincides, due to orthogonality, with the direction of the semi-mean axis of the ellipsoid constructed for the point O'. The moment of inertia about the axis defined by the root λ_i is

$$J_i' = M(\xi^2 + \eta^2 + \zeta^2) - \lambda_i \quad (i = 1, 2, 3).$$

This solution for the distribution of the principal moments of inertia in space admits of a compact geometric representation. We introduce new designations

$$A = Ma^2, \quad B = Mb^2, \quad C = Mc^2, \quad \lambda = M\varrho,$$

where a, b, c are the radii of inertia with respect to the principal axes of the central ellipsoid of inertia. In these designations equation (5.5) is written as

$$f = \frac{\xi^2}{a^2 + \varrho} + \frac{\eta^2}{b^2 + \varrho} + \frac{\zeta^2}{c^2 + \varrho} - 1 = 0.$$

Three surfaces pass through the point $O'(\xi, \eta, \zeta)$ which correspond to the values $\varrho = \varrho_i$ $(i = 1, 2, 3)$ and are confocal with a *gyrational ellipsoid*

$$\frac{\xi^2}{a^2} + \frac{\eta^2}{b^2} + \frac{\zeta^2}{c^2} = 1.$$

The normals to these surfaces coincide with the directions of the principal axes of inertia of the system for the point O' since by virtue of (5.3) we have

$$\frac{\partial f/\partial \xi}{\alpha_i} = \frac{\partial f/\partial \eta}{\beta_i} = \frac{\partial f/\partial \zeta}{\gamma_i}.$$

Thus the principal axes of inertia of a system in space are normals to the surfaces confocal with the gyrational ellipsoid.

114. Umbilical points. A point O' about which an ellipsoid of inertia is a sphere is called an *umbilical*, or *spherical*, point. Let us define the conditions under which a mechanical system has umbilical, or spherical, points and find these points. To simplify the analysis, we draw coordinate axes $x'y'z'$ through the point $O'(\xi, \eta, \zeta)$ parallel to the principal axes xyz of the central ellipsoid of inertia. The coordinates of the points of the mechanical system in these axes are related as

$$x = x' + \xi, \quad y = y' + \eta, \quad z = z' + \zeta.$$

If the point O' is umbilical, then the $x'y'z'$ axes are the principal axes of inertia and, consequently

$$\sum my'z' = 0, \quad \sum mz'x' = 0, \quad \sum mx'y' = 0.$$

In these formulas the summation is over all points of the system. Using the translation formulas, we have

$$0 = \sum my'z' = \sum myz - \eta \sum mz - \zeta \sum my + M\eta\zeta; \quad M = \sum m.$$

The sum Σmyz is zero since the xyz axes are the principal axes of the central ellipsoid of inertia; the sums Σmz, Σmy are zero since the centre of mass O of the system lies at the origin of coordinates xyz. Hence

$$\eta\zeta = 0$$

and, similarly,

$$\zeta\xi = 0, \quad \xi\eta = 0.$$

The last three equations are simultaneously satisfied if two coordinates of the point O' vanish. In other words, umbilical points may lie on the principal axes of a central ellipsoid of inertia. Without losing generality, we assume that $\xi = 0$, $\eta = 0$. Then, all the moments of inertia A', B', C' about the $x'y'z'$ axes must be equal for the umbilical point O' and the moments can be determined with the use of Steiner's theorem:

$$A' = A + M\zeta^2 = B' = B + M\zeta^2 = C' = C.$$

Hence the equality $A = B$, i.e. if there is an umbilical point on the z-axis, then the central ellipsoid of inertia must be an ellipsoid of revolution about the z-axis. Then for ζ we get an equation

$$\zeta^2 = \frac{C - A}{M},$$

which yields an inequality $C > A$. If there is an umbilical point, then the central ellipsoid of inertia must be contracted along the rotation axis, and then two umbilical points lie on the rotation axis symmetrically about the centre of mass O at the distance $\sqrt{\dfrac{C - A}{M}}$.

5.2 The Euler-Lagrange Principle

115. The fundamental principle of dynamics dates back to Johann Bernoulli's followers Hermann and Euler, who were the first academicians of the St. Petersburg Academy of Sciences. In *Phoronomia* (1716) Hermann solved the problem of a compound pendulum proceeding from the principle that if the "moving forces" are opposite in direction, they must be in equilibrium with the forces of gravity.

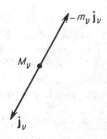

Figure 108

Euler generalized this principle and applied it to the definition of the oscillations of flexible bodies (1740) and to solutions of many other mechanics problems.

Lagrange made this principle fully general. In 1788 he published his famous *Mécanique analytiqiue*. Having thoroughly analysed the propositions and the problems that had been solved, Lagrange isolated, for the first time, the idea used by Hermann and Euler and developed it. Their idea can be explained as follows. Assume that M_ν are particles of a mechanical system, m_ν are their masses, r_ν are their radii vectors, and F_ν are the vectors of the forces acting on them. The system is assumed to be restricted by ideal constraints. Assume, furthermore, that under these constraints a particle M_ν has an acceleration j_ν at a particular moment (Fig. 108). If we apply one more force equal to $-m_\nu j_\nu$ to the particle M_ν, this force will stop the variation of the velocity. The point M_ν will be at rest or in uniform motion in a straight line since if the particle M_ν had no constraint, the force $m_\nu j_\nu$ would be sufficient to impart the acceleration j_ν. This is valid for every particle ($\nu = 1, ..., n$).

"... if we assume that we impart to every body a motion in the opposite direction to which it must acquire, then it is clear that the system will be reduced to the state of rest ... thus an inquilibrium must exist ... between the forces which can cause them (*Ed.*, these motions)."* Hence

$$\sum(F_\nu - m_\nu j_\nu)\delta r_\nu = 0.$$

Lagrange, who generalized the idea of Hermann and Euler, named this principle after D'Alembert, although D'Alembert practised complicated and laborious decomposition of motions to find the constraint forces. We shall call it the Euler-Lagrange principle**.

---

* J.L. Lagrange, *Méchanique analytique* (Paris: 1788).
** In literature it is usually called the D'Alembert principle. — *Ed.*

Some people forget the essence of the proposition of Euler and Lagrange, forget that we imagine that the forces $-m_\nu \mathbf{j}_\nu$ are added to the particles and, having called this imaginary force $-m_\nu \mathbf{j}_\nu$ an inertial force, begin discussing it as if it were a real force.

116. Linear constraints. Imagine a fixed (Euclidean) space with rectangular axes xyz. Assume that a mechanical system consists of n particles with masses m_ν and coordinates x_ν, y_ν, z_ν ($\nu = 1, \ldots, n$), respectively, and that the system's motion is restricted by linear constraints in the sense that the velocities x_ν', y_ν', z_ν' satisfy the linear relations

$$\sum (a_{j\nu} x_\nu' + b_{j\nu} y_\nu' + c_{j\nu} z_\nu') + l_j = 0 \quad (j = 1, \ldots, m), \tag{5.6}$$

where a, b, c, l are functions of time t and the coordinates x_ν, y_ν, z_ν ($\nu = 1, \ldots, n$). The primes denote time derivatives, or, in other words, x_ν', y_ν', z_ν' are equal to the projections of the velocity of the particle m_ν on the coordinate axes.

The *virtual displacements* δx_ν, δy_ν, δz_ν of the particle m_ν are defined by the system of equations

$$\sum (a_{j\nu} \delta x_\nu + b_{j\nu} \delta y_\nu + c_{j\nu} \delta z_\nu) = 0 \quad (j = 1, \ldots, m). \tag{5.7}$$

If l_j are nonzero, then the true displacements dx_ν, dy_ν, dz_ν do not satisfy equations (5.7) and are not among the virtual displacements. If the system of differential equations of constraints (5.6) is integrable in the sense that it reduces to a system $df_j (t, x_1, y_1, z_1, \ldots, x_n, y_n, z_n) = 0$ $(j = 1, \ldots, m)$, then the constraints are *holonomic*, and if the equations (5.6) are nonintegrable, then the constraints are *nonholonomic*.

If we multiply equations (5.6) by dt, we obtain

$$\sum (a_{j\nu} dx_\nu + b_{j\nu} dy_\nu + c_{j\nu} dz_\nu) + l_j dt = 0 \quad (j = 1, \ldots, m),$$

where dx_ν, dy_ν, dz_ν are the projections on the axes of the true displacement of the particle m_ν in time dt. We see that the virtual displacements δx_ν, δy_ν, δz_ν defined by equations (5.7) appear to be elementary displacements for "frozen" constraints for which t is fixed and dt is set to be zero in the last relation.

It should be pointed out that the virtual displacements δx_ν, δy_ν, δz_ν are not, in general, displacements between two infinitely close states of the mechanical system which do not violate the "frozen" constraints if their equations are not integrable for a fixed t, i.e.

$$\sum (a_{j\nu}^0 dx_\nu + b_{j\nu}^0 dy_\nu + c_{j\nu}^0 dz_\nu) = 0 \quad (j = 1, 2, \ldots, m).$$

Here $a_{j\nu}^0$, $b_{j\nu}^0$, $c_{j\nu}^0$ imply that the value of t is fixed in the functions $a_{j\nu}$, $b_{j\nu}$, $c_{j\nu}$.

If the constraint equations (5.6) are integrable (the constraints are holonomic), then after integration they can be reduced to the form

$$f_j(t, x_1, y_1, z_1, ..., x_n, y_n, z_n) = 0 \quad (j = 1, 2, ..., m). \tag{5.8}$$

In this case, by definition (according to (5.7)), the equations for virtual displacemets assume the form

$$\sum \left(\frac{\partial f_j}{\partial x_\nu} \delta x_\nu + \frac{\partial f_j}{\partial y_\nu} \delta y_\nu + \frac{\partial f_j}{\partial z_\nu} \delta z_\nu \right) = 0 \quad (j = 1, ..., m).$$

It follows that for systems restricted by holonomic constraints the virtual displacements δx_ν, δy_ν, δz_ν are not only elementary displacements for the "frozen" constraints for a fixed t, but also displacements between infinitely close states of the mechanical system which do not disturb the constraints "frozen" for the fixed t*.

After differentiating with respect to time the constraint equations (5.8), which are identities for motion, we have the following equations for the true displacements dx_ν, dy_ν, dz_ν:

$$\frac{\partial f_j}{\partial t} dt + \sum \left(\frac{\partial f_j}{\partial x_\nu} dx_\nu + \frac{\partial f_j}{\partial y_\nu} dy_\nu + \frac{\partial f_j}{\partial z_\nu} dz_\nu \right) = 0 \quad (j = 1, 2, ..., m).$$

Hence it follows that true displacemets for constraints dependent on time cannot be found among virtual displacements. If the constraints are holonomic and do not depend on time, $\frac{\partial f_j}{\partial t} = 0$ $(j = 1, ..., m)$, then some virtual displacements will be true displacements.

117. Ideal constraints. Let us imagine that acceleration forces with projections X_ν, Y_ν, Z_ν on the coordinate axes xyz act upon the particles m_ν of the mechanical system. Let us assume that the system is under linear constraints and use δx_ν, δy_ν, δz_ν to denote displacements of the particle m_ν, virtual for these constraints.

For the constraints we shall deal with we adopt the axiom stating that the action of the constraints imposed on a system can be replaced by the reaction forces acting on the particles m_ν of the system.

We use $R_{\nu x}$, $R_{\nu y}$, $R_{\nu z}$ to denote the projections of the reaction force acting on the particle m_ν. In other words, if we add reaction forces to the given forces, then we can regard the mechanical system as a system of particles free of constraints, with masses m_ν subjected to the forces $X_\nu + R_{\nu x}$, $Y_\nu + R_{\nu y}$, $Z_\nu + R_{\nu z}$ respectively. We shall deal with *ideal constraints*, for

---

* This is a reflection of an idea, often repeated by the author, that in the case of nonholonomic constraints two infinitely close states of a system permitted by the constraints cannot always be connected by an infinitely small curve which does not disturb the constraints imposed on the system. By contrast, this is always the case for holonomic constraints. — *Ed.*

which we adopt the axiom stating that the sum of the works of the reactions of ideal constraints over virtual displacements is zero, i.e.

$$\sum (R_{\nu x}\delta x_\nu + R_{\nu y}\delta y_\nu + R_{\nu z}\delta z_\nu) = 0.$$

If nonideal constraints must be considered, then we shall include the components of their reactions, which do not satisfy the condition of ideality in the given forces.

118. The Euler-Lagrange principle. The axiom of ideal constraints directly yields the fundamental principle of dynamics. Indeed, if we replace the constraints by reactions, then we can imagine the particle m_ν to be without constraints and subjected to the action of the forces X_ν, Y_ν, Z_ν and the reaction forces $R_{\nu x}$, $R_{\nu y}$, $R_{\nu z}$. Particles without constraints are defined by Newton's equations of motion

$$m_\nu \frac{d^2 x_\nu}{dt^2} = X_\nu + R_{\nu x}, \quad m_\nu \frac{d^2 y_\nu}{dt^2} = Y_\nu + R_{\nu y},$$

$$m_\nu \frac{d^2 z_\nu}{dt^2} = Z_\nu + R_{\nu z}, \quad (\nu = 1, \ldots, n). \tag{5.9}$$

Determining $R_{\nu x}$, $R_{\nu y}$, $R_{\nu z}$ from these relations and substituting the results into the expression for the axiom of ideal constraints, we get

$$\sum \left[\left(m_\nu \frac{d^2 x_\nu}{dt^2} - X_\nu \right) \delta x_\nu + \left(m_\nu \frac{d^2 y_\nu}{dt^2} - Y_\nu \right) \delta y_\nu \right.$$
$$\left. + \left(m_\nu \frac{d^2 z_\nu}{dt^2} - Z_\nu \right) \delta z_\nu \right] = 0. \tag{5.10}$$

In other words, the accelerations $d^2 x_\nu/dt^2$, $d^2 y_\nu/dt^2$, $d^2 z_\nu/dt^2$ of the particles m_ν in the true motion of a mechanical system, which is under ideal constraints, must necessarily satisfy relations (5.10) for all virtual displacements δx_ν, δy_ν, δz_ν compatible with the constraints.

Let us prove the sufficiency of (5.10) for true accelerations. We regard the system as a system of particles without constraints subjected to the forces $X_\nu + R_{\nu x}$, $Y_\nu + R_{\nu y}$, $Z_\nu + R_{\nu z}$. We can do this by virtue of the axiom stating that constraints can be replaced by reactions. We designate the virtual displacements for particles without constraints as $\hat{\delta} x_\nu$, $\hat{\delta} y_\nu$, $\hat{\delta} z_\nu$. These are arbitrary and independent quantities. The general formula (5.10) for this consideration necessarily yields

$$\sum \left[\left(m_\nu \frac{d^2 x_\nu}{dt^2} - X_\nu - R_{\nu x} \right) \hat{\delta} x_\nu + \left(m_\nu \frac{d^2 y_\nu}{dt^2} - Y_\nu - R_{\nu y} \right) \hat{\delta} y_\nu \right.$$
$$\left. + \left(m_\nu \frac{d^2 z_\nu}{dt^2} - Z_\nu - R_{\nu z} \right) \hat{\delta} z_\nu \right] = 0.$$

Since $\hat{\delta x_\nu}$, $\hat{\delta y_\nu}$, $\hat{\delta z_\nu}$ are arbitrary, we find from the last relation, on the basis of Lagrange's theorem, that

$$m_\nu \frac{d^2 x_\nu}{dt^2} = X_\nu + R_{\nu x}, \quad m_\nu \frac{d^2 y_\nu}{dt^2} = Y_\nu + R_{\nu y},$$

$$m_\nu \frac{d^2 z_\nu}{dt^2} = Z_\nu + R_{\nu z} \quad (\nu = 1, \ldots, n).$$

But we obtained (5.9) and thus proved the sufficiency of (5.10). Consequently, (5.10) is a principle sufficient to solve problems of dynamics. We shall call it the *Euler-Lagrange principle*.

5.3 General Theorems

119. The reactions $R_{\nu x}$, $R_{\nu y}$, $R_{\nu z}$ due to the constraints are absent in the Euler-Lagrange principle (5.9). This means that the equations directly derived from this principle are differential equations of motion which do not include any unknown reactions due to constraint.

If we decompose the set of virtual displacements of a system into their independent components and substitute the latter into the formula for the Euler-Lagrange principle, we obtain a complete system of independent differential equations of motion.

Now if we take one of the virtual displacements of the set and substitute it into the formula for the Euler-Lagrange principle, then we get a relation which is either a differential equation of motion or a corollary of the motion equations. We now present some general propositions to develop this idea.

120. Theorem on the motion of the centre of mass. Assume that the virtual displacements of a mechanical system as a rigid body include a translation along the x-axis

$$\delta x_\nu = a, \quad \delta y_\nu = 0, \quad \delta z_\nu = 0 \quad (\nu = 1, \ldots, m).$$

Substituting these virtual displacements δx_ν, δy_ν, δz_ν into the formula for the Euler-Lagrange principle, we get, after cancelling out a,

$$\sum \left(m_\nu \frac{d^2 x_\nu}{dt^2} - X_\nu \right) = 0.$$

But

$$\sum m_\nu x_\nu = M\xi,$$

where $M = \sum m_\nu$ is the net mass of the mechanical system and ξ is the coordinate of its centre of mass along the x-axis. We can now transform the last formula into

$$M \frac{d^2\xi}{dt^2} = \sum X_\nu,$$

i.e. if a mechanical system under ideal constraints can be translated as a rigid body along the x-axis, then the system's centre of mass will move in the direction of the x-axis as if all the forces $\sum X_\nu$ are applied to it and the total mass M of the system is concentrated in it.

We can simplify this theorem.

We divide the forces X_ν, Y_ν, Z_ν into external forces X_ν^e, Y_ν^e, Z_ν^e and internal forces X_ν^i, Y_ν^i, Z_ν^i and have

$$X_\nu = X_\nu^e + X_\nu^i, \quad Y_\nu = Y_\nu^e + Y_\nu^i, \quad Z_\nu = Z_\nu^e + Z_\nu^i.$$

We assume the internal forces to be divisible into pairs of forces which are equal in magnitude, opposite in direction, and lie on the same straight line, but are applied at different points of the system (interaction forces).

If a mechanical system is translated virtually along the x-axis as a rigid body, then the theorem on the motion of the centre of mass has the form

$$M \frac{d^2\xi}{dt^2} = \sum X_\nu^e,$$

since for internal forces we have

$$\sum X_\nu^i = 0.$$

If, in addition, $\sum X_\nu^e = 0$, then, after integrating the last equation, we get

$$\xi = \left(\frac{d\xi}{dt}\right)_0 t + \xi_0,$$

where ξ_0 and $(d\xi/dt)_0$ are the values of coordinate ξ and velocity $d\xi/dt$ respectively at the moment $t = 0$.

Let us consider some corollaries of this general statement.

If a mechanical system can be translated, as a rigid body along the x, y, z axes, then the equations of motion of the centre of mass are valid with respect to these axes,

$$M \frac{d^2\xi}{dt^2} = \sum X_\nu, \quad M \frac{d^2\eta}{dt^2} = \sum Y_\nu, \quad M \frac{d^2\zeta}{dt^2} = \sum Z_\nu,$$

i.e. the centre of mass of such a system with coordinates ξ, η, ζ moves in space as if all the given active forces are applied to it and the whole mass M is concentrated in it.

Remark. This proposition concerning the motion of the centre of mass of a mechanical system is the justification for the study of the dynamics of a particle.

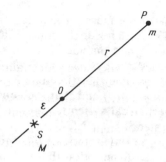

Figure 109

Example. *The two-body problem.* Assume that S is the sun, P is a planet, and O is the centre of mass of the sun and the planet (Fig. 109). We have an equation for the motion of the centre of mass of the sun and the planet. The interaction forces will cancel out in the equations of the motion of the centre of mass since they are internal forces. This means that the centre of mass O moves uniformly in a straight line. We choose an internal system of coordinates with origin at O since we can assume O to be a fixed point. Let M be the mass of the sun and m be the mass of the planet. The planet is subjected to Newton's attractive force

$$F = \frac{Mm}{SP^2}$$

if we choose a unit of measurement such that the attraction constant $f = 1$. Since $M\varepsilon = mr$, we have

$$F = \frac{Mm}{(r + \varepsilon)^2} = \frac{Mm}{(1 + m/M)^2} \frac{1}{r^2} = \frac{\mu m}{r^2} \quad \left(\mu = \frac{M}{(1 + m/M)^2} \right).$$

We can therefore assume that the planet is attracted to the fixed point O, and this is why we can consider the motion of planets as the motion of a particle attracted by a fixed pole.

Example. As an application of the general theorem on the motion of the centre of mass of a mechanical system we shall derive an equation of motion for a body of a variable mass, which is known as Meshchersky's equation.

Let us consider a mechanical system subjected to the action of forces $\mathbf{F}_\nu$. We consider a control surface S_t and a mass $m(t)$ in its interior and use $\xi(t)$ to denote the radius vector of the centre of mass $m(t)$, which is inside the control surface S_t, relative to the origin of a fixed coordinate system.

At the moment $t + dt$ the mass $m(t)$, which was previously inside the control surface S_i, will be both inside the control surface S_{t+dt}, corresponding to the moment $t + dt$, and outside of it. We use $m + dm$ to label the mass which is inside the surface S_{t+dt} and $-dm$ to label the mass which is outside of S_{t+dt}. We assume that the mass leaves the surface S due to forces interior for the mass $m(t)$ and that the constraints admit of virtual translations of the mechanical system as a rigid body.

We consider the system of particles, enveloped by the control surface S_t at the moment t, in the time interval from t to $t + dt$, to be a system with constant mass on the assumption that the mass leaves the surface S and designate the constant mass of this system as M.

On these assumptions, we have from the general theorem on the motion of the centre of mass

$$d \left(M \frac{d\xi}{dt} \right) = \sum \mathbf{F}_\nu dt.$$

Since M is a constant, we have

$$M \frac{d\xi}{dt} \bigg|_{t+dt} - M \frac{d\xi}{dt} \bigg|_t = \sum \mathbf{F}_\nu dt.$$

But

$$M \frac{d\xi}{dt} \bigg|_t = m\mathbf{v},$$

where $m(t)$ is the mass in the interior of the conrol surface S_t and $\mathbf{v}$ is the absolute velocity of the centre of mass $m(t)$, and

$$M \frac{d\xi}{dt} \bigg|_{t+dt} = (m + dm)(\mathbf{v} + d\mathbf{v}) + (-dm)\mathbf{u},$$

where $m + dm$ is the mass inside the surface S_{t+dt}, $\mathbf{v} + d\mathbf{v}$ is the absolute velocity of the centre of mass, and $\mathbf{u}$ is the absolute velocity of the mass $(-dm)$ that has left the surface S_{t+dt}. Hence we have

$$m\mathbf{v} + m\,d\mathbf{v} + dm \cdot \mathbf{v} + dm\,d\mathbf{v} - \mathbf{u}\,dm - m\mathbf{v} = \sum \mathbf{F}_\nu dt.$$

Neglecting the second-order term $dm\,d\mathbf{v}$, we get

$$m\,d\mathbf{v} + dm(\mathbf{v} - \mathbf{u}) = \sum \mathbf{F}_\nu dt,$$

and the final result is

$$m \frac{d\mathbf{v}}{dt} = (\mathbf{u} - \mathbf{v}) \frac{dm}{dt} + \sum \mathbf{F}_\nu.$$

This equation, derived by Meshchersky, is very significant in rocketry.

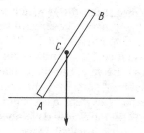

Figure 110

From the derivation of the equation we infer that the term $(\mathbf{u} - \mathbf{v})\dfrac{dm}{dt}$ is the accelerating reactive force. The vector $\mathbf{u} - \mathbf{v} = \mathbf{v}_r$ is the relative velocity of the masses that leave the control surface.

The derivation of Meshchersky's equation for the case when some masses leave the control surface and some masses enter it is similar.

This theorem also explains the phenomenon of a recoil in shooting, the motion of the centre of mass of a shrapnel exploding in vacuum, etc.

Problem. A heavy homogeneous rod AB of length $2l$ stands on one end A on a smooth horizontal table (Fig. 110) and is initially at rest at an angle α with the table. Then it begins to fall under the force of gravity. Determine the trajectory of the end B.

121. Theorem on angular momentum. Assume that the virtual displacements of a mechanical system include its rotation about a fixed z-axis as a rigid body. We designate the element of virtual revolution of the system about the z-axis as $\delta\varphi$. From Euler's theorem we have

$$\begin{pmatrix} \delta x_\nu \\ \delta y_\nu \\ \delta z_\nu \end{pmatrix} = \begin{Vmatrix} 0 & 0 & \delta\varphi \\ x_\nu & y_\nu & z_\nu \end{Vmatrix} = \begin{pmatrix} -y_\nu \ \delta\varphi \\ +x_\nu \ \delta\varphi \\ 0 \end{pmatrix} \quad (\nu = 1, \ldots, n).$$

Substituting these virtual displacements into the formula for the Euler-Lagrange principle, we obtain, after simple transformation, an equation

$$\frac{d}{dt} \sum m_\nu \left(x_\nu \frac{dy_\nu}{dt} - y_\nu \frac{dx_\nu}{dt} \right) = \sum (x_\nu Y_\nu - y_\nu X_\nu).$$

Consequently, if a system restricted by ideal constraints can rotate about an axis as a rigid body, the rate of variation of the angular momentum about the z-axis

$$K_z = \sum m_\nu \left(x_\nu \frac{dy_\nu}{dt} - y_\nu \frac{dx_\nu}{dt} \right)$$

is equal to the resultant moment of the given forces about the z-axis

$$M_z = \sum (x_\nu Y_\nu - y_\nu X_\nu).$$

In other words,

$$\frac{dK_z}{dt} = M_z.$$

We can simplify this expression still further.

We divide the forces X_ν, Y_ν, Z_ν into external X_ν^e, Y_ν^e, Z_ν^e and internal X_ν^i, Y_ν^i, Z_ν^i forces, i.e.

$$X_\nu = X_\nu^e + X_\nu^i, \quad Y_\nu = Y_\nu^e + Y_\nu^i, \quad Z_\nu = Z_\nu^e + Z_\nu^i.$$

The vector of the moment of the internal forces about any point is zero since the moments of the opposite forces $\mathbf{f}$ and $-\mathbf{f}$ about O are equal, lie on the same straight line, and are opposite in direction. Therefore

$$\sum (x_\nu Y_\nu^i - y_\nu X_\nu^i) = 0$$

and, consequently,

$$\frac{dK_z}{dt} = \sum (x_\nu Y_\nu^e - y_\nu X_\nu^e).$$

If $M_z = 0$, then the integration of this equation results in the first integral

$$K_z = \text{const},$$

which is an area integral.

If, in addition, the system not only can rotate but actually rotates about the z-axis with the angular velocity ω, then

$$\left(\begin{array}{c} \dfrac{dx_\nu}{dt} \\[2mm] \dfrac{dy_\nu}{dt} \\[2mm] \dfrac{dz_\nu}{dt} \end{array}\right) = \left\|\begin{array}{ccc} 0 & 0 & \omega \\ x_\nu & y_\nu & z_\nu \end{array}\right\| = \left(\begin{array}{c} -y_\nu\omega \\ x_\nu\omega \\ 0 \end{array}\right) \quad (\nu = 1, \ldots, n).$$

Hence

$$K_z = \omega \sum m_\nu (x_\nu^2 + y_\nu^2) = J_z\omega,$$

where J_z is the moment of inertia of the system about the z-axis. Then the theorem on the angular momentum reduces to the form

$$\frac{d}{dt} J_z\omega = M_z. \tag{5.11}$$

Corollary. If a mechanical system can rotate about x, y, and z axes as a rigid body, then

$$\frac{d\mathbf{K}}{dt} = \mathbf{M},$$

i.e. the velocity of the terminus of the angular momentum vector is equal to the moment of the given forces.

Remark. It is assumed in the theorem on the angular momentum that the virtual displacements include a *rotation* of the system, as a rigid body, *about the fixed z-axis*. The fact that the axis is fixed was used when the relation was reduced to its final form.

Since the variables x_ν, y_ν in the formulas for virtual displacements $\delta x_\nu = -y_\nu\delta\varphi$, $\delta y_\nu = x_\nu\delta\varphi$ are in the basic relation, and are differentiated with respect to t, the formula for the virtual displacement must refer to some time interval, even if small, rather than to an instantaneous state. Therefore when discussing the virtual rotation about a fixed axis we talked about the possibility of rotation about an axis, fixed during some finite, if small, time interval rather than about an instantaneous state (say, an instantaneous rotation axis).

The additional assumption that the system actually rotates about the *fixed z-axis* with the angular velocity ω refers to the state of rotation of the system as a rigid body about the fixed z-axis during some, perhaps small, time interval, since then the product $J_z\omega$ is differentiated by the formula

$$\frac{d}{dt} J_z\omega = M_z \quad (K_z = J_z\omega).$$

Example 1. A rope is passed over a weightless block of radius R which can rotate in a vertical plane about its fixed axis O. The two ends of the rope a and a' long, hang down from the block (Fig. 111). Two men with masses m and m' respectively simultaneously grasp the ends of the rope and begin climbing so that they simultaneously reach the horizontal passing through the axis of the block. The two parts of the rope are assumed to remain vertical. When will this occur?

The system can rotate as a rigid body about the axis of the block. Therefore the theorem under consideration is valid here for the system's rotation about the axis of the block, i.e.

$$\frac{dK_0}{dt} = mgR - m'gR = Rg(m - m'),$$

but

$$K_0 = mvR - m'v'R$$

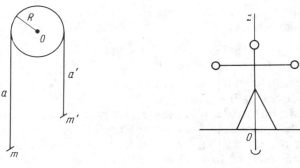

Figure 111   Figure 112

if v and v' are vertical velocities (the positive direction is downwards) of men with masses m and m' respectively. Hence

$$(mv - m'v') = g(m - m')t.$$

Carrying out the integration for the initial data of the problem, we obtain

$$-(ma - m'a') = g(m - m') \frac{T^2}{2}.$$

Example 2. A man stands at the centre of a horizontal platform which can rotate about a vertical z-axis and holds a weight of mass m in each extended hand (Fig. 112). The man and the platform gain speed and rotate about the z-axis with the angular velocity ω_1. With what angular velocity ω_2 will the man and the platform rotate if he moves his arms inwards?

Assume that the moment of inertia of the man and the platform about the z-axis is J, his arms are weightless, r_1 and r_2 are the distances from the z-axis of the weights for the extended and indrawn arms respectively. Since the system can rotate as a rigid body about the z-axis, we have, by virtue of the theorem on the angular momentum, a relation

$$\frac{dK_z}{dt} = 0$$

since the internal forces and the forces of gravity do not yield a moment about the z-axis. Integrating, we have

$$(J + 2mr_1^2)\omega_1 = (J + 2mr_2^2)\omega_2.$$

Example 3. A man stands at the centre of a horizontal platform which can rotate about a fixed vertical z-axis and holds a weight of mass m in one extended hand (Fig. 113a). The man moves the weight so that the projection of its trajectory is as shown in Fig. 113b.

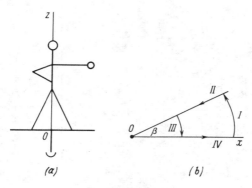

Figure 113

Through what angle α will the platform rotate?

Assume that the moment of inertia of the man and the platform about the z-axis is J, r_1 and r_2 are the distances at which the mass m is when the arm is extended and indrawn respectively. The angle β of rotation of the arm is measured from a line Ox drawn on the platform and the angle α of rotation of the platform is measured from some fixed vertical plane which passes through the z-axis. At the initial moment $\alpha_0 = 0$ and $\beta_0 = 0$. Since the system can rotate about the z-axis as a rigid body, we have for the section I, by virtue of the theorem on the angular momentum, a relation

$$J\alpha' + mr_1^2(\alpha' + \beta') = 0,$$

or, after integration, a relation

$$J\alpha_1 + mr_1^2(\alpha_1 + \beta_1) = 0.$$

On sections II ans IV there are no changes in the angles α and β. On section III we have

$$J\alpha_2 + mr_2^2(\alpha_2 - \beta_2) = 0.$$

Since $\beta_1 = \beta_2 = \beta$, we have

$$\alpha = \alpha_1 + \alpha_2 = -\beta \, \frac{Jm(r_1^2 - r_2^2)}{(J + mr_1^2)(J + mr_2^2)} \, .$$

Example 4. *Zhukovsky's problem.* We have a vertical perfectly smooth wall and a perfectly smooth horizontal floor. A beam AB, whose weight concentrated at the middle is P, slides with its ends A and B butted up against the wall and the floor (Fig. 114). An animal whose weight is Q runs along the beam, its motion is described by the function $s = s(t)$. How must the animal run so that the beam does not fall?

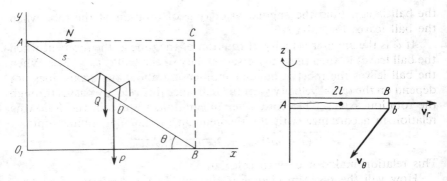

Figure 114   Figure 115

For the beam not to fall, the virtual displacements of the system must include a rotation as a rigid body, about a fixed point C of the intersection of the perpendiculars drawn from A and B, respectively, to the wall and the floor. Consequently, in accordance with the theorem on the angular momentum of the system about the point C we have

$$\frac{d}{dt}\left[\frac{Q}{g}\frac{ds}{dt}(2a\cos\theta\sin\theta)\right] = Pa\cos\theta + Q(2a - s)\cos\theta;$$

where $2a$ is the length of the beam AB, θ is the acute angle between the beam and the floor, and s is the distance from the end A of the beam to the animal, which is considered to be a particle. Since the interaction forces of the animal and the beam are internal, they do not enter into the equation for the angular momentum. We have

$$\frac{d^2s}{dt^2} = \frac{-g}{2a\sin\theta}s + \frac{(P + 2Q)g}{2Q\sin\theta},$$

or

$$\frac{d^2\sigma}{dt^2} = -\frac{g}{2a\sin\theta}\sigma,$$

where $\sigma = s - \dfrac{P + 2Q}{Q}a$. Integrating, we obtain

$$\sigma = A\cos\sqrt{\frac{g}{2a\sin\theta}}\,t + B\sin\sqrt{\frac{g}{2a\sin\theta}}\,t.$$

Example 5. A horizontal, perfectly smooth, homogeneous tube AB with mass M and length $2l$ rotates with angular velocity ω_0 about a vertical z-axis fastened at the end A (Fig. 115). A ball of mass m, fastened by a weightless thread to the end A, is inside the tube. At some instant the thread holding the ball is cut. Find the angular velocity ω of rotation of the tube when the ball leaves the tube.

If ω is the angular velocity of rotation of the tube at the moment when the ball leaves it, then the transversal velocity of the ball is $v_\theta = 2l\omega$. When the ball leaves the tube its angular momentum about the z-axis does not depend on the radial velocity $\mathbf{v}_r$ of the ball since this velocity passes through the rotation axis z. Therefore, after integration, we obtain the following relation in accordance with the theorem on the angular momentum:

$$(J + ml^2)\omega_0 = J\omega + m(2l\omega)2l.$$

This relation makes it easy to calculate ω.

How will the problem change if the tube is not perfectly smooth?

122. The work-kinetic energy theorem. If the true displacements of a mechanical system are among the virtual displacements

$$\delta x_\nu = dx_\nu, \quad \delta y_\nu = dy_\nu, \quad \delta z_\nu = dz_\nu \quad (\nu = 1, \ldots, n),$$

then

$$\sum \left[\left(m_\nu \frac{d^2 x_\nu}{dt^2} - X_\nu \right) dx_\nu + \left(m_\nu \frac{d^2 y_\nu}{dt^2} - Y_\nu \right) dy_\nu \right.$$
$$\left. + \left(m_\nu \frac{d^2 z_\nu}{dt^2} - Z_\nu \right) dz_\nu \right] = 0,$$

or

$$dT = \sum (X_\nu dx_\nu + Y_\nu dy_\nu + Z_\nu dz_\nu).$$

To put it another way, if the constraints are ideal and the real displacements of the mechanical system are among its virtual displacements, then the kinetic energy differential $T = \sum \dfrac{m_\nu}{2} (x_\nu'^2 + y_\nu'^2 + z_\nu'^2)$ is equal to the work of the active forces X_ν, Y_ν, Z_ν on the true displacements dx_ν, dy_ν, dz_ν. This proposition is known as the *work-kinetic energy theorem*.

If the additional forces X_ν, Y_ν, Z_ν have a force function $U(x_1, y_1, z_1, \ldots, x_n, y_n, z_n)$ independent of time,

$$X_\nu = \frac{\partial U}{\partial x_\nu}, \quad Y_\nu = \frac{\partial U}{\partial y_\nu}, \quad Z_\nu = \frac{\partial U}{\partial z_\nu} \quad (\nu = 1, \ldots, n),$$

then, in accordance with the equality

$$\sum (X_\nu dx_\nu + Y_\nu dy_\nu + Z_\nu dz_\nu) = dU$$

the work-kinetic energy theorem assumes an easily integrable form

$$dT = dU.$$

We get the kinetic energy integral

$$T = U + h,$$

where the constant h is the kinetic energy constant.

If a rigid body rotates about the z-axis with the angular velocity ω, then the magnitude of the velocity of a particle m_ν which is at the distance r_ν from the z-axis is $v_\nu = r_\nu \omega$. The kinetic energy of such a body is

$$T = \sum \frac{m_\nu v_\nu^2}{2} = \sum \frac{m_\nu r_\nu^2 \omega^2}{2} = J_z \frac{\omega^2}{2} .$$

Example. A homogeneous cylinder of radius r and mass m is placed, initially at rest, on an inclined plane so that its generatrix is horizontal and then so that its endface stands on the plane (Fig. 116). In the first case the cylinder rolls down without sliding and in the second case it slides without friction. The constraints are assumed to be ideal. The angle between the inclined and horizontal planes is α. Find the ratio of the vertical distances the cylinder travels in the two cases in the same time interval.

In both cases the virtual displacements include the true displacements of the cylinder and, therefore, we can solve the problem proceeding from the work-kinetic energy theorem to avoid involving the reaction forces. In the second case we can use the theorem on the motion of the centre of mass along an inclined plane since in this case the virtual displacements of the cylinder include a translation in the direction of the inclined plane.

Let us calculate the kinetic energy of the cylinder for the first case. We designate the angular velocity of the rolling cylinder as ω. Where the cylinder and the inclined plane touch is an instantaneous centre of rotation of the cylinder. The kinetic energy of the cylinder is $J_0 \dfrac{\omega^2}{2}$, where J_0 is the moment of inertia of the cylinder about the point of tangency. But from Steiner's theorem we have

$$J_0 = mr^2 + J,$$

where J is the moment of inertia of the cylinder about its centre of gravity.

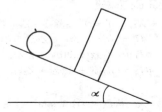

Figure 116

In the first case the work-kinetic energy theorem yields

$$d\,\frac{(J + mr^2)}{2}\,\omega^2 = mg\,ds_1 \cdot \sin\alpha,$$

where ds_1 is a true displacement of the centre of gravity of the cylinder.

When the cylinder rolls along an inclined plane we have $\omega = \dfrac{1}{r}\dfrac{ds_1}{dt}$ and, therefore, after elementary transformations we get an equation

$$\frac{(J + mr^2)}{r^2}\,\frac{d^2 s_1}{dt^2} = mg\sin\alpha,$$

or, since $J = mr^2/2$, and equation

$$\frac{3}{2}\,m\,\frac{d^2 s_1}{dt^2} = mg\sin\alpha. \tag{5.12}$$

In the second case, from the theorem on the motion of the centre of mass we get an equation

$$m\,\frac{d^2 s_2}{dt^2} = mg\sin\alpha. \tag{5.13}$$

Comparing (5.12) and (5.13), we obtain

$$\frac{3}{2}\,\frac{d^2 s_1}{dt^2} = \frac{d^2 s_2}{dt^2}.$$

Integrating twice with the initial data $v_1^0 = v_2^0 = 0$ and $s_1^0 = s_2^0 = 0$, we obtain $\dfrac{3}{2}\,s_1 = s_2$. Since the vertical distances the centre of mass of the cylinder travels in the two cases are proportional to the paths they traverse, we get

$$\frac{h_1}{h_2} = \frac{s_1}{s_2} = \frac{2}{3}.$$

Problem. Sometimes the work-kinetic energy theorem is incorrectly formulated in textbooks. To indicate the mistake in the formulation, we cite the following erroneous calculation. We consider a particle without constraints

$$X = X_1 + X_2, \quad Y = Y_1 + Y_2, \quad Z = Z_1 + Z_2,$$
$$dT = X\,dx + Y\,dy + Z\,dz,$$
$$dT_1 = X_1\,dx + Y_1\,dy + Z_1\,dz,$$
$$dT_2 = X_2\,dx + Y_2\,dy + Z_2\,dz,$$

whence

$$d(T_1 + T_2) = dT.$$

Find the error in the reasoning which leads to this obvious absurdity.

5.4 Motion About the Centre of Mass

123. König's formulas. We introduce into consideration a centre of mass G with coordinates

$$\xi = \frac{\sum m_\nu x_\nu}{M}, \quad \eta = \frac{\sum m_\nu y_\nu}{M}, \quad \zeta = \frac{\sum m_\nu z_\nu}{M},$$

where $M = \sum m_\nu$ is the net mass of the system. Let us consider the moving coordinate axes $x^* y^* z^*$ whose origin is at the centre of mass G of the mechanical system and which are parallel to the fixed axes xyz respectively. These axes are known as König's axes (Fig. 117). The motion of the mechanical system with respect to the axes $x^* y^* z^*$ is called the motion about the centre of mass. In these two systems the coordinates of the particle m are related as

$$x_\nu = x_\nu^* + \xi, \quad y_\nu = y_\nu^* + \eta, \quad z_\nu = z_\nu^* + \zeta.$$

We can prove König's formula, namely,

$$K_z = \sum m_\nu \left(x_\nu^* \frac{dy_\nu^*}{dt} - y_\nu^* \frac{dx_\nu^*}{dt} \right) + M \left(\xi \frac{d\eta}{dt} - \eta \frac{d\xi}{dt} \right).$$

The angular momentum K_z about the z-axis is equal to the angular momentum about the z^*-axis

$$K_{z^*}^* = \sum m_\nu \left(x_\nu^* \frac{dy_\nu^*}{dt} - y_\nu^* \frac{dx_\nu^*}{dt} \right)$$

plus the angular momentum of the centre of mass G about the z-axis if the total mass M is concentrated there.

The expression for the kinetic energy

$$T = \sum \frac{m_\nu}{2} \left(x_\nu'^2 + y_\nu'^2 + z_\nu'^2 \right)$$

can be transformed by analogy:

$$T = \sum \frac{m_\nu}{2} \left[\left(\frac{d(x_\nu^* + \xi)}{dt} \right)^2 + \left(\frac{d(y_\nu^* + \eta)}{dt} \right)^2 + \left(\frac{d(z_\nu^* + \zeta)}{dt} \right)^2 \right]$$

$$= \sum \frac{m_\nu}{2} \left[\left(\frac{dx_\nu^*}{dt} \right)^2 + \left(\frac{dy_\nu^*}{dt} \right)^2 + \left(\frac{dz_\nu^*}{dt} \right)^2 \right]$$

$$+ \sum \frac{m_\nu}{2} \left[\left(\frac{d\xi}{dt} \right)^2 + \left(\frac{d\eta}{dt} \right)^2 + \left(\frac{d\zeta}{dt} \right)^2 \right]$$

$$+ \sum m_\nu \left(\frac{dx_\nu^*}{dt} \frac{d\xi}{dt} + \frac{dy_\nu^*}{dt} \frac{d\eta}{dt} + \frac{dz_\nu^*}{dt} \frac{d\zeta}{dt} \right).$$

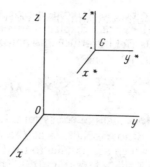

Figure 117

But the last sum can be reduced to the form

$$\frac{d\xi}{dt}\left(\frac{d}{dt}\sum m_\nu x_\nu^*\right) + \frac{d\eta}{dt}\left(\frac{d}{dt}\sum m_\nu y_\nu^*\right) + \frac{d\zeta}{dt}\left(\frac{d}{dt}\sum m_\nu z_\nu^*\right).$$

Since the centre of mass G of the system is always at the origin of the moving axes $x^* y^* z^*$, it follows that $\sum m_\nu x_\nu^* = 0$, $\sum m_\nu y_\nu^* = 0$, $\sum m_\nu z_\nu^* = 0$, and, consequently,

$$T = T_G + M \frac{v_G^2}{2},$$

i.e., the kinetic energy T of the system is equal to the kinetic energy of the system as it moves with respect to König's axes $Gx^* y^* z^*$

$$T_G = \sum \frac{m_\nu}{2}\left[\left(\frac{dx_\nu^*}{dt}\right)^2 + \left(\frac{dy_\nu^*}{dt}\right)^2 + \left(\frac{dz_\nu^*}{dt}\right)^2\right]$$

plus the kinetic energy $M \dfrac{v_G^2}{2}$ of the centre of mass if the total mass M is concentrated in it, where v_G is the velocity of the centre of mass G with respect to the fixed axes.

In many cases König's formulas make it easier to calculate the angular momentum and the kinetic energy of a mechanical system.

124. General theorems on the motion of a mechanical system about the centre of mass. There are some general theorems for certain cases on the motion of a mechanical system with respect to König's axes $Gx^* y^* z^*$, which are parallel to fixed axes and pass through the centre of mass G. These theorems do not contain any unknown reactions of ideal constraints.

Assume that the system can rotate about a fixed vertical z-axis as a rigid body, or, to put it otherwise, the virtual displacements of the mechanical system include a rotation about the z-axis as a rigid body. Assume furthermore that the system can be translated as a rigid body in the x and y axes, or, to put it otherwise, the virtual displacements of the mechanical system include translations, as a rigid body, in the direction of the x and y axes. These propositions yield relations

$$\frac{dK_z}{dt} = M_z, \quad M \frac{d^2\xi}{dt^2} = \sum X_\nu, \quad M \frac{d^2\eta}{dt^2} = \sum Y_\nu.$$

We replace K_z in accordance with König's formula and explicitly write out M_z. We obtain

$$\frac{dK_{z^*}^*}{dt} + M \left(\xi \frac{d^2\eta}{dt^2} - \eta \frac{d^2\xi}{dt^2} \right) = \sum (x_\nu^* Y_\nu - y_\nu^* X_\nu) + \xi \sum Y_\nu - \eta \sum X_\nu$$

using the last relations. The final result is

$$\frac{dK_{z^*}^*}{dt} = M_{z^*}.$$

Consequently, if a system can rotate about the z-axis as a rigid body and if the system can be translated along the x and y axes as a rigid body, then the rate of change of the angular momentum about the z^*-axis is equal to the moment of the active forces with respect to the z^*-axis. If $M_{z^*} = 0$, then $K_{z^*}^* = \text{const.}$

If the true displacements of the system are among the virtual displacements and if the system can be translated along the x, y, z axes as a rigid body, then the work-kinetic energy theorem is valid for the motion about the centre of mass:

$$dT_G = \sum (X_\nu dx_\nu^* + Y_\nu dy_\nu^* + Z_\nu dz_\nu^*).$$

The proof is similar to the proof of the preceding theorem.

Example. *Somersault.* When an acrobat leaps, he imparts to his body an angular momentum about a horizontal axis during the motion about his centre of mass. To neglect the air drag, we assume that he leaps in a vacuum. The virtual displacements of the acrobat will then include translations as a rigid body in all directions and a rotation as a rigid body about horizontal axes. Consequently, the theorem on angular momentum about a horizontal axis of fixed direction passing through the centre of mass can be applied to the motion of the acrobat about its centre of mass. Since internal forces do not enter into the theorem on the angular momentum and the moment of the force of gravity about a centre of mass is always zero, we can integrate the expression of the indicated theorem and make an inference concerning the conservation of the angular momentum of the acrobat about the horizontal axis of fixed direction in his motion about the centre of mass, i.e.

$$J\omega = \text{const.}$$

Here J is the moment of inertia of the acrobat about the indicated axis and ω is the angular velocity of rotation.

We can see from the last formula that the angular velocity ω of the acrobat increases when his moment of inertia J decreases and the latter phenomenon occurs if the acrobat pulls his body towards his centre of mass, and vice versa. Therefore, after an acrobat has turned a somersault, he considerably increases J by straightening his body, in order to decrease ω and land on his feet.

Example. A spinning top is a heavy rigid body which is shaped as a body of revolution and whose centre of mass C lies on its axis of rotation fairly high above the thin bearing point A (Fig. 118). When the top's angular velocity is high enough, it will, if started on a rough surface, assume a vertical position. In order to explain this phenomenon using the general theorems on the relative motion, we add the reaction of the rough supporting horizontal plane, i.e. the normal reaction **N** and the force of friction **T**, to the given forces acting on the top. We can regard the top as a rigid body without constraints and then all the conditions of the preceding theorem on the angular momentum in the motion about the centre of mass will be satisfied. The velocity of the terminus of the vector of the angular momentum of the relative motion is equal to the angular momentum about the centre of mass of the forces acting on the top. The moment of the force of gravity on the top about the centre of mass is zero, and therefore it does not immediately cause any precession about the centre of mass. The normal reaction **N** about the centre of mass has a moment equal to $\overrightarrow{CA} \times \mathbf{N}$ and, consequently, the reaction **N** causes a precession during which the terminus of the vector of the angular momentum of the relative motion of the top has a horizontal component of the velocity $\overrightarrow{CA} \times \mathbf{N}$. The force of friction **T**, which is opposite in direction to the velocity of the supporting plane at the tangency point, has a moment $\overrightarrow{CA} \times \mathbf{T}$ about the centre of mass C of the top, and therefore during the precession caused by the friction **T** the terminus of the vector of the moment of the relative motion of the top has a velocity component $\overrightarrow{CA} \times \mathbf{T}$. When the centre of mass C is high, this component has a positive vertical projection. Therefore, the terminus of the vector of the angular momentum of the relative motion moves upwards and, consequently, when the angular velocity of rotation of the top is high for the vector of the angular momentum to lie close to the axis of the top, the top will assume a vertical position.

Example. *A Chinese spinning top.* A Chinese top is a light spherical shell with a segment cut out of it. On the cross section of the segment

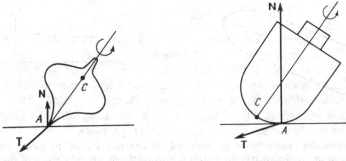

Figure 118   Figure 119

there is a small handle to start the top (Fig. 119). The centre of mass C of this top is on the axis of rotation close to the point where the axis of the top cuts the spherical surface. This is done by placing a small heavy weight at a vertex of the hemisphere. When the angular velocity is high enough, the Chinese top, set rotating on a rough horizontal surface, will turn upside down and stand on its handle.

To explain the motion of a Chinese top qualitatively, we use the theorem on the angular momentum about a centre of mass. We add the reaction forces of the rough horizontal surface, i.e. the normal reaction **N** and the force of friction **T**, to the given forces and assume that the top moves without constraints. The relative motion of the top is composed of the precessions caused by the reactions **N** and **T**. The velocity of the terminus of the angular momentum vector of the relative motion will be

$$\frac{d\mathbf{K}}{dt} = \vec{CA} \times (\mathbf{N} + \mathbf{T}).$$

When the angular velocity of rotation of the top about its axis is sufficiently high, the direction of the angular momentum vector will differ but little from the direction of the axis of the top. The velocity vector $\vec{CA} \times \mathbf{N}$ is horizontal and the velocity vector $\vec{CA} \times \mathbf{T}$ has a vertical component, and whatever the rotation imparted to the top, the handle of the top will move downwards because of this vertical component. The top will turn upside down and stand on its handle.

Problem 1. Consider the motion of an artillery shell in the air rotating about its centre of mass and, taking into account the forces with which the air acts on the shell, i.e. the tilting force and the resultant of the friction forces, explain the tendency of the axis of the rotating shell to turn in the direction of the velocity of the shell (Fig. 120).

Problem 2. Explain how an acrobat can cycle without steering.

Problem 3. A saddle and a motor driving a heavy disc DD' in the direction shown in Fig. 121 are attached to a horizontal bar AB. The bar can rotate freely in its bearings. A

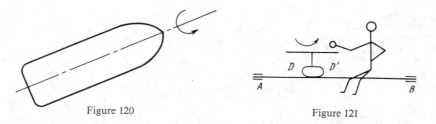

Figure 120   Figure 121

man mounts the saddle. When the man and the common axis of the motor and the disc DD' are in the vertical plane, the system is in equilibrium. On which side of the disc must the man press at D' in order to right himself from falling away from the reader?

125. The Euler-Lagrange principle for the motion about the centre of mass. Assume that the virtul displacements of a mechanical system include a translation as a rigid body in the direction of the fixed xyz axes. The equations of motion of the centre of mass along any of the three fixed coordinate axes are valid:

$$M \frac{d^2 \xi}{dt^2} = \sum X_\nu, \quad M \frac{d^2 \eta}{dt^2} = \sum Y_\nu, \quad M \frac{d^2 \zeta}{dt^2} = \sum Z_\nu. \qquad (5.14)$$

The Euler-Lagrange principle allows us to change from the fixed coordinate system to the coordinates with respect to König's axes:

$$\sum \left[\left(m_\nu \frac{d^2(x_\nu^* + \xi)}{dt^2} - X_\nu \right) \delta(x_\nu^* + \xi) + \dots \right] = 0,$$

or

$$\sum \left[\left(m_\nu \frac{d^2 x_\nu^*}{dt^2} - X_\nu \right) \delta x_\nu^* + \dots \right] + \sum \left[m_\nu \frac{d^2 x_\nu^*}{dt^2} \delta \xi + \dots \right]$$
$$+ \sum \left[m_\nu \frac{d^2 \xi}{dt^2} \delta x_\nu^* + \dots \right] + \sum \left[\left(m_\nu \frac{d^2 \xi}{dt^2} - X_\nu \right) \delta \xi + \dots \right] = 0.$$

But

$$\sum m_\nu \frac{d^2 x_\nu^*}{dt^2} \delta \xi = \delta \xi \frac{d^2}{dt^2} \sum m_\nu x_\nu^* = 0,$$

$$\sum m_\nu \frac{d^2 \xi}{dt^2} \delta x_\nu^* = \frac{d^2 \xi}{dt^2} \delta \sum m_\nu x_\nu^* = 0.$$

Similar relations hold true when the coordinates x_ν^* are replaced by y_ν^* and z_ν^*. The last sum is eliminated by virtue of (5.14). Consequently, we have the following for the motion of the system with respect to König's axes

$x^* y^* z^*$:

$$\sum \left[\left(m_\nu \frac{d^2 x_\nu^*}{dt^2} - X_\nu \right) \delta x_\nu^* + \left(m_\nu \frac{d^2 y_\nu^*}{dt^2} - Y_\nu \right) \delta y_\nu^* \right.$$
$$\left. + \left(m_\nu \frac{d^2 z_\nu^*}{dt^2} - Z_\nu \right) \delta z_\nu^* \right] = 0$$

holds true if the system can be translated as a rigid body in the direction of the $Oxyz$ axes.

5.5 Lagrange's Equations

126. Consider a system of particles with masses m_ν and coordinates x_ν, y_ν, z_ν in a fixed system of rectangular Cartesian coordinates. Assume that the particles m_ν are subject to the active forces X_ν, Y_ν, Z_ν and restricted by the holonomic constraints defined by the relations

$$x_\nu = x_\nu(t, q_1, ..., q_k), \quad y_\nu = y_\nu(t, q_1, ..., q_k),$$

$$z_\nu = z_\nu(t, q_1, ..., q_k) \quad (\nu = 1, ..., n), \tag{5.15}$$

in which $q_1, ..., q_k$ are real independent holonomic (defining) coordinates of the system and k is the number of its degrees of freedom. The virtual displacements are then defined by the relations

$$\delta x_\nu = \sum \frac{\partial x_\nu}{\partial q_s} \delta q_s, \quad \delta y_\nu = \sum \frac{\partial y_\nu}{\partial q_s} \delta q_s, \quad \delta z_\nu = \sum \frac{\partial z_\nu}{\partial q_s} \delta q_s,$$

where $\delta q_1, ..., \delta q_k$ are small arbitrary variations of the variables q_s. These formulas divide the whole set of virtual displacements of the system into k basic displacements, in each of which only q_s varies (for a fixed s) and all the other variations of Lagrange's coordinates are set to be zero ($s = 1$, ..., k). The displacement of the whole system is regarded as a set of basic displacements.

To obtain the equations of motion, we substitute δx_ν, δy_ν, δz_ν into the relation of the Euler-Lagrange principle and get

$$\sum_s \delta q_s \left\{ \sum_\nu m_\nu \left(x_\nu'' \frac{\partial x_\nu}{\partial q_s} + y_\nu'' \frac{\partial y_\nu}{\partial q_s} + z_\nu'' \frac{\partial z_\nu}{\partial q_s} \right) \right.$$
$$\left. - \sum_\nu \left(X_\nu \frac{\partial x_\nu}{\partial q_s} + Y_\nu \frac{\partial y_\nu}{\partial q_s} + Z_\nu \frac{\partial z_\nu}{\partial q_s} \right) \right\} = 0$$

The quantities

$$Q_s = \sum_\nu \left(X_\nu \frac{\partial x_\nu}{\partial q_s} + Y_\nu \frac{\partial y_\nu}{\partial q_s} + Z_\nu \frac{\partial z_\nu}{\partial q_s} \right) \quad (s = 1, ..., k)$$

also appear in analytical statics. They are generalized forces. We have

$$x_\nu'' \frac{\partial x_\nu}{\partial q_s} = \frac{d}{dt} \left(x_\nu' \frac{\partial x_\nu}{\partial q_s} \right) - x_\nu' \frac{d \frac{\partial x_\nu}{\partial q_s}}{dt},$$

but

$$x_\nu' = \frac{\partial x_\nu}{\partial t} + \sum \frac{\partial x_\nu}{\partial q_s} q_s',$$

and, consequently,

$$\frac{\partial x_\nu'}{\partial q_s'} = \frac{\partial x_\nu}{\partial q_s}, \qquad \frac{\partial x_\nu'}{\partial q_s} = \frac{\partial \frac{\partial x_\nu}{\partial q_s}}{\partial t} + \sum \frac{\partial \frac{\partial x_\nu}{\partial q_s}}{\partial q_j} q_j' = \frac{d \frac{\partial x_\nu}{\partial q_s}}{dt}.$$

Therefore

$$x_\nu'' \frac{\partial x_\nu}{\partial q_s} = \frac{d}{dt} \left(x_\nu' \frac{\partial x_\nu'}{\partial q_s'} \right) - x_\nu' \frac{\partial x_\nu'}{\partial q_s} = \frac{d}{dt} \left(\frac{\partial \frac{x_\nu'^2}{2}}{\partial q_s'} \right) - \frac{\partial \frac{x_\nu'^2}{2}}{\partial q_s}.$$

We can derive similar formulas for y_ν, z_ν. Consequently,

$$\sum m_\nu \left(x_\nu'' \frac{\partial x_\nu}{\partial q_s} + y_\nu'' \frac{\partial y_\nu}{\partial q_s} + z_\nu'' \frac{\partial z_\nu}{\partial q_s} \right)$$

$$= \frac{d}{dt} \frac{\partial}{\partial q_s'} \sum \frac{m_\nu}{2} \left(x_\nu'^2 + y_\nu'^2 + z_\nu'^2 \right) - \frac{\partial}{\partial q_s} \sum \frac{m_\nu}{2} \left(x_\nu'^2 + y_\nu'^2 + z_\nu'^2 \right).$$

The Euler-Lagrange principle can be reduced to the form

$$\sum_s \delta q_s \left[\frac{d}{dt} \left(\frac{\partial T}{\partial q_s'} \right) - \frac{\partial T}{\partial q_s} - Q_s \right] = 0, \tag{5.16}$$

where

$$T = \sum \frac{m_\nu}{2} \left(x_\nu'^2 + y_\nu'^2 + z_\nu'^2 \right)$$

is the kinetic energy of the system.

Relation (5.16) must hold for all the different values of δq_s possible. If the variables q_s ($s = 1, ..., k$) are such that the values of δq_s may have arbitrary signs, then relation (5.16) yields differential equations of motion for the mechanical system, i.e.,

$$\frac{d}{dt} \left(\frac{\partial T}{\partial q_s'} \right) - \frac{\partial T}{\partial q_s} = Q_s \quad (s = 1, ..., k). \tag{5.17}$$

These are Lagrange's equations.

127. The generalized forces Q_s are associated with the expression for the virtual work of the forces X_ν, Y_ν, Z_ν in producing a virtual displacement of the system

$$A = \sum (X_\nu \delta x_\nu + Y_\nu \delta y_\nu + Z_\nu \delta z_\nu) = \sum Q_s \delta q_s.$$

If we regard the work A_s of the given forces in producing a displacement in which one variable q_s varies by δq_s and the other variables do not vary, $\delta q_j = 0$ $(j = 1, ..., s - 1, s + 1, ..., n)$, then we can find the generalized force Q_s from the relation

$$A_s = Q_s \delta q_s.$$

If the given forces have a force function $U(t, x_1, y_1, z_1, ..., x_n, y_n, z_n)$ dependent of the position of the mechanical system and, maybe, on time, then

$$X_\nu = \frac{\partial U}{\partial x_\nu}, \quad Y_\nu = \frac{\partial U}{\partial y_\nu}, \quad Z_\nu = \frac{\partial U}{\partial z_\nu} \quad (\nu = 1, ..., n),$$

and, consequently,

$$Q_s = \sum_\nu \left(X_\nu \frac{\partial x_\nu}{\partial q_s} + Y_\nu \frac{\partial y_\nu}{\partial q_s} + Z_\nu \frac{\partial z_\nu}{\partial q_s} \right) = \frac{\partial U}{\partial q_s}.$$

In this case Lagrange's equations

$$\frac{d}{dt} \left(\frac{\partial T}{\partial q_s'} \right) - \frac{\partial T}{\partial q_s} = \frac{\partial U}{\partial q_s}$$

can be written as

$$\frac{d}{dt} \left(\frac{\partial L}{\partial q_s'} \right) - \frac{\partial L}{\partial q_s} = 0,$$

where L is the Lagrangian function

$$L = T + U.$$

128. Lagrange's equations are very convenient since they do not include unknown reactions of the constraints imposed on the system, the number of equations of motion is the minimum possible, and we only have to calculate the kinetic energy T of the mechanical system and the generalized forces Q_s in order to write Lagrange's equations.

When deriving Lagrange's equations, we have to make sure that the coordinates $q_1, ..., q_k$ are holonomic, or, in other words, that the Cartesian coordinates of the particles of the system x_ν, y_ν, z_ν can be explicitly expressed before the equations of motion are derived in terms of real variables q_s $(s = 1, 2, ...)$ which have a geometrical meaning of their own.

Example. Let us again consider Zhukovsky's problem (see Fig. 114). The position of the beam is defined by the angle $ABO_1 = \theta$ it makes with the horizontal floor. Consequently, θ is a holonomic coordinate of the system, and therefore the equation of motion in terms of the variable θ will be a Lagrange equation.

The kinetic energy of the beam can be found from König's formula

$$T_b = \frac{P}{2g}\,a^2\theta'^2 + \frac{P}{2g}\,\frac{a^2}{3}\,\theta'^2 = \frac{2}{3}\,\frac{P}{g}\,a^2\theta'^2.$$

since the moment of inertia of the beam at the centre of gravity is equal to $\dfrac{P}{g}\dfrac{a^2}{3}$.

The coordinates of the animal are

$$x = s\cos\theta, \quad y = (2a - s)\sin\theta,$$

whence the kinetic energy of the animal, regarded as a particle, can be easily calculated:

$$T_{an} = \frac{Q}{2g}\,[s'^2 - 4as'\theta'\sin\theta\cos\theta + [4a(a - s)\cos^2\theta + s^2]\theta'^2].$$

The generalized force can be found from the expression for the elementary virtual work

$$-P\delta(a\sin\theta) - Q\delta[(2a - s)\sin\theta] = Q_\theta\delta\theta,$$

whence we have

$$Q_\theta = -[Pa + Q(2a - s)]\cos\theta.$$

Consequently, the required equation of motion has the form

$$\frac{d}{dt}\left\{\frac{P}{g}a^2\frac{4}{3}\theta' + \frac{Q}{g}[4a(a - s)\cos^2\theta + s^2]\theta' - \frac{Q}{g}2as'\sin\theta\cos\theta\right\}$$

$$- \left\{-4a(a - s)\frac{Q}{g}\theta'^2\sin\theta\cos\theta - 2\frac{Q}{g}s'\theta'\cos2\theta\right\}$$

$$= -[Pa + Q(2a - s)]\cos\theta.$$

We can solve Zhukovsky's problem using this equation, i.e. find the equation of the animal's motion $s(t)$ such that the beam does not fall. It is $\theta' = 0$. Substituting this into the last equation, we have

$$-\frac{Q}{g}2as''\sin\theta\cos\theta = -[Pa + Q(2a - s)]\cos\theta.$$

Disregarding the solution $\cos\theta = 0$, which is of no interest to us, we have

(see p. 170)

$$s'' = -\frac{gs}{2a\sin\theta} + \frac{(P + 2Q)g}{2Q\sin\theta}.$$

129. Lagrange's equations retain their form

$$\frac{d}{dt}\left(\frac{\partial T}{\partial q_s'}\right) - \frac{\partial T}{\partial q_s} = Q_s$$

whatever the independent holonomic variables q_s we use to write the equations of motion. A change in the variables causes a change in T and Q, but the form of the equations remains the same. This is very significant.

Let us consider in a plane some fixed oblique coordinates with an angle ψ between them (Fig. 122). The position of the particle m can be defined by the contravariant (ξ^1, ξ^2), covariant (ξ_1, ξ_2) or mixed (ξ_1, ξ^2) coordinates. When the forces are given it is easy to write the equations of motion of the particle m in the plane using any one of these definitions of the position of the particle m.

130. Cyclic coordinates. Assume that the variables $q_{s+1}, ..., q_k$ $(s < k)$ possess the property

$$\frac{\partial L}{\partial q_\alpha} = 0 \quad (\alpha = s + 1, ..., k).$$

Coordinates q_α of this kind are called cyclic coordinates. Lagrange's equations for these coordinates have the form

$$\frac{d}{dt}\left(\frac{\partial L}{\partial q_\alpha'}\right) = 0.$$

The integrals of these equations are

$$\frac{\partial L}{\partial q_\alpha'} = \beta_\alpha \quad (\alpha = s + 1, ..., k), \qquad (5.18)$$

where β_α are arbitrary integration constants.

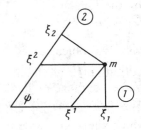

Figure 122

We introduce Routh's function R, which is defined by the relation

$$R = L - \sum q'_\alpha \beta_\alpha.$$

We replace q'_α ($\alpha = s + 1, \ldots, h$) in the function R by their expressions resulting from the first integrals (5.18) written out, and then R becomes a function of $t, q_1, \ldots, q_s, q'_1, \ldots, q'_s, \beta_{s+1}, \ldots, \beta_k$. The variation of the function R is

$$\delta R = \sum_j \frac{\partial R}{\partial q_j} \delta q_j + \sum_j \frac{\partial R}{\partial q_j'} \delta q_j' + \sum_\alpha \frac{\partial R}{\partial \beta_\alpha} \delta \beta_\alpha,$$

where $j = 1, \ldots, s$, $\alpha = s + 1, \ldots, k$.

The variation of the function R according to the formula defining it is

$$\delta R = \sum_j \frac{\partial L}{\partial q_j} \delta q_j + \sum_j \frac{\partial L}{\partial q_j'} \delta q_j' - \sum_\alpha q'_\alpha \delta \beta_\alpha.$$

Comparing the coefficients of the same variations in these two expressions, we obtain relations

$$\frac{\partial L}{\partial q_j} = \frac{\partial R}{\partial q_j}, \quad \frac{\partial L}{\partial q_j'} = \frac{\partial R}{\partial q_j'}, \quad -q'_\alpha = \frac{\partial R}{\partial \beta_\alpha},$$

in accordance to which Lagrange's equations for the noncyclic coordinates q_j ($j = 1, \ldots, s$) assume the form

$$\frac{d}{dt}\left(\frac{\partial R}{\partial q_j'}\right) - \frac{\partial R}{\partial q_j} = 0 \quad (j = 1, \ldots, s).$$

The function R does not depend on the cyclic coordinates q_α or their velocities.

The last equations "ignore" the cyclic coordinates and reduce the dynamics problem to a problem on the motion of a mechanical system with a new Lagrangian function R and with fewer number of the degrees of freedom ($s < k$). After the last equations are integrated, the cyclic coordinates q_α are defined by the integrals

$$q_\alpha = - \int \frac{\partial R}{\partial \beta_\alpha} \, dt \quad (\alpha = s + 1, \ldots, k).$$

When we analyse the motions of a mechanical system, it is best to use the variables q_i which make Lagrange's equations easier to integrate. In many cases, in order to find a cyclic coordinate, it is necessary to consider displacements δq_α on which the total work of the given forces $\dfrac{\partial U}{\partial q_\alpha}$ is zero and for which the kinetic energy of the system does not vary,

$$\frac{\partial T}{\partial q_\alpha} = 0.$$

The equations of motion of the centre of gravity and the area integral are special cases of cyclic integrals.

Example. A particle m is acted on by a central force with a force function $U(r)$. Hence the cyclic coordinates, provided that they exist, are related to displacements orthogonal to the radius Om (Fig. 123). We introduce spherical coordinates. The displacements, for which only the angles θ, φ vary are orthogonal to r,

$$\frac{\partial U}{\partial \theta} = 0, \quad \frac{\partial U}{\partial \varphi} = 0,$$

the angle φ being a cyclic coordinate. The angle θ is a noncyclic coordinate since

$$T = \frac{m}{2} (r'^2 + r^2\theta'^2 + r^2 \sin^2 \theta \varphi'^2).$$

Consequently

$$\frac{\partial T}{\partial \varphi} = 0, \quad \frac{\partial T}{\partial \theta} \neq 0.$$

The integral associated with the cyclic coordinate φ has the form

$$\frac{\partial L}{\partial \varphi'} = mr^2 \sin^2 \theta \varphi' = \beta,$$

where β is an integration constant.

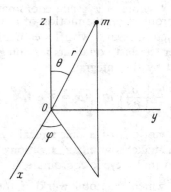

Figure 123

The Routh function has the form

$$R = T + U - \beta \varphi' = \frac{m}{2}(r'^2 + r^2\theta'^2) + U - \frac{m}{2}r^2\sin^2\theta\varphi'^2$$

$$= \frac{m}{2}(r'^2 + r^2\theta'^2) + U - \frac{\beta^2}{2mr^2\sin^2\theta}.$$

The equations of motion for noncyclic coordinates are those of a particle m in a plane in polar coordinates with kinetic energy

$$T^* = \frac{m}{2}(r'^2 + r^2\theta'^2)$$

when acted upon by forces with the force function

$$U^* = U - \frac{\beta^2}{2mr\sin^2\theta}$$

since

$$R = T^* + U^*.$$

131. Generalization of the kinetic energy integral. It is possible to find from Lagrange's equations of motion the kinetic energy integral in a form more general than the one we encountered in our presentation of the general theorems of dynamics. After differentiating equations (5.15) of the holonomic constraints imposed on the mechanical system in question, in Lagrange's holonomic coordinates, we obtain

$$x_\nu' = \frac{\partial x_\nu}{\partial t} + \sum \frac{\partial x_\nu}{\partial q_s}q_s', \quad y_\nu' = \frac{\partial y_\nu}{\partial t} + \sum \frac{\partial y_\nu}{\partial q_s}q_s'.$$

$$z_\nu' = \frac{\partial z_\nu}{\partial t} + \sum \frac{\partial z_\nu}{\partial q_s}q_s' \quad (\nu = 1, \ldots, n).$$

Hence

$$T = \sum \frac{m}{2}(x_\nu'^2 + y_\nu'^2 + z_\nu'^2) = T_2 + T_1 + T_0,$$

where

$$T_2 = \sum_\nu \frac{m_\nu}{2}\left[\left(\sum \frac{\partial x_\nu}{\partial q_s}q_s'\right)^2 + \left(\sum \frac{\partial y_\nu}{\partial q_s}q_s'\right)^2 + \left(\sum \frac{\partial z_\nu}{\partial q_s}q_s'\right)^2\right],$$

$$T_1 = \sum_\nu m_\nu\left(\frac{\partial x_\nu}{\partial t}\sum \frac{\partial x_\nu}{\partial q_s}q_s' + \frac{\partial y_\nu}{\partial t}\sum \frac{\partial y_\nu}{\partial q_s}q_s' + \frac{\partial z_\nu}{\partial t}\sum \frac{\partial z}{\partial q_s}q_s'\right],$$

$$T_0 = \sum_\nu \frac{m_\nu}{2}\left[\left(\frac{\partial x_\nu}{\partial t}\right)^2 + \left(\frac{\partial y_\nu}{\partial t}\right)^2 + \left(\frac{\partial z_\nu}{\partial t}\right)^2\right],$$

where T_2 is a homogeneous quadratic form of velocities $q_1', \ldots, q_k'$; T_1 is a homogeneous linear form of these velocities; and T_0 does not depend on velocities.

Considering the expressions for T_1, T_2, T_0, we note that if the equations of constraints do not depend on time, i.e. if

$$\frac{\partial x_\nu}{\partial t} = 0, \quad \frac{\partial y_\nu}{\partial t} = 0, \quad \frac{\partial z_\nu}{\partial t} = 0 \quad (\nu = 1, \ldots, n),$$

then $T_0 = 0$, $T_1 = 0$ and, consequently, in this case, $T = T_2$.

Assume that the forces which act on a mechanical system have a force function U. On this assumption $Q_s = \partial U / \partial q_s$ and Lagrange's equations of motion have the form

$$\frac{d}{dt}\left(\frac{\partial T}{\partial q_s'}\right) - \frac{\partial T}{\partial q_s} = \frac{\partial U}{\partial q_s} \quad (s = 1, \ldots, k).$$

We multiply Lagrange's equations by the corresponding q_s' and add the results:

$$\sum \frac{d}{dt}\left(\frac{\partial T}{\partial q_s'}\right) q_s' - \sum \frac{\partial T}{\partial q_s} q_s' = \sum \frac{\partial U}{\partial q_s} q_s'.$$

We assume, in addition, that the expressions for the kinetic energy T and the force function U in the system of variables $t, q_1, \ldots, q_k, q_1', \ldots, q_k'$ do not explicitly depend on time, i.e.

$$\frac{\partial U}{\partial t} = 0, \quad \frac{\partial T}{\partial t} = 0.$$

On this assumption we can write the last relation as

$$\frac{d}{dt}\left(\sum_s \frac{\partial T}{\partial q_s'} q_s'\right) - \left(\sum_s \frac{\partial T}{\partial q_s'} q_s'' + \sum_s \frac{\partial T}{\partial q_s} q_s'\right) = \sum_s \frac{\partial U}{\partial q_s} q_s',$$

or, if, after the substitution $T = T_2 + T_1 + T_0$, we use Euler's theorem on homogeneous functions, we get

$$\frac{d}{dt}(2T_2 + T_1) - \frac{d}{dt}(T_2 + T_1 + T_0) = \frac{dU}{dt},$$

or

$$d(T_2 - T_0) = dU.$$

By integrating this equation, we get the generalized kinetic energy integral

$$T_2 - T_0 = U + h.$$

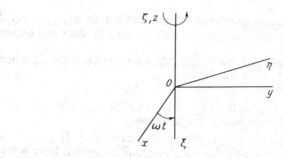

Figure 124

If on these assumptions the constraints do not explicitly depend on time, then $T_1 = 0$, $T_0 = 0$, $T = T_2$ and the last integral coincides with the kinetic energy integral we have found:

$$T = T_2 = U + h.$$

Example. Consider mutually orthogonal axes $\xi\eta\zeta$ with origin at O in a fixed coordinate system $Oxyz$ (Fig. 124). The ζ-axis coincides with the fixed z-axis and the ξ, η axes rotate about the z-axis with a constant angular velocity ω. We choose the initial moment t to be when the x and ξ axes coincide, and then $\angle xO\xi = \omega t$.

We designate the coordinates of a particle of mass m in the systems being considered as x, y, z and ξ, η, ζ respectively. The transition formulas for these coordinates are

$$x = \xi \cos \omega t - \eta \sin \omega t,$$
$$y = \xi \sin \omega t + \eta \cos \omega t, \tag{5.19}$$
$$z = \zeta.$$

We infer that the variables ξ, η, ζ are holonomic ones defining the position of the particle. In these variables, the equations of motion of the particle m are Lagrange's equations. To calculate the kinetic energy of the particle m, it is sufficient to note that the projections of the absolute velocity of the particle m on the moving axes $\xi' - \eta\omega$, $\eta' + \xi\omega$, ζ' are composed of the projections of the relative velocity ξ', η', ζ' and the projections of the transportation velocity $-\eta\omega$, $\xi\omega$, 0. Hence

$$T = \frac{m}{2}[(\xi' - \eta\omega)^2 + (\eta' + \xi\omega)^2 + \zeta'^2].$$

Although t explicitly enters into (5.19), the expression of the kinetic energy T does not explicitly depend on time. If the forces have a force function

$U(\xi, \eta, \zeta)$, then, in accordance with what we have said, we have an integral

$$\frac{m}{2}(\xi'^2 + \eta'^2 + \zeta'^2) - \frac{m\omega^2}{2}(\xi^2 + \eta^2) = U + h.$$

5.6 Determining the Reaction Forces

132. The Euler-Lagrange principle can be used to find the reaction forces of constraints. Indeed, if we add all the reaction forces of constraints to the forces acting on a mechanical system, then from the Euler-Lagrange principle we get Newton's equations for an unconstrained system. However, the method of determining separate forces of reactions is more interesting for practical purposes. Basically the given active forces are complemented by the one reaction force of interest to us. The system is then assumed to be without constraints and generating one reaction which is of interest to us. For the mechanical system thus freed from constraints and having one degree of freedom more an additional holonomic coordinate q is defined whose variation yields a displacement free of constraints in the system. The new T^* are calculated together with the generalized force Q_q^* for the unconstrained motion, and the values of the variables q_s for the true motion are substituted into the Lagrange equation

$$\frac{d}{dt}\left(\frac{\partial T^*}{\partial q'}\right) - \left(\frac{\partial T^*}{\partial q}\right) = Q_q^*.$$

The required reaction force is thus determined which enters into the expression for Q_q^*. To avoid general arguments concerning various possible cases, we give examples to explain this technique.

133. Example 1. Assume that in Zhukovsky's problem we are interested in the reaction of the vertical wall **N** (see Fig. 114). We shall try to find immediately the reaction **N**.

We add the reaction **N** to the given forces. When the other constraints were retained, the reaction force **N** prevented the translation of the system along the x-axis. Therefore the addition of the force of reaction **N** to the given active forces eliminates the link between the beam and the wall and complements the coordinate θ by a variable u associated with the free motion and equal, say, to the coordinate x of the endpoint A.

The kinetic energy T^* of the system moving without constraints consists of the kinetic energy of the animal T_{an}^*, which, for simplicity, is assumed to be a particle Λ, and the kinetic energy of the beam T_b^*.

The midpoint O of the beam has coordinates

$$x = u + a \cos \theta, \quad y = a \sin \theta$$

and hence

$$x' = u' - a \sin \theta \cdot \theta', \quad y' = a \cos \theta \cdot \theta'.$$

Consequently, the square of the velocity of the point O

$$v_0^2 = u'^2 - 2au' \theta' \sin \theta + a^2 \theta'^2,$$

and this means that

$$T_b^* = \frac{P}{2g} (u'^2 - 2au'\theta \sin \theta + a^2 \theta'^2) + \frac{P}{2g} \frac{a^3}{3} \theta'^2.$$

The quantities

$$x = u + s \cos \theta, \quad y = (2a - s) \sin \theta$$

are the coordinates of the animal Λ, and so we have

$$x' = u' + s' \cos \theta - s\theta' \sin \theta,$$
$$y' = -s' \sin \theta + (2a - s)\theta' \cos \theta;$$

and consequently

$$v_{an}^2 = (u' + s' \cos \theta - s\theta' \sin \theta)^2 + (-s' \sin \theta + (2a - s)\theta' \cos \theta)^2,$$

and this means that

$$T_{an}^* = \frac{Q}{2g} [(u' + s' \cos \theta - s\theta' \sin \theta)^2 + (-s' \sin \theta + (2a-s)\theta' \cos \theta)^2].$$

The generalized forces, which are the conjugates of u and θ, can be found from the expression of the virtual elementary work

$$N\delta u - P\delta(a \sin \theta) - Q\delta(2a - s) \sin \theta = N\delta u + \ldots.$$

Hence the expression for the force of interest to us is

$$Q_u = N.$$

We are interested in a new Lagrange equation in the variable u,

$$\frac{d}{dt} \left(\frac{\partial T^*}{\partial u'} \right) - \frac{\partial T^*}{\partial u} = Q_u,$$

but

$$\frac{\partial T^*}{\partial u'} = \frac{P}{g} (u' - a\theta'\sin \theta) + \frac{Q}{g}(u' + s' \cos \theta - s\theta' \sin \theta'),$$

$$\frac{\partial T^*}{\partial u} = 0,$$

and, consequently,

$$\frac{P}{dt}\left[\frac{P}{g}(u' - a\theta'\sin\theta) + \frac{Q}{g}(u' + s'\cos\theta - s\theta'\sin\theta)\right] = N.$$

To find N, we must substitute into this formula the variable $u = 0$, which obtains when the system is constrained. Then we get the value N' for the case when the beam falls, i.e.

$$N' = \frac{d}{dt}\left[\frac{P}{g}(-a\theta'\sin\theta) + \frac{Q}{g}(s'\cos\theta - s\theta'\sin\theta)\right].$$

For Zhukovsky's problem we must also substitute into this formula the variables

$$u' = 0, \quad \theta' = 0,$$

which correspond to a fixed beam, and then we shall have

$$N = \frac{Q}{g}s''\cos\theta.$$

By analogy we can find the force of reaction at the end B.

It is more interesting to find the reaction $\mathbf{R}$ of link between the animal Λ and the beam. We disregard this link (Fig. 125). We add the reaction of the constraint $\mathbf{R}$ imposed on the animal Λ and $\mathbf{R}'$ ($\mathbf{R}' = -\mathbf{R}$) acting on a point C of the beam to the given active forces, i.e. to the weight of the animal and the beam. The coordinate s is holonomic and such that δs expresses the motion of the animal alone without constraint. We need not calculate the kinetic energy T^* anew, it is sufficient to substitute $u' = 0$ into the preceding relation. In the unconstrained motion we are considering, the generalized forces can be found from the expression for the work

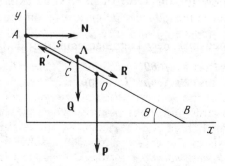

Figure 125

of the active forces and the forces of reactions, which we added when we eliminated the constraints, during an elementary virtual displacement of the motion, i.e.

$$R\delta s - P\delta(a \sin\theta) - Q\delta(2a - s)\sin\theta = Q_s\delta s + \ldots .$$

Hence

$$Q_s = R + Q \sin\theta.$$

Lagrange's equation from which the reaction force R can be found is

$$\frac{d}{dt}\left(\frac{\partial T^*}{\partial s'}\right) - \frac{\partial T^*}{\partial s} = Q_s,$$

where $(u' = 0)$

$$\frac{\partial T^*}{\partial s'} = \frac{Q}{g}\{ (s' \cos\theta - s\theta' \sin\theta)\cos\theta$$

$$-[-s' \sin\theta + (2a - s)\theta' \cos\theta]\sin\theta \},$$

$$\frac{\partial T^*}{\partial s} = \frac{Q}{g}\{ -(s' \cos\theta - s\theta' \sin\theta)\theta' \sin\theta$$

$$-[-s' \sin\theta + (2a - s)\theta' \cos\theta]\theta' \cos\theta \}.$$

Consequently, when the beam falls we have

$$\frac{d}{dt}\frac{Q}{g}\{ (s' \cos\theta - s\theta' \sin\theta)\cos\theta - [-s' \sin\theta$$

$$+ (2a - s)\theta' \cos\theta]\sin\theta \} + \frac{Q}{g}\{ (s' \cos\theta - s\theta' \sin\theta)\theta' \sin\theta$$

$$+[-s' \sin\theta + (2a - s)\theta' \cos\theta]\theta' \cos\theta \} = R + Q \sin\theta.$$

For Zhukovsky's problem, in which $\theta' = 0$ we have

$$\frac{Q}{g}(s'' \cos^2\theta + s'' \sin^2\theta) = R + Q \sin\theta,$$

or

$$R = \frac{Q}{g}s'' - Q \sin\theta.$$

Example 2. A homogeneous heavy rod AB of length $2l$ and mass m slides in a vertical plane with its ends moving over the interior surface of a smooth hemispherical bowl of radius r (Fig. 126). Find the force of reaction **R** at the end B.

We add **R** to the active forces and free the point B of constraints. We assume that the angle $OAB = \varphi$ is a new variable associated with the un-

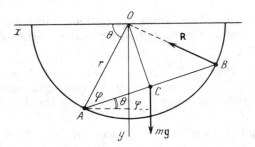

Figure 126

constrained motion. When the motion is constrained, this angle has a constant value $\varphi = \alpha$. Let C be the midpoint of the rod AB. The kinetic energy T^* of the rod moving without constraint can be found from König's theorem. To do this, we must determine the velocity of the centre of gravity C of the rod AB.

The coordinates of the point C in the coordinate system Oxy, where x is a horizontal axis directed leftwards from the centre O of the bowl and y is a vertical axis directed downwards, can be expressed as

$$x = r \cos \theta - l \cos (\theta - \varphi), \quad y = r \sin \theta - l \sin (\theta - \varphi).$$

Differentiating, we obtain x', y' which allow us to determine the square of the velocity v of the point C:

$$v^2 = x'^2 + y'^2 = r^2\theta'^2 + l^2(\theta' - \varphi')^2$$
$$- 2rl\theta'(\theta' - \varphi')[\sin(\theta - \varphi)\sin \theta + \cos(\theta - \varphi)\cos \theta].$$

If J_C is the moment of inertia of the rod AB about its centre of gravity C, then König's formula yields the following expression for the kinetic energy T^* of the system without constraints:

$$T^* = \frac{mv^2}{2} + J_C \frac{(\theta' - \varphi')^2}{2}.$$

The generalized force Q_φ can be easily found from the expression for the work of the active forces and the reaction force R during the displacement for which one variable φ varies by $\delta\varphi$. Since for $\delta\varphi > 0$ the rod rotates clockwise about the point A, Q_φ is equal to the moment of the active forces and the reaction force $\mathbf{R}$, taken with the opposite sign, about the point A, i.e.

$$Q_\varphi = mgl \cos (\theta - \varphi) - R \, 2l \sin \varphi.$$

Then we write out the Lagrange equation

$$\frac{d}{dt}\left(\frac{\partial T^*}{\partial \varphi'}\right) - \frac{\partial T^*}{\partial \varphi} = Q_\varphi$$

which assumes the following form if we take into account the value $J_C = ml^2/3$ and substitute into it the value of φ for the true constrained motion $\varphi = \alpha$, $\varphi' = 0$, viz.,

$$\frac{d}{dt}\left(\frac{ml^2\theta'}{3}\right) + mrl\theta'^2 \sin\alpha = -mgl\cos(\theta - \alpha) + 2Rl\sin\alpha.$$

The last relation immediately yields the reaction R due to the constraints.

It is interesting to consider this problem from another point of view. When we add the reaction forces **R** to the active forces, the end B of the rod being considered free, we can assume that the rotation of the rod about the other end A is a motion without constraints. This rotation does not violate the constraints on the point A and is a rotation about a fixed point for a fixed variable θ. To find the reaction force R, the virtual displacements for the motion without constraints can be substituted directly into the expression for the Euler-Lagrange principle extended by the addition of the reaction to the active forces. The acceleration is assumed to be for a true motion.

To write the formulas, we introduce a rectangular coordinate system x^*y^* with origin at A. After this substitution, the Euler-Lagrange principle yields

$$\sum m_\nu\left(x_\nu^* \frac{d^2 y_\nu}{dt^2} - y_\nu^* \frac{d^2 x_\nu}{dt^2}\right) = \sum (x_\nu^* Y_\nu - y_\nu^* X_\nu) + x_B^* R_y - y_B^* R_x.$$

The left-hand side of this expression is the sum of the masses multiplied by the moment of the accelerations about A, and the right-hand side is the moment of the active forces and the reaction force **R** about A. It is easier to derive these expressions directly from Fig. 127. We isolate an element ds at the point P of the rod which is at the distance of s from A.

The element ds contains the mass $\frac{m}{2l} ds$. In a true motion along a circle about a centre O, the point P has a centripetal acceleration $\varrho\theta'^2$ directed along PO and a tangential acceleration $\varrho\theta''$ which is orthogonal to OP. For the sake of brevity, we designate the distance OP as ϱ ($\varrho = |OP|$). The moment of the centripetal acceleration of the point P about the endpoint A is

$$\varrho\theta'^2 s\sin\psi = \theta'^2 sr\sin\alpha,$$

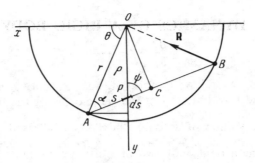

Figure 127

and the moment of the tangential acceleration of the point P about the endpoint A is

$$- \varrho \theta'' s \cos \psi = - \theta'' s(l - s).$$

Consequently, the preceding formula assumes the form

$$\frac{m}{2l} \int [\theta'^2 sr \sin \alpha + \theta'' s(s - l)] \, ds = - mgl \cos(\theta - \alpha) + 2Rl \sin \alpha,$$

or

$$\frac{ml^2}{3} \theta'' + \theta'^2 mrl \sin \alpha = - mgl \cos(\theta - \alpha) + 2Rl \sin \alpha.$$

We can solve this problem if, when adding the reaction force R to the active forces, we assume the rotation about any point of the straight line OA, except for the point O, to be an unconstrained motion.

CHAPTER 6

DYNAMICS OF A RIGID BODY

6.1 Motion of a Rigid Body About a Fixed Axis

134. Imagine a rigid body two of whose points O and O' are fixed. This rigid body can only rotate about the axis which passes through the two points (Fig. 128). We can name the axis passing through O and O' the z-axis and take two straight lines orthogonal to the z-axis and passing through the point O to be the x and y axes.

Next we assume that forces with projections X_ν, Y_ν, Z_ν act on the particles of the body with masses m_ν and coordinates x_ν, y_ν, z_ν ($\nu = 1, \ldots, n$). We use ω to designate the magnitude of the angular velocity of the rigid body about the z-axis. According to Euler's theorem, the projections of the velocity of a particle (x_ν, y_ν, z_ν) of the body on the fixed axes are

$$\frac{dx_\nu}{dt} = -\omega y_\nu, \quad \frac{dy_\nu}{dt} = \omega x_\nu, \quad \frac{dz_\nu}{dt} = 0.$$

We have established that the angular momentum of the body about the z-axis and the kinetic energy of the body are expressed by the formulas

$$K_z = J\omega, \quad T = J\frac{\omega^2}{2},$$

where J is the moment of inertia of the body about the z-axis.

When the rigid body rotates about the z-axis, it has one degree of freedom and, therefore, only one equation of motion. In this case the virtual motions of the body are rotations about the fixed z-axis and, therefore, we have here the theorem on the angular momentum about the z-axis

$$J\frac{d\omega}{dt} = N,$$

where N is the resultant moment about the z-axis of the forces which act on the body.

135. The problem of determining reactions at supports has practical importance. Assume that X, Y, Z and X', Y', Z' are projections of the reactions of the fixed points O and O' on the coordinate axes respectively. We add these reactions to the given forces and then imagine that the rigid body is not fixed at the points O and O'. From the general theorems on the

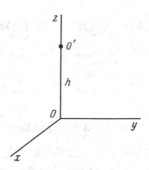

Figure 128

motion of the centre of mass and on the angular momentum, we can write the following relations for a body without constraints:

$$\sum m_\nu \frac{d^2 x_\nu}{dt^2} = \sum X_\nu + X + X',$$

$$\sum m_\nu \frac{d^2 y_\nu}{dt^2} = \sum Y_\nu + Y + Y',$$

$$\sum m_\nu \frac{d^2 z_\nu}{dt^2} = \sum Z_\nu + Z + Z', \tag{6.1}$$

$$\frac{d}{dt} \sum m_\nu \left(y_\nu \frac{dz_\nu}{dt} - z_\nu \frac{dy_\nu}{dt} \right) = L - hY',$$

$$\frac{d}{dt} \sum m_\nu \left(z_\nu \frac{dx_\nu}{dt} - x_\nu \frac{dz_\nu}{dt} \right) = M + hX',$$

where h is the distance between the points O and O' and L, M, N are the projections of the moment of the forces about the origin O on the coordinate axes.

These five relations cannot yield all six unknown support reactions X, Y, Z, X', Y', Z' and therefore the problem is indeterminate. We can explain the indeterminacy here in the same way as in statics, in that the two reactions Z and Z' simultaneously prevent the body from being translated along the z-axis and, therefore, they cannot be found separately from the above expressions.

136. Permanent and free (natural) rotation axes. We can transform (6.1) using

$$\frac{d^2 x_\nu}{dt^2} = -\omega^2 x_\nu - \frac{d\omega}{dt} y_\nu, \quad \frac{d^2 y_\nu}{dt^2} = -\omega^2 y_\nu + \frac{d\omega}{dt} x_\nu, \quad \frac{d^2 z_\nu}{dt^2} = 0,$$

to obtain

$$-\omega^2 \sum m_\nu x_\nu - \frac{d\omega}{dt} \sum m_\nu y_\nu = \sum X_\nu + X + X',$$

$$-\omega^2 \sum m_\nu y_\nu + \frac{d\omega}{dt} \sum m_\nu x_\nu = \sum Y_\nu + Y + Y',$$

$$0 = \sum Z_\nu + Z + Z', \qquad (6.2)$$

$$\omega^2 \sum m_\nu z_\nu y_\nu - \frac{d\omega}{dt} \sum m_\nu z_\nu x_\nu = L - hY',$$

$$-\omega^2 \sum m_\nu z_\nu x_\nu - \frac{d\omega}{dt} \sum m_\nu z_\nu y_\nu = M + hX'.$$

If z is the principal axis of the central ellipsoid of inertia

$$\sum m_\nu x_\nu = 0, \quad \sum m_\nu y_\nu = 0, \quad \sum m_\nu z_\nu x_\nu = 0, \quad \sum m_\nu z_\nu y_\nu = 0,$$

then the reactions at the supports are determined in the same way as in statics since all the terms on the left-hand sides of (6.2) are eliminated. When this is not the case, the reactions depend on the angular velocity of the rigid body ω and the derivative $d\omega/dt$. When the numerical values of ω and $d\omega/dt$ are large, we get large numerical values of the reactions at the supports. In order to get sufficiently small numerical values, we direct the rotation axis in centrifuges along the principal axis of the central ellipsoid of inertia.

Let $L = 0$, $M = 0$, $N = 0$. Then the body will rotate about the z-axis with a constant angular velocity ω. Let us find the conditions under which there are no reactions at the point $O'(X' = 0, Y' = 0, Z' = 0)$. Then from the last relations (6.2) we have

$$\sum m_\nu z_\nu y_\nu = 0, \quad \sum m_\nu z_\nu x_\nu = 0.$$

To put it otherwise, the z-axis must be the principal axis of the ellipsoid of inertia constructed for the rigid body with respect to the point O. Axes of this kind are known as the *permanent axes* of rotation of a rigid body.

Assume that a body is not subjected to external forces $X_\nu = 0$, $Y_\nu = 0$, $Z_\nu = 0$. Then the body will rotate with a constant angular velocity about the z-axis without support reactions at O and O' if

$$\sum m_\nu x_\nu = 0, \quad \sum m_\nu y_\nu = 0, \quad \sum m_\nu z_\nu x_\nu = 0, \quad \sum m_\nu z_\nu y_\nu = 0,$$

or, in other words, if the z-axis is the principal axis of the central ellipsoid of inertia. Axes of this kind are known as *free* (or *natural*) *axes* of rotation.

137. Compound pendulum. A compound pendulum is a heavy rigid body suspended from a fixed horizontal axis (Fig. 129). We call this the

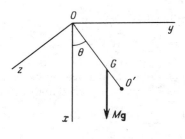

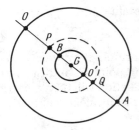

Figure 129   Figure 130

z-axis and make the vertical x-axis directed downwards. We choose the origin O such that the centre of gravity G of the compound pendulum is in the xy-plane and designate the angle between the vertical x and OG as θ.

The pendulum can oscillate about the fixed axis and therefore we can apply the theorem on the angular momentum about the z-axis.

$$Mk^2 \, \frac{d^2\theta}{dt^2} = - Mgl \sin \theta,$$

where M is the mass of the pendulum, k is the radius of gyration of the pendulum about the z-axis ($J_z = Mk^2$) and $l = OG$.

Let us compare the equation

$$\frac{d^2\theta}{dt^2} = - \frac{lg}{k^2} \sin \theta$$

with the equation of motion for a simple pendulum of length l'

$$\frac{d^2\theta}{dt^2} = - \frac{g}{l'} \sin \theta.$$

We see that a simple pendulum of length $l' = k^2/l$ oscillates as a compound pendulum. Therefore l' is called an *equivalent length of a compound pendulum*.

From the point of suspension of the compound pendulum on the straight line OG we lay off a segment OO' equal to the equivalent length of the compound pendulum $OO' = l'$. The point O' is the *centre of oscillation*.

138. Huygens' principle. If a compound pendulum is suspended by its centre of oscillation, it will oscillate with the same period.

Indeed, let ϱ be the central radius of gyration with respect to the axis which is parallel to the z-axis and passes through the centre of gravity G. According to Steiner's theorem

$$Mk^2 = M\varrho^2 + Ml^2.$$

Hence $l' = \varrho^2/l + l$, or $(l' - l)l = \varrho^2$, or, else,

$$OG \cdot O'G = \varrho^2.$$

This formula is symmetric about the centre of oscillation O' and the suspension point O, and this proves that a pendulum suspended by its centre of oscillation will oscillate with the same period. We can immediately calculate its equivalent length l^*:

$$l^* = \frac{\varrho^2}{(l' - l)} + (l' - l) = l'.$$

If follows that the two circles with centres at the point G and radii OG and $O'G$ are the locus of suspension points of a compound pendulum for which the period of oscillation of the pendulum is the same (Fig. 130). If $l > \varrho$, we have a *pendulum* and if $l < \varrho$ we have a *rocker*.
  The formula

$$\frac{dl'}{dl} = 1 - \frac{\varrho^2}{l^2}$$

shows that when we make the length l of the pendulum smaller ($l > \varrho$), the oscillation period decreases $\left(\dfrac{dl'}{dl} > 0 \right)$, and when we make the length l shorter ($l < \varrho$), the oscillation period increases $\left(\dfrac{dl'}{dl} < 0 \right)$.

  Example. *Kater's reversible pendulum.* This pendulum consists of a ruler with two prismatic fulcra attached to it. The edges of the prisms yield two suspension points O and O'. A large disc is placed at one end of the ruler so that the centre of mass of the system is not the midpoint of the distance between the edges of the prisms. An adjuster easily changes the position of the centre of mass between the points O and O'. We can make the pendulum oscillate alternately about each fulcrum (the points O and O') and, by changing the centre of mass, force it to oscillate with the same period. All the time we must ensure that the distances between the fulcra and the centre of mass are not equal (i.e. to have $OG \neq O'G$). In addition, the fulcra must be on different sides of the centre of gravity. Then $O'O = l'$ will be the equivalent length of the pendulum.
  We draw a circle of radius ϱ such that

$$OG \cdot O'G = \varrho^2.$$

When this circle meets the straight line connecting the fulcra and the centre of oscillation at two points P and Q.
  We construct a point O' by drawing a circle of radius ϱ with centre at G. Its intersection with the straight line connecting the points O and

G yields two points P and Q. We draw tangents to this circle from O and connect the points of tangency by a straight line which intersects OG at a point B. From the point G as the centre we draw a circle of radius BG which yields a point O'.

From this construction we see that

$$OG \cdot BG = \varrho^2, \quad BG = GO',$$

i.e.

$$OG \cdot O'G = \varrho^2.$$

6.2 Motion of a Rigid Body About a Fixed Point

139. Consider a rigid body whose particle O is fixed. Assume that the x, y, z axes are rectangular orthogonal axes associated with the body (Fig. 131). The projections of the vector of the instantaneous angular velocity ω of the body on the x, y, z axes are p, q, r respectively, the coordinates of the point of the body are x, y, z, and its mass is m. Let v be the vector of the absolute velocity of the particle m. According to Euler's theorem the projections of the vector on the x, y, z axes are

$$\begin{pmatrix} v_x \\ v_y \\ v_z \end{pmatrix} = \left\| \begin{matrix} p & q & r \\ x & y & z \end{matrix} \right\|,$$

or

$$v_x = qz - ry, \quad v_y = rx - pz, \quad v_z = py - qx.$$

The kinetic energy of the body is

$$T = \sum \frac{mv^2}{2} = \frac{1}{2}(Ap^2 + Bq^2 + Cr^2 - 2Dqr - 2Erp - 2Fpq),$$

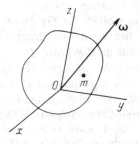

Figure 131

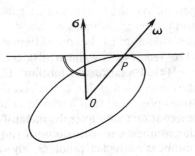

Figure 132

where A, B, C are the moments of inertia of the body about the x, y, z axes respectively, and D, E, F are the products of inertia about the same axes.

140. The quantities σ_x, σ_y, σ_z are the projections of the angular momentum σ of the rigid body on the x, y, z axes:

$$\begin{pmatrix} \sigma_x \\ \sigma_y \\ \sigma_z \end{pmatrix} = \sum \left\| \begin{matrix} x & y & z \\ mv_x & mv_y & mv_z \end{matrix} \right\|,$$

hence

$$\sigma_x = Ap - Fq - Er = \frac{\partial T}{\partial p},$$

$$\sigma_y = -Fp + Bq - Dr = \frac{\partial T}{\partial q},$$

$$\sigma_z = -Ep - Dq + Cr = \frac{\partial T}{\partial r}.$$

The last relations establish an interesting geometric connection between the direction of the instantaneous angular velocity vector ω and the angular momentum vector σ. Let $P(x, y, z)$ be the point of intersection of the vector ω and the ellipsoid of inertia $\frac{x}{p} = \frac{y}{q} = \frac{z}{r} = \varkappa$ (Fig. 132). The equation of the ellipsoid of inertia is

$$f = Ax^2 + By^2 + Cr^2 - 2Dyz - 2Ezx - 2Fxy = 1$$

and so

$$\frac{\partial f}{\partial x} = 2(Ax - Ez - Fy) = 2\varkappa(Ap - Er - Fq) = 2\varkappa \frac{\partial T}{\partial p} = 2\varkappa\sigma_x,$$

then

$$\frac{\sigma_x}{\frac{\partial f}{\partial x}} = \frac{\sigma_y}{\frac{\partial f}{\partial y}} = \frac{\sigma_z}{\frac{\partial f}{\partial z}}.$$

The vector σ is parallel to grad f constructed at the point P.

141. Equations of motion. The virtual motions of a rigid body which has one fixed point O are rotations about any axes passing through the fixed point and thus a rotation about fixed mutually orthogonal axes which meet at O. Consequently, the absolute velocity of the terminus of the angular momentum vector σ with respect to the fixed point O is equal to the moment of the active forces. This proposition can also be written in moving axes.

The absolute velocity of the terminus of the angular momentum vector σ (σ_x, σ_y, σ_z) is equal to the sum of the relative velocity ($d\sigma_x/dt$, $d\sigma_y/dt$, $d\sigma_z/dt$) and the transportation velocity ($q\sigma_z - r\sigma_y$, $r\sigma_x - p\sigma_z$, $p\sigma_y - q\sigma_x$). Hence

$$\frac{d\sigma_x}{dt} + q\sigma_z - r\sigma_y = L, \quad \frac{d\sigma_y}{dt} + r\sigma_x - p\sigma_z = M, \quad \frac{d\sigma_z}{dt} + p\sigma_y - q\sigma_x = N,$$

where L, M, N are the projections of the resultant moment of the active forces about O.

142. The equations of motion of a rigid body about a fixed point which we have obtained are especially simple when the principal axes of the ellipsoid of inertia constructed for a fixed point O are assumed to be the moving axes x, y, z. In this case

$$2T = Ap^2 + Bq^2 + Cr^2$$

and, consequently,

$$\sigma_x = Ap, \quad \sigma_y = Bq, \quad \sigma_z = Cr.$$

Substituting these expressions into the equations of motion, we obtain

$$A\,\frac{dp}{dt} = (B - C)qr + L, \quad B\,\frac{dq}{dt} = (C - A)rq + M,$$

$$C\,\frac{dr}{dt} = (A - B)pq + N.$$

These equations of motion of a rigid body with one fixed point were first derived by Euler and are therefore known as *Euler's equations*.

143. Equations of motion in moving axes not associated with the body. Consider a system of Cartesian rectangular coordinate axes x, y, z with origin at a fixed point O of the body (Fig. 133). Assume that the axes move, and the vector Ω (with projections P, Q, R on these axes) is the vector of an instantaneous absolute angular velocity of the moving system of coordinates. Let p, q, r be the projections of the vector ω of the absolute angular velocity of the rigid body on the x, y, z axes respectively.

According to Euler's theorem the velocity $\mathbf{v}$ of the particle $M(x, y, z)$ of the rigid body is defined by the formula $\mathbf{v} = \omega \times \mathbf{r}$ or the projections of this velocity are defined by the matrix

$$\begin{pmatrix} v_x \\ v_y \\ v_z \end{pmatrix} = \left\| \begin{matrix} p & q & r \\ x & y & z \end{matrix} \right\|,$$

and, consequently,

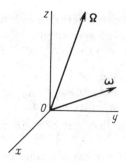

Figure 133

$$v_x = qz - ry, \quad v_y = rx - pz, \quad v_z = py - qx.$$

Hence

$$2T = Ap^2 + Bq^2 + Cr^2 - 2Dqr - 2Erp - 2Fqp,$$

where A, B, C are the moments of inertia of the rigid body about the x, y, z axes and D, E, F are the corresponding products of inertia. Since the body moves relative to the axes x, y, z, the quantities A, B, C, D, E, F are, in general, variables.

The projections of the vector σ of the angular momentum (absolute) of the body on the coordinate axes satisfy the relations

$$\sigma_x = \frac{\partial T}{\partial p}, \quad \sigma_y = \frac{\partial T}{\partial q}, \quad \sigma_z = \frac{\partial T}{\partial r}.$$

The equations of motion are

$$\frac{d\sigma_x}{dt} + Q\sigma_z - R\sigma_y = L, \quad \frac{d\sigma_y}{dt} + R\sigma_x - P\sigma_z = M, \quad \frac{d\sigma_z}{dt} + P\sigma_y - Q\sigma_x = N,$$

where L, M, N are the resultant moments of the active forces with respect to the x, y, z axes.

6.3 Euler's Case

144. Euler considered the motion of a rigid body with one fixed point when the forces acting on the body can be reduced to a resultant force which passes through the fixed point. In this case the resultant moments of the acting forces are zero, i.e. $L = 0$, $M = 0$, $N = 0$, and, consequently, the equations of motion of the body in the principal axes of the ellipsoid of inertia of the rigid body about the point O have the form

$$A \frac{dp}{dt} = (B - C)qr, \quad B \frac{dq}{dt} = (C - A)rp, \quad C \frac{dr}{dt} = (A - B)pq.$$

The equations of motion can be integrated in two steps: we first integrate the equations of motion and obtain p, q, r as functions of t and then use the results to calculate the Euler angles ψ, θ, φ as functions of t (see p. 212). The equations of motion have two first integrals. Multiplying the first equation by p, the second by q, and the third by r and adding the results, we get a relation

$$Ap \frac{dp}{dt} + Bq \frac{dq}{dt} + Cr \frac{dr}{dt} = 0.$$

Integrating this, we obtain

$$Ap^2 + Bq^2 + Cr^2 = h,$$

where h is an integration constant.

Multiplying the equations of motion by Ap, Bq, Cr respectively and adding the results, we obtain

$$A^2p \frac{dp}{dt} + B^2q \frac{dq}{dt} + C^2r \frac{dr}{dt} = 0.$$

Designating the integration constant as l^2 we get

$$A^2p^2 + B^2q^2 + C^2r^2 = l^2.$$

These integrals can be understood from general theorems. The first integral is the kinetic energy integral, the second is the angular momentum integral. Virtual displacements include the true displacements of a rigid body with one fixed point. The work done by active forces which can be reduced to one resultant force, passing through the fixed point, during the true displacement is zero, and, consequently, we have here the kinetic energy integral $2T = h$. This rigid body can rotate about any fixed axis which passes through the fixed point O. The resultant moment of the active forces about the fixed point is zero and therefore it follows from the general theorem on the angular momentum that the angular momentun σ of the rigid body is a vector which is fixed in a fixed space. It follows that its length $\sigma^2 = l^2$ is constant. From the general theorem we have got more than the second first integral, namely, we have found that σ not only has a constant length but also a constant direction in the fixed space.

145. Poinsot gave an elegant geometric interpretation of the motion of a rigid body with one fixed point in Euler's case.

Consider an ellipsoid of inertia of a rigid body relative to a fixed point O and let x, y, z be the principal axes of this ellipsoid (Fig. 134). If A, B, C are the moments of inertia of the body about the x, y, z axes respectively, then the equation of the ellipsoid of inertia is

$$Ax^2 + By^2 + Cz^2 = 1.$$

Assume, for definiteness, that $A \geqslant B \geqslant C$.

At the moment t the instantaneous angular velocity of the body ω (p, q, r) cuts the surface of the ellipsoid of inertia at a point P, which Poinsot called a *pole*. Let x, y, z be the coordinates of the pole and then

$$\frac{p}{x} = \frac{q}{y} = \frac{r}{z} = \frac{\omega}{\Delta} \quad (\Delta = OP).$$

Poinsot formulated the following theorems.

1°. *The ratio ω/Δ is a constant quantity.*

Let us replace p, q, r in the kinetic energy integral by the expressions in the ratio. We get

$$\frac{\omega^2}{\Delta^2} = h,$$

and this proves the assertion.

2°. *The plane π tangential to the ellipsoid of inertia at the pole P is orthogonal to the angular momentum σ.*

The equation of the plane π tangential to the ellipsoid of inertia at the point P with coordinates x, y, z is

$$AxX + ByY + CzZ = 1,$$

where X, Y, Z are the running coordinates, or

$$\frac{\Delta}{\omega}(ApX + BqY + CrZ) = 1.$$

But this is the equation of the plane orthogonal to the vector of the angular momentum $\sigma_x = Ap$, $\sigma_y = Bq$, $\sigma_z = Cr$.

3°. *The distance between the plane π and the fixed point O is constant and equal to $\sqrt{h}/l$.*

This distance is equal to the constant term divided by the square root of the sum of the squares of the coefficients:

$$\delta = \frac{\omega/\Delta}{\sqrt{A^2p^2 + B^2q^2 + C^2r^2}} = \frac{\sqrt{h}}{l}.$$

Hence the plane π tangential at P is a fixed plane in a fixed space since it is orthogonal to the invariable direction of σ and is a constant distance

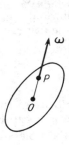

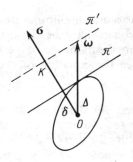

Figure 134   Figure 135

δ from the fixed point O. The ellipsoid of inertia of the body rolls over the fixed plane π without sliding relative to O.

The projection OK of the instantaneous angular velocity vector ω on the direction of the angular momentum vector σ is a constant quantity. Indeed (Fig. 135)

$$\frac{OK}{\delta} = \frac{\omega}{\Delta} = \sqrt{h}.$$

Hence

$$OK = \delta\sqrt{h} = \frac{h}{l}.$$

In other words, the terminus of the instantaneous angular velocity vector lies in the plane π' which is parallel to π, and lies at the distance h/l from the fixed point.

Corollaries. 1. The principal axes of inertia are constant axes of rotation of a rigid body.

2. If an ellipsoid of inertia is an ellipsoid of rotation, then the body precesses regularly: herpolhodes* are circles and axoides are circular cones.

146. The stability of the constant rotations of a body. Let

$$D = \frac{l^2}{h} = \frac{1}{\delta^2}.$$

The first integrals yield relations

$$Ax^2 + By^2 + Cz^2 = 1,$$
$$A^2x^2 + B^2y^2 + C^2z^2 = D.$$

---

* A herpolhode is a plane curve described by a pole on a fixed plane π. — *Ed.*

From this we have

$$A(A - D)x^2 + B(B - D)y^2 + C(C - D)x^2 = 0.$$

This is an equation of a moving axoide. For this cone to be real, it is necessary that

$$A \geqslant D \geqslant C.$$

But this is an obvious condition under which the distance δ between a fixed plane π and a fixed point must not be smaller than the semi-minor axis of the ellipsoid of inertia and must not be larger than the semi-major axis of the ellipsoid of inertia.

If $D = A$, then the cone degenerates into the x-axis ($y = 0$, $z = 0$), if $D = C$, then the cone degenerates into the z-axis ($x = 0$, $y = 0$), and if $D = B$, then the cone degenerates into a pair of planes

$$[\sqrt{A(A - D)}x + \sqrt{C(D - C)}z][\sqrt{A(A - D)}x - \sqrt{C(D - C)}z] = 0.$$

Small deviations of D from these critical values give a clear idea of the stability of constant rotations about the semi-major and semi-minor axes of the ellipsoid of inertia and the instability of constant rotations about its semi-mean axis (Fig. 136).

147. Let us consider how to determine the reaction of a fixed point. When we were interested in the first integrals or integration, we did not use vectors, but for the reaction of a fixed point it is more convenient to use vector notation.

We assume that the principal axes of the ellipsoid of inertia of a body constructed for a fixed point O are the xyz axes, **K** is the angular momen-

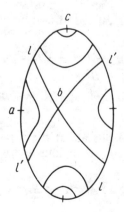

Figure 136

tum of the body, ω is instantaneous angular velocity of the body, $\mathbf{F}_\nu$ are the active forces acting on the body, and $\mathbf{R}$ is the reaction of the fixed point. We designate the radii vectors of the particles of the body as $\mathbf{r}$, their masses as m and the radius vector of the centre of gravity of the body as ξ. The velocity of a particle of the body is $\omega \times \mathbf{r}$, and hence the angular momentum vector $\mathbf{K}$ is defined by the relations

$$\mathbf{K} = \sum m\omega \times \mathbf{r} = \omega \times \sum m\mathbf{r} = \omega \times M\xi = M(\omega \times \xi),$$

where $M = \Sigma m$ is the total mass of the body.

148. Theorem on the angular momentum of a body. The absolute velocity of the terminus of the angular momentum vector $\mathbf{K}$ is equal to the sum of all active forces $\Sigma \mathbf{F}$ and the reaction force $\mathbf{R}$. On the other hand, this absolute velocity is equal to the sum of the relative velocity $\left(\dfrac{d\mathbf{K}}{dt}\right)_{x,y,z}$ relative to the system xyz and the transportation velocity $\omega \times \mathbf{K}$ since the instantaneous angular velocity ω of the system xyz associated with the body is the same as that of the body. Hence the theorem on the angular momentym yields a relation

$$\left(\frac{d\mathbf{K}}{dt}\right)_{xyz} + \omega \times \mathbf{K} = \sum \mathbf{F} + \mathbf{R}.$$

If we substitute the value $\mathbf{K} = M(\omega \times \xi)$ we have found into this formula, we obtain

$$\left(\frac{d\omega}{dt}\right)_{xyz} \times \xi + \omega \times (\omega \times \xi) = \frac{\Sigma \mathbf{F} + \mathbf{R}}{M},$$

since the vector ξ in the xyz system is constant. From this we can find $\mathbf{R}$.

149. Integration of the equations of motion. We have the following system of equations:

$$Ap^2 + Bq^2 + Cr^2 = h,$$
$$A^2p^2 + B^2q^2 + C^2r^2 = Dh,$$
$$B\frac{dq}{dt} = (C - A)rp, \quad A < B < C.$$

The cases of constant rotations about the principal axes of the ellipsoid of inertia about a fixed point are obvious and we shall not consider them. Therefore we assume that

$$A < D < B.$$

From the first two equations we have

$$p^2 = \frac{\begin{vmatrix} h - Bq^2 & C \\ Dh - B^2q^2 & C^2 \end{vmatrix}}{\begin{vmatrix} A & C \\ A^2 & C^2 \end{vmatrix}} = \frac{B(C-B)}{A(C-A)}(f^2 - q^2),$$

where

$$f^2 = \frac{h(C-D)}{B(C-B)} > 0.$$

Similarly,

$$r^2 = \frac{\begin{vmatrix} A & h - Bq^2 \\ A^2 & Dh - B^2q^2 \end{vmatrix}}{\begin{vmatrix} A & C \\ A^2 & C^2 \end{vmatrix}} = \frac{B(B-A)}{C(C-A)}(g^2 - q^2), \qquad (6.3)$$

where

$$g^2 = \frac{h(D-A)}{B(B-A)} > 0.$$

Since

$$f^2 - g^2 = \frac{h}{B} \frac{(B-D)(C-A)}{(C-B)(B-A)} > 0,$$

we have

$$g^2 < f^2.$$

It follows that for formula (6.3) to yield nonnegative values for r^2, q may vary on the interval $[-g, +g]$, $r^2 = 0$ when $q^2 = g^2$. The quantity p always retains its sign and does not vanish. Let $p > 0$. From Euler's second equation we have

$$\frac{dq}{dt} = \sqrt{\frac{(C-B)(B-A)}{AC}} \cdot \sqrt{(f^2 - q^2)(g^2 - q^2)}.$$

The substitution $q = gs$, $k^2 = g^2/f^2 < 1$ yields

$$\frac{ds}{dt} = n\sqrt{(1 - s^2)(1 - k^2 s^2)},$$

where

$$n = f\sqrt{\frac{(C-B)(B-A)}{AC}}.$$

Hence

$$n(t - t_0) = \int_0^s \frac{ds}{\sqrt{(1 - s^2)(1 - k^2 s^2)}}.$$

14*

Thus, in Euler's case, the integration of the equations of motion reduces to elliptic integrals.

150. The *case of regular precession* occurs when an ellipsoid of inertia is a body of revolution. Then the equations of motion are integrated with the use of elementary functions.

Let $A = B$. In this case, we choose the direction of the angular momentum vector σ, constant in a fixed space, to be the z_1-axis and some fixed axes, which are orthogonal to z_1 and to each other and pass through a fixed point O, to be x_1 and y_1 axes. We shall define the position of the principal x, y, z axes of the ellipsoid of inertia relative to the fixed axes x_1, y_1, z_1 by Euler's angles ψ, θ, φ (Fig. 137). We have

$$p = \psi' \sin\theta \sin\varphi + \theta' \cos\varphi,$$
$$q = \psi' \sin\theta \cos\varphi - \theta' \sin\varphi,$$
$$r = \psi' \cos\theta + \varphi'.$$

In this case we have

$$\sigma_x = Ap = \sigma \sin\theta \sin\varphi, \quad \sigma_y = Bq = \sigma \sin\theta \cos\varphi, \quad \sigma_z = Cr = \sigma \cos\theta.$$

For $A = B$ we obtain the following first integral from Euler's third equation:

$$r = r_0.$$

The quantity σ is constant and $\cos\theta = Cr_0/\sigma$ is also constant. Hence $\theta = \theta_0$ and

$$Ap = A\psi' \sin\theta \sin\varphi = \sigma \sin\theta \sin\varphi,$$
$$Aq = A\psi' \sin\theta \cos\varphi = \sigma \sin\theta \cos\varphi.$$

For the angle ψ the equations yield

$$A\psi' = \sigma.$$

Integration yields

$$\psi = \frac{\sigma}{A} t + \text{const.}$$

From the integral $r = r_0$ we obtain, in the Euler angles, a relation

$$r_0 = \psi' \cos\theta_0 + \varphi',$$

which makes it possible to derive a differential equation for the angle φ

$$\varphi' = \left(r_0 - \frac{\sigma}{A} \cos\theta_0\right).$$

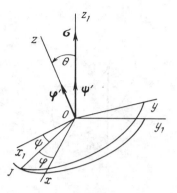

Figure 137

Integration yields

$$\varphi = \left(r_0 - \frac{\sigma}{A} \cos \theta_0\right) t + \text{const.}$$

6.4 Lagrange's Case

151. For Lagrange's case the ellipsoid of inertia is an ellipsoid of revolution with respect to the z-axis and the centre of gravity of the rigid body lies on the axis of dynamic symmetry at the distance ζ from a fixed point O. We have $A = B$, $\xi = 0$, $\eta = 0$, $\zeta > 0$. The body is heavy.

The first integrals;

1. *The kinetic energy integral.* The virtual displacements include true displacements and the forces have a force function $U = -Mg\zeta_1$. Here ζ_1 is the distance from the fixed point O to the projection of the centre of gravity on the fixed vertical Oz_1 (Fig. 138). The work-kinetic energy theorem is

$$d\frac{1}{2}(Ap^2 + Aq^2 + Cr^2) = -Mgd\zeta_1,$$

whence it follows that

$$Ap^2 + Aq^2 + Cr^2 = h - 2Mg\zeta \cos \theta.$$

2. *The area integral.* The body can rotate about a fixed z_1-axis and the moment of the active forces is zero about the z_1-axis ($N_1 = 0$) because the force of gravity is vertical and hence parallel to z_1. From this we get the

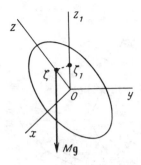

Figure 138

area integral

$$\frac{d\sigma_{z_1}}{dt} = 0,$$

i.e.

$$\sigma_{z_1} = Ap\gamma + Aq\gamma' + Cr\gamma'' = k,$$

where k is the integration constant and γ, γ', γ'' are the direction cosines of the vertical z_1 in the system xyz, viz.,

$$\gamma = \sin\theta\sin\varphi, \quad \gamma' = \sin\theta\cos\varphi, \quad \gamma'' = \cos\theta.$$

3. *Lagrange's integral.* For Lagrange's case $A = B$, $N = 0$, Euler's third equation

$$C\frac{dr}{dt} = (A - B)pq + N$$

yields $\qquad\qquad\qquad r = r_0.$

152. Transformations. We shall write the first integral as

$$r = r_0, \quad p^2 + q^2 = \alpha - a\cos\theta,$$
$$\sin\theta(p\sin\varphi + q\cos\varphi) = \beta - br_0\cos\theta.$$

Here

$$\alpha = \frac{h - Cr_0^2}{A}, \quad a = \frac{2Mg\zeta}{A} > 0, \quad \beta = \frac{k}{A}, \quad b = \frac{C}{A} > 0.$$

Substituting the values of p, q, r expressed in terms of Euler's angles into the first integrals

$$p = \psi'\sin\theta\sin\varphi + \theta'\cos\varphi,$$
$$q = \psi'\sin\theta\cos\varphi - \theta'\sin\varphi,$$
$$r = \psi'\cos\theta + \varphi',$$

we obtain

$$\sin^2 \theta \left(\frac{d\psi}{dt} \right)^2 + \left(\frac{d\theta}{dt} \right)^2 = \alpha - a \cos \theta,$$

$$\sin^2 \theta \left(\frac{d\psi}{dt} \right) = \beta - br_0 \cos \theta,$$

$$\frac{d\psi}{dt} \cos \theta + \frac{d\varphi}{dt} = r_0,$$

We multiply the first equation by $\sin^2 \theta$

$$\left(\sin^2 \theta \, \frac{d\psi}{dt} \right)^2 + \left(\sin \theta \, \frac{d\theta}{dt} \right)^2 = (\alpha - a \cos \theta) \sin^2 \theta$$

and use the second equation. We obtain

$$(\beta - br_0 \cos \theta)^2 + \left(\frac{d \cos \theta}{dt} \right)^2 = (\alpha - a \cos \theta)(1 - \cos^2 \theta).$$

The substitution $u = \cos \theta$ yields

$$\left(\frac{du}{dt} \right)^2 = (\alpha - au)(1 - u^2) - (\beta - br_0 u)^2 = f(u).$$

This equation leads to an elliptic integral. Then the angles ψ and φ can be found by integrating the equations

$$\frac{d\psi}{dt} = \frac{\beta - br_0 u}{1 - u^2}, \quad \frac{d\varphi}{dt} = r_0 - u \frac{\beta - br_0 u}{1 - u^2}.$$

In a mechanics problem the polynomial $f(u) = (\alpha - au)(1 - u^2) - (\beta - br_0 u)^2$ has three real roots. Indeed, for the initial moment the value $u_0 = \cos \theta_0$ lies in the interval $(-1, +1)$ and we have $f(u_0) = (du/dt)_0^2 \geqslant 0$. We assume that $f(u_0) > 0$. The case when $f(u_0) = 0$ is much simpler. For $u = \pm 1$ we have $f(\pm 1) = -(\beta \mp br_0)^2 \leqslant 0$. We assume that $f(\pm 1) < 0$. As $u \to +\infty$, the sign of the polynomial $f(u)$ is defined by its leading term

$$f(u) = au^3 + \ldots > 0$$

since $a > 0$.

Descartes' rule of signs ascertains that the polynomial $f(u)$ has three real roots u_1, u_2 and u' (Fig. 139). Since in motion the value $u = \cos \theta$ must lie in the interval $[-1, +1]$ where $f(u) \geqslant 0$, u can only vary in the interval

$$u_1 \leqslant u \leqslant u_2$$

and oscillates between u_1 and u_2.

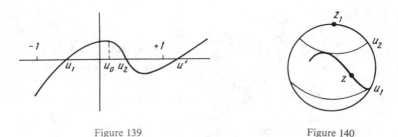

Figure 139   Figure 140

153. Consider a sphere of unit radius with origin at a fixed point (Fig. 140). Two parallels u and u_2 correspond to the roots u_1 and u_2. The z-axis describes a curve on the sphere between the two parallels since u varies in the closed interval $[u_1, u_2]$. Since u is defined by en elliptic integral, z will necessarily touch the parallels.

Let us find the types of curve the z-axis can describe on the sphere. We express the tangent of the angle ϑ that the trace of the z-axis makes on the sphere with the meridian $z_1 z_2$ at the moment $t + dt$. In Fig. 141 ZK is the arc of the circle of radius $\sin\theta$. Obviously,

$$\widehat{ZK} = \sin\theta\, d\psi, \quad \widehat{KZ'} = d\theta$$

so that

$$\tan\vartheta = \frac{\sin\theta\, d\psi}{d\theta} = \frac{\sin^2\theta\, d\psi}{\sin\theta\, d\theta} = -\frac{(1 - u^2)d\psi}{du} = -\frac{\beta - br_0 u}{\sqrt{f(u)}}.$$

Let us determine how the trace of the z-axis approaches the circle u_j ($j = 1, 2$). If, for the circle u_j, the expression in the numerator is nonzero, i.e. $\beta - br_0 u_j \neq 0$, then the curve (z) touches the circle since $\tan\vartheta = \infty$.

If $\beta - br_0 u_j = 0$ for the circle u_j, then $\tan\vartheta$ of the curve (z) has an indeterminate value $0/0$. Using L'Hospital's rule, we find the value of the indeterminate form:

$$\tan\vartheta = \frac{br_0}{(f'(u)/(2\sqrt{f(u)}))_{u=u_j}} = 0.$$

Consequently, in this case the curve (z) approaches the parallel at right angles.

We introduce into consideration a parallel $\bar{u}$ defined by the root of the expression $\beta - br_0\bar{u} = 0$. Here $d\psi/dt = 0$. The following cases are possible here.

1°. $\bar{u}$ lies outside the interval (u_1, u_2). The curve (z) touches the parallels u_1 and u_2, and $d\psi/dt$ retains sign (Fig. 140).

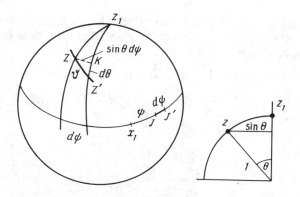

Figure 141

2°. $\bar{u}$ is the interval (u_1, u_2), $d\psi/dt$ changes sign on the parallel $\bar{u}$, and the curve (z) touches the parallels u_1 and u_2 and cuts the parallel $\bar{u}$ at right angles (Fig. 142a).

3°. $\bar{u}$ coincides with u_j $(j = 1, 2)$. We shall prove that $\bar{u}$ cannot coincide with u_1. We have

$$f(\bar{u}) = (\alpha - a\bar{u})(1 - \bar{u}^2) = 0$$

whence it follows that

$$\alpha - a\bar{u} = 0, \quad \delta f = -a(1 - \bar{u}^2)\delta u - 2\bar{u}(\alpha - a\bar{u})\delta u = -a(1 - \bar{u}^2)\delta u.$$

Hence $\delta f < 0$ for $\delta u > 0$. This means that $\bar{u}$ can only coincide with u_2 (see Fig. 139).

For $\bar{u} = u_0$ the curve (z) has cuspidal points on the upper parallel u_2 (Fig. 142b).

Corollary. The case shown in Fig. 142c is impossible. This can be derived from the continuous dependence of the solutions on the constants α and β and the impossibility of $\bar{u} = u_1$.

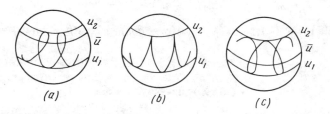

Figure 142

Let us find the conditions under which the roots of $f(u)$ are greater than $1 - \delta$, where $\delta > 0$ is a small constant. For the ideal case $\theta_0 = 0$, $\theta_0' = 0$, $\psi_0' = 0$, $\alpha - a = 0$, $\beta = br_0$,

$$f(u) = (1 - u)^2[a(1 + u) - b^2r_0^2], \quad a(1 + u) - b^2r_0^2 = 0,$$

$$u = \frac{b^2r_0^2}{a} - 1 > 1 - \delta, \quad b^2r_0^2 - 2a > -a\delta,$$

for an arbitrarily small δ we have

$$b^2r_0^2 - 2a > 0.$$

We have the stability condition for the Lagrangian sleeping top.

154. Gyroscope. Let us consider a special case when r_0 is very large and $p_0 = q_0 = 0$. Then $\theta_0' = 0$, $\psi_0' = 0$, or, for $1 - u_0 \neq 0$,

$$\bar{u} = u_0, \quad \alpha - au_0 = 0, \quad \beta - br_0u_0 = 0$$

which corresponds to case $3°$.

$$\left(\frac{du}{dt}\right)^2 = (u_0 - u)[a(1 - u^2) - b^2r_0^2(u_0 - u)],$$

one root is $u_2 = u_0 = \bar{u}$. Hence

$$a(1 - u_1^2) - b^2r_0^2(u_0 - u_1) = 0,$$

or

$$u_0 - u_1 = \frac{a(1 - u_1^2)}{b^2r_0^2}.$$

If r_0 is very large, then u_1 is very close to $u_0 = u_2$,

$$\frac{d\psi}{dt} = \frac{br_0(u_0 - u)}{1 - u^2}.$$

For $r_0 > 0$ we have $\dfrac{d\psi}{dt} \gtrless 0$. This is a pseudoregular precession (Fig. 142b).

The condition for regular precession. The roots u_1 and u_2 coincide:

$$f(u_0) = (\alpha - au_0)(1 - u_0^2) - (\beta - br_0u_0)^2 = 0,$$

$$\left(\frac{df}{du}\right)_{u_0} = -a(1 - u_0^2) - 2u_0(\alpha - au_0) + 2br_0(\beta - br_0u_0) = 0.$$

6.5 Kovalevskaya's Case

155. An ellipsoid of inertia is a symmetric ellipsoid extended along the z-axis, $A = B = 2C$. The centre of mass of a rigid body lies in the equatori-

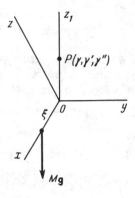

Figure 143

al plane, which is perpendicular to the z-axis (Fig. 143), say, on the x-axis. Then

$$\xi > 0, \quad \eta = 0, \quad \zeta = 0.$$

i°. **Euler's equations.** The moment of the weight with respect to the coordinate axes associated with the rigid body is defined by the matrix

$$\left\| \begin{array}{ccc} \xi & 0 & 0 \\ -mg\gamma & -mg\gamma' & -mg\gamma'' \end{array} \right\|,$$

$$2\frac{dp}{dt} = qr, \quad 2\frac{dq}{dt} = -rp + c\gamma'', \quad \frac{dr}{dt} = -c\gamma',$$

$$c = \frac{mg\xi}{C}.$$

2°. **Poisson's equations** (immobility of the particle $P(\gamma, \gamma', \gamma'')$):

$$\frac{d\gamma}{dt} + q\gamma'' - r\gamma' = 0, \quad \frac{d\gamma'}{dt} + r\gamma - p\gamma'' = 0, \quad \frac{d\gamma''}{dt} + p\gamma' - q\gamma = 0.$$

Without losing generality we can assume that $c = 1$, since Poisson's equations are homogeneous with respect to $\gamma, \gamma', \gamma''$ and in Euler's equations $\gamma, \gamma', \gamma''$ are with constant multipliers c.

156. First integrals. 1°. *The kinetic energy integral.* The virtual displacements include true displacements and the force of gravity has a force function independent of time,

$$d\frac{1}{2}[2(p^2 + q^2) + r^2] = -d\gamma,$$

or

$$p^2 + q^2 + \frac{r^2}{2} = -\gamma + h.$$

2°. *The integral of the angular momentum with respect to the vertical* z_1. A body can rotate about the vertical z_1 and the resultant moment of the forces of gravity with respect to the z_1-axis is zero

$$2(p\gamma + q\gamma') + r\gamma'' = k.$$

3°. *The trivial integral*

$$\gamma^2 + \gamma'^2 + \gamma''^2 = 1.$$

4°. *Kovalevskaya's integral.* We introduce the designations

$$u = p^2 - q^2 - \gamma, \quad v = 2pq - \gamma'$$

and obtain

$$\frac{du}{dt} = 2p\,\frac{dp}{dt} - 2q\,\frac{dq}{dt} - \frac{d\gamma}{dt}$$
$$= pqr + qrp - q\gamma'' + q\gamma'' - r\gamma' = r(2pq - \gamma'),$$

i.e.

$$\frac{du}{dt} = rv,$$

$$\frac{dv}{dt} = 2p\,\frac{dq}{dt} + 2q\,\frac{dp}{dt} - \frac{d\gamma'}{dt} = -rp^2 + p\gamma'' + rq^2 + r\gamma - p\gamma''$$
$$= -r(p^2 - q^2 - \gamma).$$

i.e.

$$\frac{dv}{dt} = -ru,$$

whence it follows that

$$\frac{d}{dt}(u^2 + v^2) = 0, \quad u^2 + v^2 = \text{const.}$$

The time t does not appear explicitly in the equations of motion. Excluding t, we have five equations for which we have found four first integrals. According to the last-factor theory, the problem reduces to taking integrals. Kovalevskaya proved that except for the four known cases, those of Euler, Lagrange, the complete kinetic symmetry $A = B = C$, and her own case, there were no cases when the general solution of the equations of motion could be a meromorphic function in a complex plane of the variable t.

6.6 Gyroscope in Gimbals

The problem of the motion of a heavy symmetric gyroscope in gimbals, when the axis of the outer ring is vertical, has much in common with a well studied problem of the motion of a heavy rigid body with one fixed point in Lagrange's case. We can also consider a problem of the stability with respect to the angle of nutation.

157. Consider a gyroscope in gimbals shown schematically in Fig. 144. We have introduced the following notation: x_1, y_1, z_1 are fixed coordinate axes, z_1 is the spin axis of the outer gimbal, ψ is the angle of rotation of the outer gimbal, x is the axis of rotation of the inner gimbal, θ is the angle of rotation of the inner gimbal, x, y, z are the axes of the inner gimbal, φ is the angle of spin of the gyroscope in the inner gibmal.

The projections of the angular velocity of the inner gimbal on the x, y, z axes are

$$p^0 = \theta', \quad q^0 = \psi' \sin\theta, \quad r^0 = \psi' \cos\theta.$$

The projections of the angular velocity of the gyroscope on the axes of the inner gimbal are

$$p = \theta', \quad q = \psi' \sin\theta, \quad r = \psi' \cos\theta + \varphi'.$$

The kinetic energy of the outer gimbal is

$$\frac{1}{2} J \psi'^2.$$

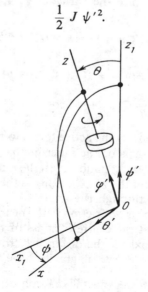

Figure 144

where J is the moment of inertia of the outer gimbal about the z_1-axis.
The kinetic energy of the inner gimbal is

$$\frac{1}{2}(A^0 p^{02} + B^0 q^{02} + C^0 r^{02}).$$

where A^0, B^0, C^0 are the moments of inertia of the inner gimbal about
the x, y, z axes which we take to be the principal axes of the ellipsoid of
inertia of the inner gimbal relative to a fixed point O.

The kinetic energy of the gyroscope is

$$\frac{1}{2}(Ap^2 + Bq^2 + Cr^2),$$

where A, B, C are the moments of inertia of the gyroscope about the x,
y, z axes.

We assume that the ellipsoid of inertia of the gyroscope relative to the
point O is an ellipsoid of rotation about the z-axis, i.e. $A = B$.

The net kinetic energy has the form

$$2T = \psi'^2[J + (A + B^0)\sin^2\theta + (C + C^0)\cos^2\theta]$$
$$+ \theta'^2(A + A^0) + C\varphi'^2 + 2C\varphi'\,\psi'\cos\theta.$$

We can now write the equations of motion for the system being considered
as Lagrange's equations since the variables ψ, θ, φ are independent and
holonomic,

$$(A + A^0)\theta'' - [(A + B^0)\sin\theta\cos\theta - (C + C^0)\sin\theta\cos\theta]\psi'^2$$
$$+ C\varphi'\psi'\sin\theta = Q_\theta,$$

$$\frac{d}{dt}[\psi'[J + (A + B^0)\sin^2\theta + (C + C^0)\cos^2\theta] + C\varphi'\cos\theta] = Q_\psi,$$

$$\frac{d}{dt}C(\varphi' + \psi'\cos\theta) = Q_\varphi.$$

Here $Q_\theta\delta\theta$ is the work done by the active forces to rotate the system through
the angle $\delta\theta$ about the x-axis of the inner gimbal, $Q_\psi\delta\psi$ is the work done
by the forces to rotate the outer gimbal with the inner gimbal and gyroscope
through the angle $\delta\psi$ about the z_1-axis, $Q_\varphi\delta\varphi$ is the work done by the forces
to rotate the gyroscope through the angle $\delta\varphi$.

158. Assume that there is no friction in the bearing, that the active forces
are the gravity force, that the centre of mass of the inner gimbal and the
gyroscope lies on the z-axis at the distance ζ from the origin O and that
z_1 is a vertical axis. The forces of gravity have a force function

$$U = -mg\zeta\cos\theta.$$

Then we can use the equations of motion to derive the first integrals of the following kinds:

$1°$. $\varphi' + \psi' \cos\theta = r_0$,

$2°$. $\psi'[J + (A + B^0)\sin^2\theta + (C + C^0)\cos^2\theta] + C\varphi'\cos\theta = k$,

$3°$. $\psi'^2[J + (A + B^0)\sin^2\theta + (C + C^0)\cos^2\theta] + (A + A^0)\theta'^2$
$+ C\varphi'^2 + 2C\varphi'\psi'\cos\theta = -2mg\zeta\cos\theta + h$.

The first two integrals are associated with the cyclic coordinates φ and ψ respectively. The last integral, which is the kinetic energy integral, can also be immediately found since the virtual displacements include true ones, r_0, k, h are the corresponding constants of the first integrals, and m is the mass of the gyroscope and the inner gimbal.

159. From integrals $1°$ and $2°$ we have

$$\psi = \frac{k - Cr_0\cos\theta}{J + (A + B^0)\sin^2\theta + C^0\cos^2\theta}.$$

The denominator is a quantity strictly larger than zero. We substitute ψ' into $3°$ and, with due account of $1°$, obtain

$$\frac{(k - Cr_0\cos\theta)^2}{J + (A + B^0)\sin^2\theta + C^2\cos^2\theta} + (A + A^0)\theta'^2$$
$$= h - Cr_0^2 - 2mg\zeta\cos\theta.$$

For Euler's angles ψ, θ, φ we get the following system of differential equations:

$$\left(\frac{du}{dt}\right)^2 = \frac{(\alpha - au)(\varepsilon - eu^2) - (\beta - br_0u)^2}{\varepsilon - eu^2},$$

$$\frac{d\psi}{dt} = \frac{\beta - br_0u}{\varepsilon - eu^2}, \quad \frac{d\varphi}{dt} = r_0 - u\frac{\beta - br_0u}{\varepsilon - eu^2}.$$

Here we have accepted the notation

$$u = \cos\theta, \quad \alpha = \frac{h - Cr_0^2}{A + A^0}, \quad a = \frac{2mg\zeta}{A + A^0} > 0, \quad \varepsilon = \frac{J + A + B^0}{A + A^0} > 0,$$

$$e = \frac{A + B^0 - C^0}{A + A^0}, \quad \beta = \frac{k}{A + A^0}, \quad b = \frac{C}{A + A^0} > 0.$$

For Lagrange's case

$$\varepsilon = 1, \quad e = 1.$$

The form of the differential equations for Euler's angles assures us that in the case of a heavy gyroscope in gimbals the nutations of the gyroscope axis play the same leading role as in Lagrange's case. It is therefore natural to begin integration from the first equation, from which we have

$$\int_{u_0}^{u} \frac{(\varepsilon - eu^2)du}{\sqrt{[(\alpha - au)(\varepsilon - eu^2) - (\beta - br_0u)^2](\varepsilon - eu^2)(1 - u^2)}} = t - t_0.$$

After the inversion of this hyperelliptic integral, the calculation of ψ and φ reduces to taking integrals.

We introduce the designations

$$f(u) = (\alpha - au)(\varepsilon - eu^2) - (\beta - br_0u)^2.$$

If $e > 0$, which is the case for most of the practical problems, then in a mechanics problem the polynomial $f(u)$ has three real roots

$$-\sqrt{\frac{\varepsilon}{e}} < u_1 \leqslant u_0, \quad u_0 \leqslant u_2 < \sqrt{\frac{\varepsilon}{e}}, \quad \sqrt{\frac{\varepsilon}{e}} < u' < \infty.$$

The quantity u varies on the interval between the neighbouring points -1, $+1$, u_1, u_2 between which u_0 lies.

For the case of a pseudoregular precession defined by the initial conditions $\theta_0 \neq 0$, $p_0 = 0$, $q_0 = 0$, r_0 having a very large numerical value, we have $\beta - br_0u_0 = 0$, $\alpha - au_0 = 0$, and, consequently,

$$f(u) = (u_0 - u) \ [a(\varepsilon - eu^2) - b^2r_0^2(u_0 - u)],$$

whence it follows that

$$u_0 - u_1 = \frac{a(\varepsilon - eu_1^2)}{b^2r_0^2} > 0,$$

and, consequently, for a sufficiently large r_0, u varies on the interval $(u_1, u_2 = u_0)$, the interval being the smaller the larger r_0.

For the gyroscope in gimbals a regular precession is also possible, but then the polynomial $f(u)$ must have a multiple root $u_1 = u_2 = u_0$. This occurs under the condition that $f(u_0) = 0$, $f'(u_0) = 0$. From these relations we have

$$(A + B^0 - C^0)u_0\psi_0'^2 - Cr_0\psi_0' + mg\zeta = 0.$$

The condition that the roots of this quadratic equation must be real for ψ_0' is expressed by the inequality

$$C^2r_0^2 - 4(A + B^0 - C^0)u_0mg\zeta > 0,$$

or

$$C^2\varphi_0'^2 - 4(A - C + B^0 - C^0)u_0mg\zeta > 0.$$

This condition is analogous to Mayevsky's stability condition for a projectile.

We can derive the conditions for small deviations of the value of u from unity for the interesting practical case of $e > 0$ in the same way as we did for Lagrange's case. To do so, it is sufficient to require that the roots of the polynomial $f(1 - \delta - z)$ be negative, i.e. that the roots of the polynomial $f(u)$ lie on the right of $1 - \delta$. Here δ denotes a small positive constant. The minimum positive number satisfying the inequalities

$$b^2 r_0^2 - 2ae + e(\alpha - a) + 3ae\delta > 0,$$

$$\{b^2 r_0^2 + (\alpha - a(1 - \delta))e - 2ae(1 - \delta)\}\{2br_0(\beta - br_0(1 - \delta))$$

$$-a(\varepsilon - e(1 - \delta)^2) - 2e(1 - \delta)(\alpha - a(1 - \delta))\}$$

$$-ae\{-(\alpha - a(1 - \delta))(\varepsilon - e(1 - \delta)^2) + (\beta - br_0(1 - \delta)^2)\} > 0,$$

$$(\beta - br_0(1 - \delta))^2 - (\alpha - a(1 - \delta))(\varepsilon - e(1 - \delta)^2) > 0$$

will be such a δ. We can carry out the investigation of these equations for the interesting practical case $\theta_0 = 0$, $\theta_0'^2 > 0$, $\psi_0' = 0$ in the same way as we did in Lagrange's case.

6.7 Special Cases of Integration

160. The case of Goryachev-Chaplygin. Assume that $A = B = 4C$, $\sigma_{z_1} = 0$, z_1 is a vertical passing through a fixed point, $\xi \neq 0$, $\eta = 0$, $\zeta = 0$, i.e. the centre of mass lies in the equatorial plane.

The moment of the force of gravity about the moving axes is defined by the matrix

$$\left\| \begin{matrix} \xi & 0 & 0 \\ -mg\gamma & -mg\gamma' & -mg\gamma'' \end{matrix} \right\|.$$

We have Euler's equations

$$4C \frac{dp}{dt} = 3Cqr, \qquad 4C \frac{dq}{dt} = -3Cpr + mg\xi\gamma'', \qquad C \frac{dr}{dt} = -mg\xi\gamma'.$$

Then, for simplification, we introduce $a = mg\xi/C$.

Poisson's equations are

$$\frac{d\gamma}{dt} + q\gamma'' - r\gamma' = 0, \quad \frac{d\gamma'}{dt} + r\gamma - p\gamma'' = 0, \quad \frac{d\gamma''}{dt} + p\gamma' - q\gamma = 0.$$

The first integrals are

1°. $4(p^2 + q^2) + r^2 = -2a\gamma + h$, the kinetic energy integral.
2°. $4(p\gamma + q\gamma') + r\gamma'' = 0$, the area integral.
3°. $\gamma^2 + \gamma'^2 + \gamma''^2 = 1$, the trivial integral.

In the case of Goryachev-Chaplygin the fourth integral is

$$S = r(p^2 + q^2) - ap\gamma''.$$

Indeed, by virtue of the equations of motion,

$$\frac{dS}{dt} = -\frac{aq}{4} \; (4p\gamma + 4q\gamma' + r\gamma'').$$

Using integral 2°, we get

4°. $r(p^2 + q^2) - ap\gamma'' = $ const.

161. The case of Bobylev-Steklov. Assume that $B = 2A$, $\xi = 0$, $\eta > 0$, $\zeta = 0$, i.e. the centre of mass of the body lies on the Oy axis.

The moments of the force of gravity about the axes associated with the body are defined by the matrix

$$\left\| \begin{matrix} 0 & \eta & 0 \\ -mg\gamma & -mg\gamma' & -mg\gamma'' \end{matrix} \right\|.$$

Euler's equations are

$$A\,\frac{dp}{dt} = (2A - C)qr - mg\eta\gamma'',$$

$$2A\,\frac{dq}{dt} = (C - A)rp,$$

$$C\,\frac{dr}{dt} = -Apq + mg\eta\gamma.$$

Poisson's equations complete the system.
Bobylev and Steklov found the special solution

$$r = 0, \quad q = q_0, \quad p = \frac{mg\eta\gamma}{Aq_0} = \mu\gamma.$$

It is easy to see that this solution satisfies Euler's second and third equations.
Poisson's equations for the particular problem are

$$\frac{d\gamma}{dt} + q_0\gamma'' = 0, \quad \frac{d\gamma'}{dt} - \mu\gamma\gamma'' = 0, \quad \frac{d\gamma''}{dt} + \mu\gamma\gamma' - q_0\gamma = 0. \quad (6.4)$$

Euler's first equation reduces to the form

$$\frac{d\gamma}{dt} = -q_0\gamma'',$$

i.e. is satisfied at the same time as Poisson's equations. We multiply the first and second equations in (6.4) by $\mu\gamma$ and q_0 respectively and add the results:

$$\mu\gamma \frac{d\gamma}{dt} + q_0 \frac{d\gamma'}{dt} = 0.$$

We obtain an integral

$$\mu \frac{\gamma^2}{2} + q_0\gamma' = J = \text{const.}$$

Finding γ' from this relation, we have the following from the trivial integral:

$$\gamma'' = \sqrt{1 - \gamma^2 - \gamma'^2} = \sqrt{1 - \gamma^2 - \frac{1}{q_0^2} \left(J - \frac{\mu\gamma^2}{2} \right)^2}.$$

The first equation in (6.4) leads to an elliptic integral for γ.

162. Hess' case. Consider an ellipsoid of *gyration* which has an equation

$$\frac{x^2}{A} + \frac{y^2}{B} + \frac{z^2}{C} = 1, \quad A > B > C \tag{6.5}$$

in the principal axes of inertia of a body for the point O. Hess' case occurs when (1) the centre of gravity of the body lies on the normal at the point O to the π plane of the circular section of the ellipsoid of gyration, (2) at the initial moment angular momentum vector σ lies in the π plane.

Let us find analytic expressions for these conditions.

1. We obtain the equations of the planes of circular sections of the ellipsoid of gyration as the intersection of the surface of the ellipsoid with a sphere

$$\frac{1}{B} (x^2 + y^2 + z^2) = 1.$$

We subtract this equation from (6.5) and get, after transformations,

$$x^2(A - B)C - z^2(B - C)A = 0.$$

This equation disintegrates into two equations

$$x\sqrt{C(A - B)} - z\sqrt{A(B - C)} = 0,$$
$$x\sqrt{C(A - B)} + z\sqrt{A(B - C)} = 0.$$

Assume that the last equation defines the π plane and α, β, γ are direction cosines of the normal to this plane,

$$\frac{\alpha}{\sqrt{C(A - B)}} = \frac{\beta}{0} = \frac{\gamma}{\sqrt{A(B - C)}}, \quad \beta = 0. \tag{6.6}$$

If ξ, η, ζ are the coordinates of the centre of gravity of the body, we can write the condition as

$$\frac{\xi}{\alpha} = \frac{\eta}{\beta} = \frac{\zeta}{\gamma},$$

or

$$\eta = 0, \quad \frac{\xi}{\sqrt{C(A - B)}} = \frac{\zeta}{\sqrt{A(B - C)}}. \tag{6.7}$$

2. If σ_x, σ_y, σ_z are the projections of the angular momentum vector, we can write condition (2) as

$$(\sigma_x \alpha + \sigma_y \beta + \sigma_z \gamma)_0 = 0.$$

Since $(\sigma_x)_0 = Ap_0$, $(\sigma_z)_0 = Cr_0$, it follows from (6.6) and (6.7) that

$$Ap_0 \sqrt{C(A - B)} + Cr_0 \sqrt{A(B - C)} = 0$$

and

$$Ap_0\xi + Cr_0\zeta = 0.$$

The equations of motion are

$$A \frac{dp}{dt} = (B - C)qr + mg\zeta\gamma',$$

$$B \frac{dq}{dt} = (C - A)pr - \zeta mg\gamma + \xi mg\gamma'',$$

$$C \frac{dr}{dt} = (A - B)pq - mg\xi\gamma'.$$

Poisson's equations complete the system.

The first integrals are

1°. $Ap^2 + Bq^2 + Cr^2 = -2mg(\xi\gamma + \zeta\gamma'') + h$, the kinetic energy integral.

2°. $Ap\gamma + Bq\gamma' + Cr\gamma'' = k$, the area integral.

3°. $\gamma^2 + \gamma'^2 + \gamma''^2 = 1$.

We multiply Euler's first and third equations by ξ and ζ respectively and add the results:

$$A\xi \frac{dp}{dt} + C\zeta \frac{dr}{dt} = q[(B - C)\xi r + (A - B)\zeta p]. \tag{6.8}$$

The left-hand side is equal to $\dfrac{d}{dt} (A\xi p + C\zeta r)$. We square equation (6.7) and multiply the result by AC:

$$\frac{A\xi^2}{A - B} = \frac{C\zeta^2}{B - C} = \frac{1}{\mu}.$$

Since

$$A - B = \mu A \xi^2, \quad B - C = \mu C \zeta^2,$$

we can reduce the right-hand side of (6.8) to the form

$$\mu q \xi \zeta (Ap\xi + Cr\zeta).$$

Thus equation (6.8) has a particular solution

4°. $Ap\xi + Cr\zeta = 0.$

163. N.E. Zhukovsky gave a geometric interpretation to Hess' case. Let $C(\xi, 0, \zeta)$ be the centre of mass of the body. Then the projections u, v, w of the velocity of the particle C can be represented as

$$u = q\zeta, \quad v = r\xi - p\zeta, \quad w = -q\xi.$$

We use T^* to designate the kinetic energy of a particle of mass m (the mass of the body) which is at the centre of mass C:

$$T^* = \frac{m}{2}(u^2 + v^2 + w^2) = \frac{m}{2}(q^2\zeta^2 + r^2\xi^2 + p^2\zeta^2 + q^2\xi^2 - 2rp\xi\zeta).$$

Using Hess' particular integral, we obtain

$$-2rp\xi\zeta = r\zeta(-p\xi) + p\xi(-r\zeta) = \frac{Cr^2\zeta^2}{A} + \frac{Ap^2\xi^2}{C}$$

and

$$T^* = \frac{m}{2}\left[p^2\left(\zeta^2 + \frac{A}{C}\xi^2 \right) + q^2(\xi^2 + \zeta^2) + r^2\left(\xi^2 + \frac{G}{A}\zeta^2 \right) \right].$$

Since

$$\frac{\xi^2}{C(A - B)} = \frac{\zeta^2}{A(B - C)} = \frac{\varrho^2}{B(A - C)}, \quad \varrho^2 = \xi^2 + \zeta^2, \quad (6.9)$$

it follows that

$$T^* = \frac{m}{2}\varrho^2\left[p^2\left(\frac{A(B - C)}{B(A - C)} + \frac{AC(A - B)}{CB(A - C)} \right) \right.$$

$$+ q^2\left(\frac{C(A - B)}{B(A - C)} + \frac{A(B - C)}{B(A - C)} \right) + r^2\left(\frac{C(A - B)}{B(A - C)} + \frac{CA(B - C)}{AB(A - C)} \right) \bigg]$$

$$= \frac{m\varrho^2}{B}\cdot\frac{1}{2}(Ap^2 + Bq^2 + Cr^2),$$

i.e. $T^* = \varepsilon^2 T = \varepsilon^2(-2mgz_C + h)$, where $\varepsilon^2 = m\varrho^2/B$ and the kinetic energy integral has been used.

The projections of the angular momentum σ^* of the particle of mass m lying at the centre of mass of the body can be found from the matrix

$$\left\| \begin{array}{ccc} \xi & 0 & \varsigma \\ mq\varsigma & m(r\xi - p\varsigma) & -mq\xi \end{array} \right\|,$$

$$\sigma_x^* = -\varsigma m(r\xi - p\varsigma), \quad \sigma_y^* = mq(\xi^2 + \varsigma^2), \quad \sigma_z^* = \xi m(r\xi - p\varsigma).$$

It follows from Hess' integral and from (6.9) that

$$\sigma_x^* = mp\left(\varsigma^2 + \frac{A}{C}\xi^2\right) = m\varrho^2 p\left(\frac{A(B-C)}{B(A-C)} + \frac{A}{C}\frac{C(A-B)}{B(A-C)}\right) = \frac{m\varrho^2}{B}Ap,$$

$$\sigma_y^* = mq\varrho^2 = \frac{m\varrho^2}{B}Bq, \quad \sigma_z^* = \frac{m\varrho^2}{B}Cr,$$

i.e.

$$\sigma^* = \varepsilon^2 \sigma.$$

And since $\sigma_{z_1} = k$, it follows that $\sigma_{z_1}^* = \varepsilon^2 k$. This means that in Hess' case the centre of mass moves as would a spherical pendulum if the acceleration due to gravity were replaced by $g' = \varepsilon^2 g$.

Let us consider a particle β with coordinates $x = 0$, $y = \sqrt{B}$, $z = 0$ in the π-plane. The projections of its velocity $\mathbf{v}_\beta$ can be determined from the matrix

$$\left\| \begin{array}{ccc} p & q & r \\ 0 & \sqrt{B} & 0 \end{array} \right\|,$$

$$u_\beta = -\sqrt{B}r, \quad v_\beta = 0, \quad w_\beta = \sqrt{B}p.$$

We use θ to designate the angle that $\mathbf{v}_\beta$ makes with the π-plane. Then

$$\sin\theta = \cos(\mathbf{v}_\beta, \mathbf{n}) = \frac{u_\beta \alpha + w_\beta \gamma}{\sqrt{u_\beta^2 + w_\beta^2}}$$

$$= \frac{-r\sqrt{B}\sqrt{C(A-B)/(B(A-C))} + p\sqrt{B}\sqrt{A(B-C)/(B(A-C))}}{\sqrt{B}\sqrt{p^2 + r^2}}$$

$$= \frac{p\sqrt{A(B-C)} - r\sqrt{C(A-B)}}{\sqrt{B(A-C)}\sqrt{p^2 + r^2}}$$

$$= \frac{\sqrt{AC}}{\sqrt{B(A-C)}}\frac{p\sqrt{(B-C)/C} - r\sqrt{(A-B)/A}}{\sqrt{p^2 + r^2}}.$$

Using Hess' integral and (6.9), we find that

$$\frac{p}{r} = -\sqrt{\frac{C(B-C)}{A(A-B)}}.$$

Hence
$$\sin \theta = \frac{\sqrt{AC}}{\sqrt{B(A + C - B)}} ,$$

i.e. θ remains constant.

6.8 Equations of Motion of an Unconstrained Rigid Body

164. Consider an unconstrained rigid body moving due to forces $\mathbf{F}_\nu$ ($\nu = 1, 2, 3, \ldots$).

There are no other constraints imposed on the system of particles except for those requiring the system to be a rigid body.

Let a point G be the centre of mass of the rigid body and have coordinates ξ, η, ζ in a fixed coordinate system xyz.

The virtual displacements of the rigid body include translations along the three axes and, consequently, the equations of motion of the centre of mass along all the coordinate axes are

$$m \frac{d^2 \xi}{dt^2} = \sum X_\nu, \quad m \frac{d^2 \eta}{dt^2} = \sum Y_\nu, \quad m \frac{d^2 \zeta}{dt^2} = \sum Z_\nu. \quad (6.10)$$

The virtual displacements also include rotations of a rigid body about any axis; consequently, the theorem on the angular momentum about appropriate axes holds true. But this theorem is not convenient for investigations in such a coordinate system.

The rigid body in question can rotate about any axis and also translate along all the axes. Consequently, we have a theorem on the angular momentum with respect to all the coordinate axes during a relative motion (the motion with respect to König's axes which are drawn through the centre of mass). This theorem is also inconvenient for use.

It is more convenient to formulate these theorems in the axes associated with the central ellipsoid of inertia constructed for the point G.

We additionally construct two systems of coordinate axes whose origin coincides with the point G (the centre of mass of the rigid body): $\tilde{x}$, $\tilde{y}$, $\tilde{z}$ are axes parallel to the principal axes x, y, z during the whole motion (König's axes), $\bar{x}$, $\bar{y}$, $\bar{z}$ are axes associated with the rigid body and directed along its principal central axes of inertia.

The theorem on the angular momentum about König's axes written in the $\bar{x}$, $\bar{y}$, $\bar{z}$ axes yields Euler's dynamics equations

$$A \frac{dp}{dt} = (B - C)qr + L, \quad B \frac{dq}{dt} = (C - A)rp + M,$$

$$C \frac{dr}{dt} = (A - B)pq + N. \quad (6.11)$$

Here A, B, C are constants.

Equations (6.10) and (6.11) are independent of each other and have as many degrees of freedom as an unconstrained rigid body, i.e. six. We have chosen the quantities ξ, η, ζ, ψ, θ, φ as the variables characterizing these degrees of freedom.

Thus the motion of an unconstrained rigid body is resolved into the motion of the centre of mass (equations (6.10)) and the motion about the centre of mass as a motion about a fixed point (equations (6.11)). We studied these motions in the sections on the dynamics of a particle and on the motion of a rigid body about a fixed point.

ANALYTICAL DYNAMICS

165. Lagrange's *Mécaniqué analytique* was published in 1788. It was the result of comprehensive surveys of contemporary mathematical achievements, and in it Lagrange brought out, for the first time, and developed the fundamental principles of analytical dynamics.

Not very many new ideas have been added since Lagrange: Hamilton developed the analogy between optics and mechanics, Gauss established the principle of least constraint, and the principle of stability was spontaneously discovered by Lagrange, Laplace, Poisson, Poincaré, and Lyapunov in their studies of cosmogony.

7.1 Constraints

166. Imagine a fixed space xyz and a mechanical system in it consisting of particles P_ν ($\nu = 1, \ldots, n$). We take m_ν to be the mass of the particle P_ν and x_ν, y_ν, z_ν to be its Cartesian coordinates.

Assume furthermore that the motion of the system is constrained. If at a moment t the constraints are rigid, the elementary displacements δx_ν, δy_ν, δz_ν the particle P_ν may make at that moment are known as its *virtual displacememts*.

The displacements dx_ν, dy_ν, dz_ν of the particle P_ν which it actually has during the time dt under the action of the given forces are known as *real*, or *true, displacements* of the particle P_ν. If a mechanical system is constrained, then the true displacements of its various points cannot be arbitrary but must be limited by some relationships of the general form:

$$dx_\nu = a_\nu^1 dq_1 + \ldots + a_\nu^k dq_k + a_\nu^0 dt,$$
$$dy_\nu = b_\nu^1 dq_1 + \ldots + b_\nu^k dq_k + b_\nu^0 dt, \qquad (7.1)$$
$$dz_\nu = c_\nu^1 dq_1 + \ldots + c_\nu^k dq_k + c_\nu^0 dt$$
$$(\nu = 1, \ldots, n).$$

The coefficients a_ν^s, b_ν^s, c_ν^s are functions of the position of the particles P_ν of the system and, maybe, of time t. The auxiliary variables q_s are assumed to be mutually independent and are known as *Lagrange coordinates,* and

k is the *number of degrees of freedom*. The system of equations (7.1) is the analytical definition of the constraints imposed on the mechanical system.

The constraint relations in (7.1) compose a system of Pfaffian differential equations. If (7.1) are nonintegrable, then the constraints are *nonholonomic* and if they are integrable, then the constraints are *holonomic*.

An example of nonholonomic constraints. A ball of radius a rolls over a plane without sliding (Fig. 145). The variables are ξ, η, φ, ψ, θ, where ξ, η are the coordinates of the centre G of the ball and φ, ψ, θ are Euler's angles. The absolute velocity of a particle P is zero, $(\xi'$, η', 0) are the projections of the transportation velocity of the particle P, $\left\| \begin{matrix} p & q & r \\ 0 & 0 & -a \end{matrix} \right\|$ is the relative velocity of the particle P, and p, q, r are the projections of the instantaneous angular velocity of the ball on the $Gxyz$ axes. The equations of constraints are

$$\xi' - qa = 0, \quad \eta' + pa = 0,$$
$$p = \varphi' \sin\theta \sin\psi + \theta' \cos\psi, \quad q = -\varphi' \sin\theta \cos\psi + \theta \sin\psi.$$

The constraints are evidently nonintegrable since the first equation cannot be reduced to the form $\dfrac{d}{dt} f(\xi, \varphi, \psi, \theta) = 0$ because ψ' is absent and ψ enters into the equations. The same is true of the second equation.

When the constraints are holonomic, the equations of constraints (7.1) can be reduced to a more convenient form, viz.,

$$x_\nu = x_\nu(t, q_1, \ldots, q_k), \quad y_\nu = y_\nu(t, q_1, \ldots, q_k), \tag{7.2}$$
$$z_\nu = z_\nu(t, q_1, \ldots, q_k) \quad (\nu = 1, \ldots, n),$$

where $q_1, \ldots, q_k$ are generalized Lagrange's (holonomic) coordinates.

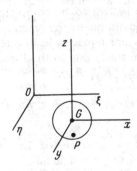

Figure 145

Differentiating (7.2), we find that

$$a_\nu^s = \frac{\partial x_\nu}{\partial q_s}, \quad b_\nu^s = \frac{\partial y_\nu}{\partial q_s}, \quad c_\nu^s = \frac{\partial z_\nu}{\partial q_s}.$$

An example of nonholonomic coordinates. The position of a particle in a plane can be defined by two independent coordinates r, σ, with $\sigma' = r^2\theta'$, where r, σ are nonholonomic coordinates since we cannot express the Cartesian coordinates of the particle in terms of them in the form (7.2).

167. By definition, the displacements δx_ν, δy_ν, δz_ν of the particles P_ν of a mechanical system are virtual (in the ordinary sense of analytical statics) at a fixed time ($dt = 0$). This immediately yields the following relations for virtual displacements for general constraints:

$$\delta x_\nu = \sum_s a_\nu^s \delta q_s, \quad \delta y_\nu = \sum_s b_\nu^s \delta q_s, \quad \delta z_\nu = \sum_s c_\nu^s \delta q_s \tag{7.3}$$

$$(\nu = 1, \ldots, n).$$

For the case of holonomic constraints we can similarly obtain the following relations from (7.2):

$$\delta x_\nu = \sum_s \frac{\partial x_\nu}{\partial q_s} \delta q_s, \quad \delta y_\nu = \sum_s \frac{\partial y_\nu}{\partial q_s} \delta q_s, \quad \delta z_\nu = \sum_s \frac{\partial z_\nu}{\partial q_s} \delta q_s. \tag{7.4}$$

In (7.3) and (7.4) the quantities δq_s are arbitrary small variations of Lagrange's independent variables q_s.[*]

168. Assume that, in addition to the constraints, several forces $\mathbf{F}_\nu$ with projections X_ν, Y_ν, Z_ν ($\nu = 1, \ldots, n$) on the rectangular axes act on the particles P_ν. When the mechanical system moves due to these forces, the constraints act on various particles P_ν of the system with some forces, which are the forces of the reaction $\mathbf{R}_\nu$ ($R_{\nu x}$, $R_{\nu y}$, $R_{\nu z}$).

To determine the reaction forces $\mathbf{R}_\nu$, we assume the mechanical action of the constraints on the system to be completely specified in the sense that we can replace the constraints by the reactions they cause and suppose that when the constraints are replaced by the reactions every particle P_ν can be considered to move freely due to the forces $\mathbf{F}_\nu$ (X_ν, Y_ν, Z_ν) and the constraint reactions $\mathbf{R}_\nu$ ($R_{\nu x}$, $R_{\nu y}$, $R_{\nu z}$).

$$m_\nu \frac{d^2 x_\nu}{dt^2} = X_\nu + R_{\nu x}, \quad m_\nu \frac{d^2 y_\nu}{dt^2} = Y_\nu + R_{\nu y},$$

$$m_\nu \frac{d^2 z_\nu}{dt^2} = Z_\nu + R_{\nu z} \quad (\nu = 1, \ldots, n). \tag{7.5}$$

---

[*] Holonomic constraints possess the property that $d\delta x_\nu/dt = \delta dx_\nu/dt$, ... For holonomic constraints δf has the sense of a differential when t does not vary.

169. The constraints imposed on a system depend on the physical nature of the mechanisms responsible for them. Therefore the characteristics of the constraints must be introduced into mechanics as an axiom that reflects real experimental relations. The following definition of ideal constraints is accepted as such an axiom:

$$\sum_{\nu} (R_{\nu x}\delta x_{\nu} + R_{\nu y}\delta y_{\nu} + R_{\nu z}\delta z_{\nu}) = 0. \tag{7.6}$$

The sum of the actual works of the constraint reactions is zero. This axiom generalizes the observations of reactions of simple smooth surfaces.

7.2 The Euler-Lagrange Principle

170. If we now find $R_{\nu x}$, $R_{\nu y}$, $R_{\nu z}$ from (7.5) and substitute the results into (7.6), we get the fundamental relation of analytical mechanics

$$\sum_{\nu} \left[\left(m_{\nu}\frac{d^2 x_{\nu}}{dt^2} - X_{\nu} \right)\delta x_{\nu} + \left(m_{\nu}\frac{d^2 y_{\nu}}{dt^2} - Y_{\nu} \right)\delta y_{\nu} \right.$$
$$\left. + \left(m_{\nu}\frac{d^2 z_{\nu}}{dt^2} - Z_{\nu} \right)\delta z_{\nu} \right] = 0. \tag{7.7}$$

This relation is valid for any virtual displacement of a system under ideal constraints and defines the relationship between the virtual displacements δx_{ν}, δy_{ν}, δz_{ν} which can exist given the constraints imposed on the system, the forces X_{ν}, Y_{ν}, Z_{ν} acting on the system, and the accelerations $\dfrac{d^2 x_{\nu}}{dt^2}$, $\dfrac{d^2 y_{\nu}}{dt^2}$, $\dfrac{d^2 z_{\nu}}{dt^2}$ caused by the given forces under the constraints imposed.

Lagrange suggested that relation (7.7) be called *d'Alembert's principle.* When all the accelerations are zero and, consequently, the system is in equilibrium, D'Alembert's principle (the principle of Euler-Lagrange) becomes Bernoulli's principle of virtual displacements, which is the fundamental principle of analytical statics.

171. Reactions of constraints are absent in the Euler-Lagrange principle. The relations which it immediately yields are differential equations of motion.

If we isolate a system of independent displacements from the virtual displacements of a system and substitute the isolated displacements into the formula for the Euler-Lagrange principle, we obtain a complete system of independent differential equations of motion.

Now if we isolate any one of the virtual displacements and substitute it into the formula for the Euler-Lagrange principle, the relation obtained will be either one of the differential equations of motion or a consequence of these equations.

We can obtain the general theorems of the dynamics of a system as important examples (see Section 5.3).

172. Poincaré established a complete relationship between the differential equations of motion and the group of virtual displacements (see Supplement 2).

7.3 Hamilton's Principle

173. Hamilton established a significant principle of mechanics, the principle of least action, for the case of ideal holonomic constraints.

We assume the real positions P_0 and P_1 of a system to be known at the moments t_0 and t_1 in a certain motion; x_ν, y_ν, z_ν are functions of time which satisfy the constraints and for $t = t_0$ and $t = t_1$ assume values corresponding to the positions P_0 and P_1. Let $x_\nu + \delta x_\nu$, $y_\nu + \delta y_\nu$, $z_\nu + \delta z_\nu$ be some functions of t which are sufficiently close to x_ν, y_ν, z_ν, satisfy the constraint equations at $t = t_0$ and $t = t_1$, and assume the same values as x_ν, y_ν, z_ν. Hence the expressions δx_ν, δy_ν, δz_ν have the sense of virtual displacements and cancel out at $t = t_0$ and $t = t_1$.

To derive Hamilton's principle, we integrate (7.7) from the Euler-Lagrange principle with respect to time in the limits from t_0 to t_1:

$$\int_{t_0}^{t_1} \sum \left[\left(m_\nu \frac{d^2 x_\nu}{dt^2} - X_\nu \right) \delta x_\nu + \left(m_\nu \frac{d^2 y_\nu}{dt^2} - Y_\nu \right) \delta y_\nu \right.$$

$$\left. + \left(m_\nu \frac{d^2 z_\nu}{dt^2} - Z_\nu \right) \delta z_\nu \right] dt = 0. \tag{7.8}$$

For holonomic constraints integration by parts yields

$$\int_{t_0}^{t_1} m_\nu \frac{d^2 x_\nu}{t^2} \delta x_\nu dt = \left[m_\nu \frac{dx_\nu}{dt} \delta x_\nu \right]_0^1 - \int_{t_0}^{t_1} \delta \frac{m_\nu}{2} \left(\frac{dx_\nu}{dt} \right)^2 dt$$

$$= - \int_{t_0}^{t_1} \delta \frac{m_\nu}{2} \left(\frac{dx_\nu}{dt} \right)^2 dt.$$

The expressions for y and z are similar. Hence expression (7.8) can be reduced to

$$\int_{t_0}^{t_1} \left[\delta T + \sum (X_\nu \delta x_\nu + Y_\nu \delta y_\nu + Z_\nu \delta z_\nu) \right] dt = 0, \tag{7.9}$$

where T is the kinetic energy of the system. Expression (7.9) is Hamilton's principle of least action for the general case of the given forces.

When the forces X_ν, Y_ν, Z_ν have a force function U the equation for Hamilton's principle (7.9) is very simple:

$$X_\nu = \frac{\partial U}{\partial x_\nu}, \quad Y_\nu = \frac{\partial U}{\partial y_\nu}, \quad Z_\nu = \frac{\partial U}{\partial z_\nu} \quad (\nu = 1, \ldots, n).$$

In this case

$$\sum (X_\nu \delta x_\nu + Y_\nu \delta y_\nu + Z_\nu \delta z_\nu) = \delta U$$

and Hamilton's principle assumes the form

$$\delta \int_{t_0}^{t_1} (T + U) dt = 0. \tag{7.10}$$

To put it another way, for the real motion $\int_{t_0}^{t_1} (T + U)\, dt$ has a stationary value. A more thorough study shows that it has a minimum if the time integral $t_1 - t_0$ is sufficiently small.

174. Hamilton's principle makes it easy to derive the differential equations of motion in Lagrange's generalized coordinates q_s. Assume, as before, that the holonomic constraints are defined by the equations

$$x_\nu = x_\nu(t, q_1, \ldots, q_k), \quad y_\nu = y_\nu(t, q_1, \ldots, q_k),$$
$$z_\nu = z_\nu(t, q_1, \ldots, q_k).$$

From this we have

$$\sum_\nu (X_\nu \delta x_\nu + Y_\nu \delta y_\nu + Z_\nu \delta z_\nu)$$

$$= \sum_s \delta q_s \left[\sum_s \left(X_\nu \frac{\partial x_\nu}{\partial q_s} + Y_\nu \frac{\partial y_\nu}{\partial q_s} + Z_\nu \frac{\partial z_\nu}{\partial q_s} \right) \right] = \sum Q_s \delta q_s,$$

where $Q_s = \sum_\nu \left(X_\nu \frac{\partial x_\nu}{\partial q_s} + Y_\nu \frac{\partial y_\nu}{\partial q_s} + Z_\nu \frac{\partial z_\nu}{\partial q_s} \right)$. If, in the expression for the kinetic energy T, we replace x_ν', y_ν', z_ν' by their values in terms of Lagrange's coordinates, we get $T = T(t, q_1, \ldots, q_k, q_1', \ldots, q_k')$.

Hamilton's principle yields

$$\int_{t_0}^{t_1} \left(\sum_s \frac{\partial T}{\partial q_s'} \delta q_s' + \sum_s \frac{\partial T}{\partial q_s} \delta q_s + \sum Q_s \delta q_s \right) dt = 0,$$

or, after integration by parts,

$$\int_{t_0}^{t_1} \sum_s \delta q_s \left[\frac{d}{dt} \left(\frac{\partial T}{\partial q_s'} \right) - \frac{\partial T}{\partial q_s} - Q_s \right] dt = 0.$$

This relation must be zero for arbitrary variations of Lagrange's independent coordinates, and consequently, the brackets must be zero:

$$\frac{d}{dt} \left(\frac{\partial T}{\partial q_s'} \right) - \frac{\partial T}{\partial q_s} = Q_s \quad (s = 1, \ldots, k). \tag{7.11}$$

The differential equations of motion (7.11) are known as *Lagrange's second-order equations.*

When the forces X_ν, Y_ν, X_ν have a force function U, we have $Q_s = \dfrac{\partial U}{\partial q_s}$ and Lagrange's equations assume the form

$$\frac{d}{dt} \left(\frac{\partial T}{\partial q_s'} \right) - \frac{\partial T}{\partial q_s} = \frac{\partial U}{\partial q_s} \quad (s = 1, \ldots, n).$$

Lagrange's equations are especially convenient when the expression $T(t, q_1, \ldots, q_k, q_1', \ldots, q_k')$ is easy to find. It must be emphasized that Lagrange's equations can only be written in the coordinates q_s $(s = 1, \ldots, k)$ in terms of which the Cartesian coordinates can be expressed in a finite form.

175. By way of example, we shall derive Euler's familiar equations for a rigid body which has one fixed point (Fig. 146). Such a body is subjected to holonomic constraints and the coordinates of its particles can be defined by Euler's angles θ, φ, ψ between the fixed coordinate axes $x_1 y_1 z_1$ and the moving xyz coordinate axes. For simplicity we shall assume that the moving axes xyz are the axes of the ellipsoid of inertia of the body with respect to the fixed point O. We know that the kinetic energy is

$$T = \frac{1}{2} (Ap^2 + Bq^2 + Cr^2),$$

where

$$p = \psi' \sin\theta \sin\varphi + \theta' \cos\varphi,$$
$$q = \psi' \sin\theta \cos\varphi - \theta' \sin\varphi,$$
$$r = \psi' \cos\theta + \varphi'.$$

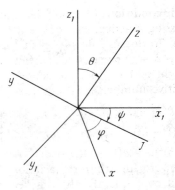

Figure 146

Choosing θ, φ, ψ as Lagrange's coordinates q_s, we have from this Lagrange's equation for the angle φ:

$$\frac{d}{dt}\left(\frac{\partial T}{\partial \varphi'}\right) - \frac{\partial T}{\partial \varphi} = Q_\varphi.$$

We have $\dfrac{\partial T}{\partial \varphi'} = \dfrac{\partial T}{\partial r}\dfrac{\partial r}{\partial \varphi'} = Cr$, $\dfrac{\partial T}{\partial \varphi} = \dfrac{\partial T}{\partial p}\dfrac{\partial p}{\partial \varphi} + \dfrac{\partial T}{\partial q}\dfrac{\partial q}{\partial \varphi} = (A - B)pq$,

since $\dfrac{\partial p}{\partial \varphi} = q$, $\dfrac{\partial q}{\partial \varphi} = -p$. Hence

$$C\frac{dr}{dt} = (A - B)pq + N.$$

We shall explain the meaning of N. Assume that only φ varies, and then

$$\begin{pmatrix} \delta x_\nu \\ \delta y_\nu \\ \delta z_\nu \end{pmatrix} = \begin{vmatrix} 0 & 0 & \delta\varphi \\ x_\nu & y_\nu & z_\nu \end{vmatrix},$$

and, composing an expression for work, we have, for $\delta\varphi \neq 0$,

$$\sum (X_\nu \delta x_\nu + \ldots) = N\delta\varphi.$$

We can obtain the other Euler equations if we change the names of the axes.

7.4 Canonical Equations

176. For the case of mechanical systems subjected to holonomic ideal constraints and acted upon by forces which have a force function U, Hamilton reduced the differential equations of motion to a very significant form. Instead of the velocities q_s' we introduce new variables

$$p_s = \frac{\partial T}{\partial q_s'} \quad (s = 1, \ldots, k).$$

The variables p_s are known as *impulses* and are *conjugates of Lagrange's coordinates* q_s. Since the highest order form with respect to q_s' in the expression for the kinetic energy T is a positive definite quadratic form, the transition from q_s' to p_s is one-to-one.

For systems subjected to ideal holonomic constraints and to the forces which have a force function U we can write Hamilton's principle

$$\delta \int_{t_0}^{t_1} (T + U)\, dt = 0$$

in the form

$$\delta \int_{t_0}^{t_1} \left[\sum_s p_s \frac{dq_s}{dt} - \left(\sum_s p_s q_s' - T - U \right) \right] dt = 0.$$

If we set

$$H = \sum_s p_s q_s' - T - U = H(t, q_1, \ldots, q_k, p_1, \ldots, p_k), \quad (7.12)$$

then

$$\delta \int_{t_0}^{t_1} \left(\sum_s p_s \frac{dq_s}{dt} - H \right) dt = 0.$$

When the limiting values of q_s are fixed*, we can get the following differential equations of motion from Hamilton's principle by the ordinary techniques of variational calculus**.

$$\frac{dq_s}{dt} = \frac{\partial H}{\partial p_s}, \quad \frac{dp_s}{dt} = -\frac{\partial H}{\partial q_s} \quad (s = 1, \ldots, k). \quad (7.13)$$

These are *Hamilton's,* or *canonical, equations.* We shall learn later on that they are theoretically very significant.

---

* This is Poincaré's form of Hamilton's principle, and (1) δp, δq are arbitrary on an interval, and (2) $\delta q = 0$, $t = t_0$, t_1.
** $q_s' = \partial H/\partial p_s$ follows from the transformations.

177. Consider a special case of the holonomic constraints being independent of time. The kinetic energy T is a positive definite quadratic form of the derivatives q_s' and the Hamiltonian function

$$H = \sum_s p_s q_s' - T - U = \sum_s \frac{\partial T}{\partial q_s'} q_s' - T - U = 2T - T - U$$

has the sense of the total energy of the system

$$H = T - U.$$

In this case H does not evidently depend on time and from the canonical equations (7.13) we get

$$\frac{dH}{dt} = \sum_s \frac{\partial H}{\partial q_s} \frac{dq_s}{dt} + \sum_s \frac{\partial H}{\partial p_s} \frac{dp_s}{dt} \equiv 0.$$

Hence the Hamiltonian function does not change value in a real motion and is the first integral

$$H \equiv h,$$

i.e. the kinetic energy integral.

178. Assume now that the constraints imposed on the system explicitly depend on time. Let us consider the function

$$V = \int_{t_0}^{t} (T + U)\, dt. \qquad (7.14)$$

This is customarily called the *action*, or *Hamilton's principal function* and the integration is along the trajectory under study, for which the coordinates q_s and impulses p_s must satisfy the canonical equations (7.13). We agree that V is a function of the variables t, q_s, q_s^0, where q_s^0 are the values of the coordinates q_s at $t = t_0$.

Let us see how the principal function V will vary when we pass from one motion to another caused by small variations of the initial data.

We write the principal function V in the form:

$$V = \int_{t_0}^{t} \left(\sum_s p_s \frac{dq_s}{dt} - H \right) dt.$$

Since the endpoints are not fixed, we have

$$\delta V = \left(\sum_s p_s \delta q_s \right)_{t_0}^{t} + \int_{t_0}^{t} \left[\sum_s \delta p_s \left(\frac{dq_s}{dt} - \frac{\partial H}{\partial p_s} \right) \right.$$

$$+ \sum_s \delta q_s \left(- \frac{dp_s}{dt} - \frac{\partial H}{\partial q_s} \right) \Bigg] dt.$$

But the coordinates q_s and impulses p_s must satisfy (7.13) since the integral in the principal function is taken with respect to the motion of the system. In the last formula the integral is lost

$$\delta V = \sum_s p_s \delta q_s - \sum_s p_s^0 \delta q_s^0.$$

The variation of the function $V(t, q_1, \ldots, q_k, q_1^0, \ldots, q_k^0)$ has the form

$$\delta V = \sum_s \frac{\partial V}{\partial q_s} \delta q_s + \sum_s \frac{\partial V}{\partial q_s^0} \delta q_s^0.$$

Two expressions for the variation δV of the same function V, which are valid for arbitrary values of δq_s and δq_s^0, yield

$$p_s = \frac{\partial V}{\partial q_s}, \quad -p_s^0 = \frac{\partial V}{\partial q_s^0} \quad (s = 1, \ldots, k). \tag{7.15}$$

These relations are very interesting. The last group of equations defines the law of motion of the system $q_s = q_s(t, q_1^0, \ldots, q_k^0, p_1^0, \ldots, p_k^0)$ and the first group yields the corresponding impulses. Consequently, it is sufficient to know the action V to solve the principal problem of mechanics using the simple formulas (7.15).

179. To determine the action V from formula (7.14), it is necessary to know the law of motion of a mechanical system. Therefore, it is no wonder that formulas (7.15) yielded so easily what we assumed at the beginning. To obviate the difficulties in finding the action V from formula (7.14), Hamilton found the partial first-order differential equation for which the action V is a complete function.

From the general form of the principal function $V = V(t, q_1, \ldots, q_k, q_1^0, \ldots, q_k^0)$ and formula (7.14) we have

$$\frac{dV}{dt} = \frac{\partial V}{\partial t} + \sum \frac{\partial V}{\partial q_s} \frac{dq_s}{dt} = T + U,$$

or, from (7.15) and (7.12), we have

$$\frac{\partial V}{\partial t} + H\left(t, q_1, \ldots, q_k, \frac{\partial V}{\partial q_1}, \ldots \frac{\partial V}{\partial q_k} \right) = 0. \tag{7.16}$$

This equation is known as the *Hamilton-Jacobi equation*.

180. Jacobi completed Hamilton's investigations by showing that it is sufficient to know any complete integral $V(t, q_1, \ldots, q_k, \alpha_1, \ldots, \alpha_k)$ of

(7.16) dependent on the constants α_s none of which is additive*, in order to obtain the general solution of (7.13) in the form

$$p_s = \frac{\partial V}{\partial q_s}, \quad \beta_s = \frac{\partial V}{\partial \alpha_s} \quad (s = 1, \ldots, k), \tag{7.17}$$

where β_s are new constants.

The proof of Jacobi's theorem can be found in a course on first-order partial differential equations.

181. In many problems of dynamics Imshenetsky's substitution makes it easier to find the complete integral of the Hamilton-Jacobi equation. If we have a first-order partial differential equation

$$\left(\Phi\left(\psi\left(x_1, \frac{\partial z}{\partial x_1} \right), x_2, \ldots, x_n, \frac{\partial z}{\partial x_2}, \ldots, \frac{\partial z}{\partial x_n} \right) \right) = 0,$$

then its complete integral is

$$z(x_1, \ldots, x_n) = f(x_1) + w(x_2, \ldots, x_n),$$

where f is the integral of the equation

$$\psi\left(x_1, \frac{df}{dx_1} \right) = \alpha_1$$

(α_1 is a constant) and w is the complete integral of the equation

$$\Phi\left(\alpha_1, x_2, \ldots, x_n, \frac{\partial z}{\partial x_2}, \ldots, \frac{\partial z}{\partial x_n} \right) = 0.$$

Example 1. If the constraints do not depend explicitly on time, then the function H does not depend explicitly on t either and the Hamilton-Jacobi equation

$$\frac{\partial V}{\partial t} + H\left(q_1, \ldots, q_k, \frac{\partial V}{\partial q_1}, \ldots, \frac{\partial V}{\partial q_k} \right) = 0$$

admits of the substitution $\dfrac{df}{dt} = -h$ or $f = -h(t - t_0)$, where h is a constant. Hence the complete integral V has the form

$$V = -h(t - t_0) + W(q_1, \ldots, q_k),$$

---

* The complete integral of equation (7.16) is a solution of the form $V(t, q_1, \ldots, q_k, \alpha_1, \ldots, \alpha_k)$, where $\alpha_1, \ldots, \alpha_k$ are arbitrary constants, and

$$\left\| \frac{\partial^2 V}{\partial q_i \partial \alpha_j} \right\| \neq 0 \quad (i, j = 1, \ldots, k).$$

where W is the complete integral of Hamilton's equation

$$H\left(q_1, \ldots, q_k, \frac{\partial W}{\partial q_1}, \ldots, \frac{\partial W}{\partial q_k}\right) = h. \tag{7.18}$$

Example 2. Assume that the Hamiltonian function H does not depend explicitly on q_1. Then the integral V of equation (7.16) has the form

$$V = f(q_1, \alpha_1) + W(t, q_2, \ldots, q_k, \alpha_1, \ldots, \alpha_k),$$

and the corresponding substitution yields one of the integrals of motion

$$p_1 = \frac{\partial V}{\partial q_1} = \frac{df}{dq_1} = \alpha_1.$$

The coordinate q_1 is *cyclic*.

Example 3. Assume that the kinetic energy T and the force function U are

$$2T = (A_1 + \ldots + A_k)(B_1 q_1'^2 + \ldots + B_k q_k'^2),$$
$$U = \frac{U_1 + \ldots + U_k}{A_1 + \ldots + A_k},$$

where A_s, B_s, U_s are functions of one variable q_s. When we pass to impulses p_s, Hamilton's equation (7.18) assumes the form

$$\sum_s \left[\frac{1}{2B_s}\left(\frac{\partial W}{\partial q_s}\right)^2 - hA_s - U_s\right] = 0.$$

The variables are separable. The complete integral W is

$$W = \sum_s V_s,$$

where V_s is a function of one variable q_s and is defined by an easily integrable equation

$$\frac{1}{2B_s}\left(\frac{\partial V_s}{\partial q_s}\right)^2 - hA_s - U_s = \alpha_s \quad (s = 1, \ldots, k).$$

The constants α_s are related as

$$\alpha_1 + \ldots + \alpha_k = 0.$$

Thus the complete integral of the Hamilton-Jacobi equation (7.16) that makes it possible to solve the problem is

$$V = -h(t - t_0) + W = -h(t - t_0) + \sum_s \int \sqrt{2B_s(hA_s + U_s + \alpha_s)}\, dq_s.$$

This case was established by Liouville.

Example 4. Let us consider the motion of a planet in space in obedience to Kepler's laws, assuming the sun to be fixed and placed at the origin. In the spherical coordinates r, θ, φ Hamilton's equation (7.18) has the form

$$\left(\frac{\partial W}{\partial r}\right)^2 + \frac{1}{r^2}\left[\left(\frac{\partial W}{\partial \theta}\right)^2 + \frac{1}{\sin^2\theta}\left(\frac{\partial W}{\partial \varphi}\right)^2\right] - \frac{2m\mu}{r} = h.$$

The following three successive integrable substitutions are obvious:

$$\frac{\partial V_\varphi}{\partial \varphi} = \alpha_1, \quad \left(\frac{dV_\theta}{d\theta}\right)^2 + \frac{\alpha_1^2}{\sin^2\theta} = \alpha_2, \quad \left(\frac{dV_r}{dr}\right)^2 + \frac{\alpha_2}{r^2} - \frac{2m\mu}{r} = h.$$

Hence the complete integral of equation (7.16) is

$$V = -h(t - t_0) + V_r + V_\theta + V_\varphi,$$

where

$$V_r = \int\sqrt{2h + \frac{2m\mu}{r} - \frac{\alpha_2}{r^2}}\, dr, \quad V_\theta = \int\sqrt{\alpha_2 - \frac{\alpha_1^2}{\sin^2\theta}}\, d\theta,$$

$$V_\varphi = \alpha_1\varphi.$$

182. The integration of the mechanics equations are not usually considered in detail in a general course. Those who are interested in these problems should read Jacobi's *Lectures on Dynamics* and Whittaker's textbook on analytical mechanics*.

183. A Coriolis force arises in the relative motion of a particle of mass m. If x', y', z' are the projections of the relative velocity of the particle m on the moving coordinate axes which rotate with the angular velocities p, q, r with respect to fixed axes, then the components of the Coriolis force are defined by the matrix

$$-2m\left\|\begin{matrix} p & q & r \\ x' & y' & z' \end{matrix}\right\|.$$

A force of the same kind arises when an electron moves in a magnetic field (H_1, H_2, H_3):

$$e\left\|\begin{matrix} H_1 & H_2 & H_3 \\ x_1' & x_2' & x_3' \end{matrix}\right\|.$$

---

* C.G.J. Jacobi, *Vorlesungen über Dynamik*. Zweite revidirte Ausgabe (Druck und Verlag von G. Reimer, Berlin: 1884).

E. Whittaker, *A Treatise on the Analytical Dynamics of Particles and Rigid Bodies* (Cambridge University Press, Cambridge: 1937).

where e is the charge of the electron and the speed of light is assumed to be unity.

In a relative motion the particle m is unconstrained, and we have here Hamilton's principle. Consequently, we can apply Hamilton's principle to the similar case of the motion of an electron in an electromagnetic field. Indeed, let $T = \dfrac{m}{2}(x_1'^2 + x_2'^2 + x_3'^2)$, where m is the mass of the electron, and

$$L = T - e\varphi + e(A_1x_1' + A_2x_2' + A_3x_3').$$

Provided that the endpoints are fixed, the variational principle

$$\delta \int_{t_0}^{t_1} L \, dt = 0$$

yields the differential equations of motion of the electron as the Euler-Lagrange equations

$$\frac{d}{dt}\left(\frac{\partial L}{\partial x_i'}\right) - \frac{\partial L}{\partial x_i} = 0 \quad (i = 1, 2, 3).$$

The electric F_1, F_2, F_3 and magnetic H_1, H_2, H_3 forces acting on the electron are defined by the quantities φ, A_1, A_2, A_3 appearing in the formulas

$$F_i = -e\left(\frac{\partial\varphi}{\partial x_i} + \frac{\partial A_i}{\partial t}\right) \quad (i = 1, 2, 3)$$

and

$$\begin{pmatrix} H_1 \\ H_2 \\ H_3 \end{pmatrix} = \left\| \begin{array}{ccc} \dfrac{\partial}{\partial x_1} & \dfrac{\partial}{\partial x_2} & \dfrac{\partial}{\partial x_3} \\ A_1 & A_2 & A_3 \end{array} \right\|.$$

The functions φ and (A_1, A_2, A_3) are known as the *scalar* and *vector* *potentials*. We introduce generalized impulses P_i instead of the ordinary impulses $p_i = \partial T/\partial x_i'$ using the formulas

$$P_i = p_i + eA_i \quad (i = 1, 2, 3), \quad P_i = \frac{\partial L}{\partial x_i'}.$$

Hamilton's principle yields

$$\delta \int_{t_0}^{t_1} \left(\sum P_i x_i' - H\right) dt = 0,$$

where $$H = \sum P_i x_i' - L.$$

The last form of Hamilton's principle immediately yields the canonical equations of motion of the electron in an electromagnetic field

$$\frac{dx_i}{dt} = \frac{\partial H}{\partial P_i}, \quad \frac{dP_i}{dt} = -\frac{\partial H}{\partial x_i} \quad (i = 1, 2, 3).$$

The explicit form of the function H is

$$H = \sum \left(\frac{\partial T}{\partial x_i'} + eA_i \right) x_i' - T + e\varphi - e\sum A_i x_i'$$

$$= T + e\varphi = \frac{1}{2m} \sum (P_i - eA_i)^2 + e\varphi.$$

And now we can write the Hamilton-Jacobi equation.

7.5 Gauss' Principle

184. Gauss established a very interesting principle in mechanics, the principle of least constraint.

If, at the moment t, a mechanical system is in a definite position, in which the coordinates x_ν, y_ν, z_ν and velocities x_ν', y_ν', z_ν' of the particles m_ν have definite values, then the constraint forces constantly acting on the system will cause small variations in the velocities dx_ν', dy_ν', dz_ν' during a very small time interval dt. Gauss suggested a principle which makes it possible to separate real motions due to these forces from the imaginary motions which occur for the same values of x_ν, y_ν, z_ν, x_ν', y_ν', z_ν' (for the moment t) and the same constraints, but are due to other forces.

We designate the variations of the velocity of the particle m during the time dt for the real motion as dx_ν', dy_ν', dz_ν' and those for the imaginary motion as $\delta x_\nu'$, $\delta y_\nu'$, $\delta z_\nu'$. Assume, for simplicity, that the holonomic constraints are

$$f(t, x_\nu, y_\nu, z_\nu) = 0 \quad (s = 1, \ldots, m).$$

Differentiating twice with respect to time, we obtain the following relations for the real and imaginary motions:

$$\sum_\nu \left(\frac{\partial f_s}{\partial x_\nu} dx_\nu' + \frac{\partial f_s}{\partial y_\nu} dy_\nu' + \frac{\partial f_s}{\partial z_\nu} dz_\nu' \right) + A_s \, dt = 0,$$

$$\sum_\nu \left(\frac{\partial f_s}{\partial x_\nu} \delta x_\nu' + \frac{\partial f_s}{\partial y_\nu} \delta y_\nu' + \frac{\partial f_s}{\partial z_\nu} \delta z_\nu' \right) + A_s \, dt = 0,$$

where the expressions for A_s depend only on time t, the coordinates x_ν, y_ν, z_ν and the velocities x_ν', y_ν', z_ν', which are constant for the values of time t being considered in the set of the imaginary motions that Gauss compared. Subtracting the second relation from the first, we get

$$\sum_\nu \left[\frac{\partial f_s}{\partial x_\nu} (dx_\nu' - \delta x_\nu') + \frac{\partial f_s}{\partial y_\nu} (dy_\nu' - \delta y_\nu') \right.$$

$$\left. + \frac{\partial f_s}{\partial z_\nu} (dz_\nu' - \delta z_\nu') \right] = 0 \quad (s = 1, \ldots, m).$$

In other words, virtual displacements include the differences $dx_\nu' - \delta x_\nu'$, $dy_\nu' - \delta y_\nu'$, $dz_\nu' - \delta z_\nu'$. The Euler-Lagrange principle yields

$$\sum_\nu [(m_\nu dx_\nu' - X_\nu dt)(dx_\nu' - \delta x_\nu')$$

$$+ (m_\nu dy_\nu' - Y_\nu dt)(dy_\nu' - \delta y_\nu') + (m_\nu dz_\nu' - Z_\nu dt)(dz_\nu' - \delta z_\nu')] = 0.$$

Let us remove some of the constraints from the system, discarding some functions f_s. Assume that $\partial x_\nu'$, $\partial y_\nu'$, $\partial z_\nu'$ denote real variations of the velocities of the particles m_ν of the system free of some of the constraints for the same external forces and the same values of coordinates and velocities for the moment t. The displacements which are virtual for all the constraints will be among the virtual displacements for the new system. Hence it follows from the Euler-Lagrange principle that for the real motion free of some of the constraints

$$\sum_\nu [(m_\nu \partial x_\nu' - X_\nu dt)(dx_\nu' - \delta x_\nu')$$

$$+ (m_\nu \partial y_\nu' - Y_\nu dt)(dy_\nu' - \delta y_\nu') + (m_\nu \partial z_\nu' - Z_\nu dt)(dz_\nu' - \delta z_\nu')] = 0.$$

Subtracting this relation from the preceding one, we have

$$A_{d\delta} + A_{d\partial} - A_{\delta\partial} = 0, \tag{7.19}$$

where

$$A_{d\delta} = \sum_\nu m_\nu [(dx_\nu' - \delta x_\nu')^2 + (dy_\nu' - \delta y_\nu')^2 + (dz_\nu' - \delta z_\nu')^2]$$

is the measure of the deviation (constraint) of the motion d from the motion δ. The formulas for $A_{d\partial}$ and $A_{\delta\partial}$ are similar.

185. Relation (7.19) yields two fundamental inequalities, namely,

$$A_{d\delta} < A_{\partial\delta},$$

which signifies that a deviation of an imaginary motion δ of the system from the real one d is always smaller than a deviation from a real free motion ∂, and

$$A_{d\partial} < A_{\delta\partial}^*,$$

which signifies that the deviation of the real motion d from the real free motion ∂ is smaller than the deviation of any of the imaginary motions δ from the real partly free motion ∂ (Mach's inequality).

186. If we remove all the constraints imposed on a system, the last inequality will express Gauss' principle of least constraint. The expression $(A \, dt^2 = A_{d\partial})$

$$A = \sum m_\nu \left[\left(x_\nu'' - \frac{X_\nu}{m_\nu} \right)^2 + \left(y_\nu'' - \frac{Y_\nu}{m_\nu} \right)^2 + \left(z_\nu'' - \frac{Z_\nu}{m_\nu} \right)^2 \right] \quad (7.20)$$

has the least value for a real motion**. We can establish Gauss' principle for linear nonholonomic constraints by complete analogy.

187. Gauss' principle easily yields Appell's equations.

Assume that $q_1, \ldots, q_k$ $(k = 3n - m)$ are independent coordinates of a mechanical system. The set of Gauss' imaginary motions includes

$$A = A(q_1'', \ldots, q_k'').$$

When only q_s'' vary, it is easy to write the condition for an extremum of A:

$$\frac{\partial A}{\partial q_s''} = 0 \quad (s = 1, \ldots, k).$$

These are Appell's equations. If we introduce the acceleration energy into consideration

$$S = \sum \frac{m_\nu}{2} (x_\nu''^2 + y_\nu''^2 + z_\nu''^2),$$

then we can write Appell's equations as***

---

* This inequality was first strictly proved by Prof. E.A. Bolotov (*Izvestia Fiz.-Mat. Obshch. Kazan Univ.*, Ser. 2: 21, No. 3 (1916)).

** According to de La Vallée Poussin: $\delta x_\nu = \frac{dt^2}{2} \delta x_\nu''$, $\delta A = 0$, $\delta^2 A > 0$.

***

$$Q_s = \sum \left(X_\nu \frac{\partial x_\nu''}{\partial q_s''} + Y_\nu \frac{\partial y_\nu''}{\partial q_s''} + Z_\nu \frac{\partial z_\nu''}{\partial q_s''} \right).$$

If the constraints are holonomic, then

$$\frac{\partial x_\nu''}{\partial q_s''} = \frac{\partial x_\nu}{\partial q_s}, \quad \frac{\partial y_\nu''}{\partial q_s''} = \frac{\partial y_\nu}{\partial q_s}, \quad \frac{\partial z_\nu''}{\partial q_s''} = \frac{\partial z_\nu}{\partial q_s},$$

and Q_s has an ordinary form

$$Q_s = \sum \left(X_\nu \frac{\partial x_\nu}{\partial q_s} + Y_\nu \frac{\partial y_\nu}{\partial q_s} + Z_\nu \frac{\partial z_\nu}{\partial q_s} \right).$$

$$\frac{\partial S}{\partial q_s''} = Q_s \quad (s = 1, \ldots, k).$$

Note that Appell's equations can be generalized to Mach's inequality if the real free motion ∂ is known *.

Gauss' principle has the same generality as the Euler-Lagrange principle for linear constraints both holonomic and nonholonomic.

7.6 The Principle of Least Action

188. If a mechanical system is under smooth holonomic constraints and is acted upon by conservative forces, and the virtual displacements include the real displacements of the system (the constraints are independent of time), then there is a kinetic energy integral

$$T - U = h.$$

The system, left to its own devices, can choose its motions from motions with the available total energy h. This makes it possible to restrict the set of comparable trajectories and regard the work-kinetic energy law as a condition and seek the principle of mechanics in the form of a conventional variational principle.

This problem was solved by Lagrange.

If the constraints are smooth, holonomic, and independent of time, if there is a kinetic energy integral $T - U = h$ for the conservative forces with a force function U, and if the finite positions of the system x_ν^0, y_ν^0, z_ν^0 and x_ν^1, y_ν^1, z_ν^1 and the initial moment t_0 are given, then

$$\delta \int_{t_0}^{t} T \, dt = 0.$$

Indeed, we compose for a conditional variational problem a Lagrange function with multiplier λ:

$$F = T + \lambda(T - U - h).$$

The condition for the sliding end for the upper bound

$$F - \sum_s q_s' \frac{\partial F}{\partial q_s'} = 0$$

---

* The generalization of Appell's equations is $\partial A_{d\partial}/\partial(dq_s') = 0$.

yields

$$T + \lambda(T - U - h) - (1 + \lambda)\sum_s q_s' \frac{\partial T}{\partial q_s'} = -(1 + 2\lambda)T = 0,$$

or $\lambda = -1/2$.

We have found that for the parameter λ Euler's equations

$$\frac{d}{dt}\left[(1 + \lambda)\frac{\partial T}{\partial q_s'}\right] - (1 + \lambda)\frac{\partial T}{\partial q_s} + \lambda\frac{\partial U}{\partial q_s} = 0$$

yield Lagrange's equations of motion

$$\frac{d}{dt}\left(\frac{\partial T}{\partial q_s'}\right) - \frac{\partial T}{\partial q_s} = \frac{\partial U}{\partial q_s} \quad (s = 1, \ldots, k).$$

189. For the same case Jacobi suggested an interesting geometric principle. Let the kinetic energy of the system be

$$2T = \sum_{i,j} a_{ij}q_i'q_j'.$$

We shall consider a configurational space defined by the linear element

$$ds^2 = \sum_{i,j} a_{ij}dq_i dq_j.$$

Assume that the finite positions of the system P_0 and P_1 are defined. For the forces and constraints being considered, Jacobi's principle is valid for the real trajectory:

$$\delta \int_{P_0}^{P_1} \sqrt{2(U + h)} \, ds = 0. \tag{7.21}$$

Indeed, assume that λ is a function of the coordinates q_s. This function defines a family of surfaces $\lambda (q_1, \ldots, q_k) = c$ which cut the required trajectory of the system as well as other curves which are very close to it and are drawn through the points P_0 and P_1 (Fig. 147). Then each of these curves can be considered to be defined by its coordinates expressed as functions of λ. Let the letter δ correspond to a transition from some point of the required trajectory to a point of a neighbouring comparable path which corresponds to the same λ. Then we can replace (7.21) by a variational problem with fixed limits λ_0, λ_1 and fixed ends P_0 and P_1:

$$\delta \int_{\lambda_0}^{\lambda_1} \sqrt{2(U + h)} \, \frac{ds}{d\lambda} \, d\lambda = 0.$$

The Euler-Lagrange equations are

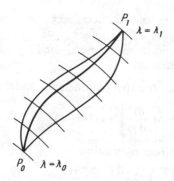

Figure 147

$$\frac{d}{d\lambda}\left[\sqrt{2(U+h)}\;\frac{\partial\sqrt{\sum_{ij}a_{ij}\dfrac{dq_i}{d\lambda}\dfrac{dq_j}{d\lambda}}}{\partial(dq_s/d\lambda)}\right]$$

$$-\frac{\partial}{\partial q_s}\left(\sqrt{2(U+h)}\;\sqrt{\sum_{ij}a_{ij}\frac{dq_i}{d\lambda}\frac{dq_i}{d\lambda}}\right)=0. \tag{7.22}$$

If dt is nonzero along the trajectory, then we can choose the time t measured on the trajectory on which the work-kinetic energy law holds true to be the parameter λ:

$$\sum_{ij}a_{ij}q_i'q_j'=2(U+h).$$

Then (7.22) becomes Lagrange's equations of motion

$$\frac{d}{dt}\left(\frac{\partial T}{\partial q_s'}\right)-\frac{\partial T}{\partial q_s}=\frac{\partial U}{\partial q_s},$$

and this proves Jacobi's principle.

Jacobi's principle is geometric in nature. Jacobi did not ascertain the connection between the sense we defined for the symbol δ, when the time t is measured along the required trajectory, and the virtual displacements of the mechanical system constrained by the condition $T = U + h$ of the work-kinetic energy law when time measurement is independent.

In mechanics, we cannot assign fixed values to the quantities t_0, t_1, P_0, P_1, and h, and, for the mechanical sense of δ to be connected with the notion of the virtual displacement of a system for $T = U + h$, the principle must be similar in form to Lagrange's principle.

190. Lagrange's equations are necessary conditions for an extremum in the variational principles of dynamics.

We know from variational calculus, that a sufficient condition for a minimum of the integral

$$\int_{t_0}^{t_1} f(x_i, \, x_i') \, dt$$

consists of two requirements.

1°. The function of ε_i

$$f(x_i, \, x_i' + \varepsilon_i) - \sum_i \varepsilon_i \frac{\partial f}{\partial x_i'} - f(x_i, \, x_i')$$

must be minimal as $\varepsilon_i \to 0$.

2°. Condition 1° is not sufficient for the integral to have a minimum if the limits t_0 and t_1 are not arbitrarily close to each other. In the general case Jacobi's condition concerning the kinetic foci* must be fulfilled.

191. Let us verify condition 1° for Hamilton's (7.10) and Jacobi's (7.21) principles. Partitioning the function $L = T + U$ into forms L_i, which are homogeneous forms of degree i with respect to the velocities q_s', we have

$$L = L_0 + L_1 + L_2,$$

where L_2 is a positive definite quadratic form. We have

$$L_1(q_i, \, q_i' + \varepsilon_i) = L_1(q_i, \, q_i') + \sum_i \varepsilon_i \frac{\partial L_1}{\partial q_i'} \, ,$$

$$L_2(q_i, \, q_i' + \varepsilon_i) = L_2(q_i, \, q_i') + \sum_i \varepsilon_i \frac{\partial L_2}{\partial q_i'} + L_2(q_i, \, \varepsilon_i),$$

and, hence,

$$L(q_i, \, q_i' + \varepsilon_i) - \sum_i \varepsilon_i \frac{\partial L}{\partial q_i'} - L(q_i, \, q_i') = L_2(q_i, \, \varepsilon_i) > 0.$$

Condition 1° is always fulfilled for Hamilton's principle. This means that Hamilton's principle actually ascertains the minimum of action.

Let us prove that condition 1° is fulfilled for Jacobi's principle as well. In Jacobi's principle, when the time is measured with the use of the trajectory, only the multiplier $\sqrt{2T(q_i, \, q_i')}$ depends on the velocities q_s'. We can

---

* E. Whittaker, *A Treatise on the Analytical Dynamics of Particles and Rigid Bodies* (Cambridge University Press, Cambridge: 1937).

expand $\sqrt{2T(q_i,\ q_i' + \varepsilon)}$ in the series of ε_i by expanding the expression $\sqrt{2T(q_i,\ q_i' + \varepsilon_i u)}$ in terms of the parameter u, and then assuming u to be unity, i.e. $u = 1$. Let

$$2T(q_i,\ q_i' + \varepsilon_i u) = au^2 + 2bu + c.$$

Since the kinetic energy is positive definite, we have

$$ac - b^2 > 0.$$

Irrespective of the values of the parameter u,

$$\frac{\partial^2}{\partial u^2}\ \sqrt{2T(q_i,\ q_i' + \varepsilon_i u)} = \frac{ac - b^2}{(au^2 + 2bu + c)^{3/2}} > 0,$$

condition 1° is satisfied, and, consequently, we deal with a minimum in Jacobi's principle.

7.7 Canonical Transformations

192. We can make the canonical transformations of Hamilton's equations using the principle of least action

$$\delta \int_{t_0}^{t_1} \left(\sum p_s dq_s - H\, dt \right) = 0.$$

Assume that instead of the old variables q_s, p_s we have new variables Q_s, P_s which satisfy the relation

$$\sum p_s \delta q_s = \sum P_s \delta Q_s + \delta W, \tag{7.23}$$

where $W(t, q_s, Q_s)$ is a differentiable function which is known as the *characteristic function* of the transformation*. In the new variables Q_s, P_s, the principle of least action has the form

$$\delta \int_{t_0}^{t_1} \left[\sum P_s dQ_s - \left(H + \frac{\partial W}{\partial t} \right) dt \right] + \delta \int_{t_0}^{t} dW = 0.$$

The variation of the last integral is zero, and this means that in the new variables Q_s and P_s the differential equations of motion also have a canoni-

---

* This function is also called a generating function of a canonical transformation. — *Ed.*

cal form. The expression

$$H^* = H + \frac{\partial W}{\partial t}$$

plays the part of the new Hamiltonian function.

The transformations which satisfy relation (7.23) are called *canonical,* or *contact, transformations*:

$$p_s = \frac{\partial W}{\partial q_s}, \quad -P_s = \frac{\partial W}{\partial Q_s}. \tag{7.24}$$

193. Let $V(t, q_s, q_s^0)$ be an action. We know that

$$\delta V = \sum p_s \delta q_s - \sum p_s^0 \delta q_s^0.$$

Comparing this relation with (7.23), we find that the variables q_s, p_s for the initial moment q_s^0, p_s^0 and for the moment t are related by a canonical transformation, and the action V plays the part of the characteristic function W.

The moments t_0 and t can be chosen arbitrarily. Consequently, we can regard the motion of a holonomic conservative mechanical system as a chain of canonical transformations of the variables q_s and p_s.

194. For the characteristic function $W(t, q_s, Q_s)$ which satisfies the relation

$$\frac{\partial W}{\partial t} + H = 0, \tag{7.25}$$

the new function H^* is zero and the canonical equations in the variables P_s, Q_s have the form

$$\frac{dP_s}{dt} = 0, \quad \frac{dQ_s}{dt} = 0,$$

i.e. P_s, Q_s are simply constants. The transformation formulas (7.24) reduce relation (7.25) to the Hamilton-Jacobi partial differential equation (7.16) and make the characteristic function $W(t, q_s, Q_s)$ a complete integral of the Hamilton-Jacobi equation, and equations (7.24) themselves acquire the contents of Jacobi's well-known theorem (7.17).

195. Consider an infinitesimal canonical transformation. Assume that the new variables differ but little from the initial ones, i.e.

$$P_s = p_s + \Delta p_s, \quad Q_s = q_s + \Delta q_s.$$

Let

$$\Delta q_s = u_s(t, q_j, p_j) \Delta t, \quad \Delta p_s = v_s(t, q_j, p_j) \Delta t.$$

Condition (7.23) for the canonical transformation yields

$$\sum[p_s\delta q_s - (p_s + \Delta p_s)\delta(q_s + \Delta q_s)] = \delta W,$$

or, with an accuracy to within the quantities of the least order of smallness,

$$\Delta t \sum(v_s\delta q_s + p_s\delta u_s) = -\delta W.$$

Assume that $\delta W = \Delta t \, \delta U$ and then

$$\sum(v_s\delta q_s - u_s\delta p_s) = -\delta\left(U + \sum p_s u_s\right) = -\delta H.$$

Comparing the coefficients of the independent variations δq_s, δp_s, we have

$$\frac{\Delta p_s}{\Delta t} = -\frac{\partial H}{\partial q_s}, \quad \frac{\Delta q_s}{\Delta t} = \frac{\partial H}{\partial p_s} \quad (s = 1, \ldots, k).$$

In the limit as $\Delta t \to 0$, these relations pass into Hamilton's canonical equations. Consequently, for mechanical systems subjected to holonomic constraints and acted upon by forces with a force function, Hamilton's canonical equations express the fact that motion is a sequence of canonical infinitely small transformations of the variables q_s, p_s which are continuous in time.

196. *A bilinear covariant* of the linear form

$$\omega_\delta = \sum X_\nu\delta x_\nu$$

is the expression

$$\Delta\omega_\delta - \delta\omega_\Delta = \sum(\Delta X_\nu\delta x_\nu - \delta X_\nu\Delta x_\nu) = \sum_{\mu,\nu}\left(\frac{\partial X_\mu}{\partial x_\nu} - \frac{\partial X_\nu}{\partial x_\mu}\right)\Delta x_\nu\delta x_\mu.$$

The bilinear covariant of a total differential is zero. The bilinear covariant of the sum of linear forms is the sum of the bilinear covariants of the linear forms.

197. The linear form (7.23) of the canonical transformation of the variables q_s, p_s and Q_s, P_s has a linear covariant

$$\sum(\Delta p_s\delta q_s - \Delta q_s\delta p_s) = \sum(\Delta P_s\delta Q_s - \Delta Q_s\delta P_s).$$

In other words, the expression $\sum(\Delta p_s\delta q_s - \Delta q_s\delta p_s)$ is an invariant under canonical transformations.

198. Using Hamilton's principal function $V(t, q_s, q_s^0)$ we have ascertained that the variables q_s^0, p_s^0 and q_s, p_s are related by a canonical transformation. Hence

$$\sum_s(\Delta p_s\delta q_s - \Delta q_s\delta p_s) = \sum_s(\Delta p_s^0\delta q_s^0 - \Delta q_s^0\delta p_s^0).$$

Assume that δ denotes a transition to a motion defined by the values $q_1^0, \ldots, q_k^0, p_1^0, \ldots, p_{r-1}^0, p_r^0 + \delta p_r^0, p_{r+1}^0, \ldots, p_k^0$ at the moment t_0 and

Δ denotes a transition to a motion defined by the values $q_1, \ldots, q_k,$ $p_1, \ldots, p_{i-1}, p_i + \Delta p_i, p_{i+1}, \ldots, p_k$ at the moment t. The preceding relation yields

$$\Delta p_i \delta q_i = -\Delta q_r^0 \delta p_r^0.$$

A change in a coordinate δq_i in an arbitrary time interval caused by a change in the impulse δp_r^0 is equal and opposite in sign to a change of the coordinate Δq_r^0 in the inverse motion in the same time interval caused by a change in the impulse Δp_i of the same magnitude (Helmholtz).

7.8 Vibrations of Mechanical Systems

199. Lagrange began studying vibrations of mechanical systems. He obtained differential equations for perturbed motions in perturbations by forces by the method of variation of arbitrary constants. Assume that a mechanical system is subjected to ideal holonomic constraints and to the action of forces with a force function. Let q_s, p_s be the system's coordinates and impulses and $H_0(t, q_s, p_s)$ be a Hamiltonian function for an unperturbed motion.

Assume, furthermore, that the system is acted upon by perturbing forces with a force function W. The Hamiltonian function for perturbed motions is $H = H_0 - W,$

$$\frac{dq_i}{dt} = \frac{\partial H}{\partial p_i}, \quad \frac{dp_i}{dt} = -\frac{\partial H}{\partial q_i} \quad (i = 1, \ldots, k).$$

If we know the solution of the unperturbed problem

$$\frac{dq_i}{dt} = \frac{\partial H_0}{\partial p_i}, \quad \frac{dp_i}{dt} = -\frac{\partial H_0}{\partial q_i} \quad (i = 1, \ldots, k)$$

or the complete integral $V(t, q_s, \alpha_s)$ of the Hamilton-Jacobi equation for unperturbed motions $\dfrac{\partial V}{\partial t} + H_0\left(t, q_s, \dfrac{\partial V}{\partial q_s}\right) = 0$, then we can ask whether the constants α_i, β_i of the unperturbed motion (H_0) can be replaced by some functions of time so that when these functions are substituted into the solutions of the unperturbed equations

$$q_i = q_i(t, \alpha_1, \ldots, \alpha_k, \beta_1, \ldots, \beta_k),$$
$$p_i = p_i(t, \alpha_1, \ldots, \alpha_k, \beta_1, \ldots, \beta_k)$$

we get solutions for the perturbed motion. This problem was solved by Lagrange. For the perturbed motion we have

$$\frac{\partial H}{\partial p_i} = \frac{dq_i}{dt} = \frac{\partial q_i}{\partial t} + \sum_s \frac{\partial q_i}{\partial \alpha_s}\frac{d\alpha_s}{dt} + \sum_s \frac{\partial q_i}{\partial \beta_s}\frac{d\beta_s}{dt}.$$

But, by the hypothesis, we have the following for the unperturbed motion

$$\frac{\partial q_i}{\partial t} = \frac{\partial H_0}{\partial p_i}$$

and, hence,

$$\sum_s \frac{\partial q_i}{\partial \alpha_s} \frac{d\alpha_s}{dt} + \sum_s \frac{\partial q_i}{\partial \beta_s} \frac{d\beta_s}{dt} = -\frac{\partial W}{\partial p_i},$$

and similarly

$$\sum_s \frac{\partial p_i}{\partial \alpha_s} \frac{d\alpha_s}{dt} + \sum_s \frac{\partial p_i}{\partial \beta_s} \frac{d\beta_s}{dt} = \frac{\partial W}{\partial q_i}.$$

From these relations $(i = 1, \ldots, k)$ we must determine the derivatives $d\alpha_\nu/dt$, $d\beta_\nu/dt$.

We introduce Lagrange's parenthesis

$$(a, b) = \sum_i \left(\frac{\partial q_i}{\partial a} \frac{\partial p_i}{\partial b} - \frac{\partial q_i}{\partial b} \frac{\partial p_i}{\partial a} \right).$$

If we multiply the first of the two preceding relations by $\partial p_i/\partial \alpha_\nu$ and the second by $-\partial q_i/\partial \alpha_\nu$ and sum up the results over all i and term-by-term, we obtain

$$\sum_s (\alpha_s, \alpha_\nu) \frac{d\alpha_s}{dt} + \sum_s (\beta_s, \alpha_\nu) \frac{d\beta_s}{dt} = -\frac{\partial W}{\partial \alpha_\nu}, \tag{7.26}$$

and similarly

$$\sum_s (\alpha_s, \beta_\nu) \frac{d\alpha_s}{dt} + \sum_s (\beta_s, \alpha_\nu) \frac{d\beta_s}{dt} = -\frac{\partial W}{\partial \beta_\nu}.$$

For the unperturbed motion the constants α_ν, β_ν are defined by Jacobi's theorem

$$\beta_\nu = \frac{\partial V}{\partial \alpha_\nu}, \quad p_\nu = \frac{\partial V}{\partial q_\nu} \quad (\nu = 1, \ldots, k).$$

These relations make it possible to find the values of Lagrange's parentheses, namely,

$$(\alpha_\nu, \beta_s) = \frac{\partial}{\partial \beta_s} \sum_i p_i \frac{\partial q_i}{\partial \alpha_\nu} - \frac{\partial}{\partial \alpha_\nu} \sum_i p_i \frac{\partial q_i}{\partial \beta_s}$$

$$= \frac{\partial}{\partial \beta_s} \sum_i \frac{\partial V}{\partial q_i} \frac{\partial q_i}{\partial \alpha_\nu} - \frac{\partial}{\partial \alpha_\nu} \sum_i \frac{\partial V}{\partial q_i} \frac{\partial q_i}{\partial \beta_s}.$$

But

$$\frac{\partial V}{\partial \alpha_\nu} = \left(\frac{\partial V}{\partial \alpha_\nu}\right) + \sum_i \frac{\partial V}{\partial q_i} \frac{\partial q_i}{\partial \alpha_\nu} = \beta_\nu + \sum_i \frac{\partial V}{\partial q_i} \frac{\partial q_i}{\partial \alpha_\nu},$$

$$\frac{\partial V}{\partial \beta_s} = \sum_i \frac{\partial V}{\partial q_i} \frac{\partial q_i}{\partial \beta_s}.$$

Hence

$$(\alpha_\nu, \beta_s) = -\frac{\partial \beta_\nu}{\partial \beta_s}.$$

The relations

$$(\alpha_s, \alpha_\nu) = 0, \quad (\beta_s, \beta_\nu) = 0$$

can be proved by analogy.

Substituting the values of Lagrange's parentheses we have found into (7.26), we finally obtain the fundamental equations of the perturbation theory

$$\frac{d\alpha_\nu}{dt} = \frac{\partial W}{\partial \beta_\nu}, \quad \frac{d\beta_\nu}{dt} = -\frac{\partial W}{\partial \alpha_\nu} \quad (\nu = 1, \ldots, k). \tag{7.27}$$

200. Poincaré derived differential equations for the motion of holonomic conservative mechanical systems when only the initial values of the coordinates q_s and impulses p_s are perturbed. Let the Hamiltonian function be $H(t, q_s, p_s)$. i.e.

$$\frac{dq_s}{dt} = \frac{\partial H}{\partial p_s}, \quad \frac{dp_s}{dt} = -\frac{\partial H}{\partial q_s} \quad (s = 1, \ldots, k).$$

For some definite (unperturbed) motion of this system $q_s = q_s(t)$ and $p_s = p_s(t)$. Let us consider a motion which is close to this unperturbed motion:

$$q_s + \xi_s, \quad p_s + \eta_s.$$

In the first approximation the deviations of the coordinates ξ_s and impulses η_s satisfy the following system of differential equations:

$$\frac{d\xi_s}{dt} = \sum_\nu \left(\frac{\partial^2 H}{\partial p_s \partial q_\nu} \xi_\nu + \frac{\partial^2 H}{\partial p_s \partial p_\nu} \eta_\nu\right),$$

$$\frac{d\eta_s}{dt} = -\sum_\nu \left(\frac{\partial^2 H}{\partial q_s \partial q_\nu} \xi_\nu + \frac{\partial^2 H}{\partial q_s \partial p_\nu} \eta_\nu\right) \quad (s = 1, \ldots, k). \tag{7.28}$$

These are Poincaré's equations in variations.

201. Poincaré established a number of interesting propositions.

If ξ_s, η_s and ξ_s', η_s' are two particular solutions of equations in variations (7.28), then

$$I = \sum (\xi_s \eta_s' - \eta_s \xi_s') = \text{const}^*. \tag{7.29}$$

Indeed,

$$\frac{d}{dt} \sum (\xi_s \eta_s' - \eta_s \xi_s')$$

$$= \sum_{s,\nu} \left(\frac{\partial^2 H}{\partial p_s \partial q_\nu} \xi_\nu \eta_s' + \frac{\partial^2 H}{\partial p_s \partial p_\nu} \eta_s' \eta_\nu - \frac{\partial^2 H}{\partial q_s \partial q_\nu} \xi_\nu' \xi_s - \frac{\partial^2 H}{\partial q_s \partial p_\nu} \eta_\nu' \xi_s \right)$$

$$+ \sum_{s,\nu} \left(\frac{\partial^2 H}{\partial q_s \partial q_\nu} \xi_\nu \xi_s' + \frac{\partial^2 H}{\partial q_s \partial p_\nu} \eta_\nu \xi_s' - \frac{\partial^2 H}{\partial p_s \partial q_\nu} \xi_\nu' \eta_s - \frac{\partial^2 H}{\partial p_s \partial p_\nu} \eta_\nu' \eta_s \right) \equiv 0.$$

We can prove by analogy that if Poincaré's equations in variations have a linear integral

$$\sum_s (A_s \xi_s + B_s \eta_s) = \text{const},$$

then these equations have a particular solution

$$\xi_s = -B_s, \quad \eta_s = A_s.$$

If $\Phi(t, q_s, p_s) = \text{const}$ is an integral of the canonical equations of motion, then

$$\sum_s \left(\frac{\partial \Phi}{\partial q_s} \xi_s + \frac{\partial \Phi}{\partial p_s} \eta_s \right) = \text{const}$$

is a line integral of Poincaré's equations in variations and, consequently, there is a particular solution

$$\xi_s = -\frac{\partial \Phi}{\partial p_s}, \quad \eta_s = \frac{\partial \Phi}{\partial q_s}$$

of the equations in variations.

Using these relations, it is easy to prove Poisson's well-known theorem. If $\Phi = a$ and $\Phi_1 = a_1$ are two first integrals of the canonical equations of motion, then the following are particular solutions of Poincaré's equations in variations:

$$\xi_s = -\frac{\partial \Phi}{\partial p_s}, \quad \eta_s = \frac{\partial \Phi}{\partial q_s}, \quad \xi_s' = -\frac{\partial \Phi_1}{\partial p_s}, \quad \eta_s' = \frac{\partial \Phi_1}{\partial q_s}.$$

---

* I is Poincaré's invariant. — *Ed.*

From Poincaré's invariant we have

$$(\Phi, \ \Phi_1) = \sum \left(\frac{\partial \Phi}{\partial q_s} \frac{\partial \Phi_1}{\partial p_s} - \frac{\partial \Phi}{\partial p_s} \frac{\partial \Phi_1}{\partial q_s} \right) = \text{const.}$$

202. Consider the motion of a mechanical system (holonomic, conservative, and subjected to constraints independent of time) close to its equilibrium position.

In the first approximation, Poincaré's equations in variations with constant coefficients are the equations of perturbed motion of such mechanical systems. The solutions of linear equations with constant coefficients have the form

$$\xi_s = A_s e^{\varkappa t}, \quad \eta_s = B_s e^{\varkappa t}$$

for every characteristic exponent $\varkappa$.

For every solution ξ_s, η_s there is always another solution ξ_s', η_s' with an exponent $\varkappa'$ for which Poincaré's invariant is nonzero. For such a solution formula (7.29) yields

$$e^{(\varkappa + \varkappa')t} \Sigma (A_s B_s' - A_s' B_s) = \text{const} \gtreqless 0.$$

In other words, for every exponent $\varkappa$ there is another exponent $\varkappa'$ such that

$$\varkappa + \varkappa' = 0. \tag{7.30}$$

If the exponents $\varkappa$ include at least one with a positive real part, then the particular solution ξ_s, η_s corresponding to it will increase indefinitely with time and, consequently, the equilibrium will be unstable.

Consequently, for the equilibrium to be stable, it is necessary, according to (7.30), that all the exponents or characteristic roots $\varkappa$ be pure imaginary. This proposition indicates that in the vicinity of a stable position of equilibrium of a mechanical system the perturbed motions are vibratory in nature.

203. Lagrange proved a very important theorem. *If a force function U has an isolated maximum at the equilibrium, then the equilibrium is stable.*

The simplest proof of Lagrange's theorem results from Lyapunov's general theorem on stability. *If the differential equations of a perturbed motion are such that it is possible to find a function of constant sign V whose derivative V', by virtue of these equations, is either a function of constant sign opposite to the sign of V or identically zero, then the unperturbed motion is stable.*

For the case we are interested in, the canonical equations of motion with the Hamiltonian functions $H = T - U$ are complete equations of perturbed motion. If $U = 0$ in the equilibrium position, then H is evidently a positive definite function of q_s and p_s. But $dH/dt = 0$ and, consequently, on the basis of Lyapunov's theorem, the equilibrium where U has an isolated maximum is stable. The converse of Lagrange's theorem is significant and difficult to prove.

204. Small vibrations about the equilibrium. Assume that $q_s = 0$ in the position of equilibrium for all s, $U(0, \ldots, 0) = 0$ and let $2T = \sum a_{ij}q_i'q_j'$ and $2U = \sum b_{ij}q_iq_j$ to within lowest order terms and that $a_{ij} = a_{ji}$, $b_{ij} = b_{ji}$ are constants. The equations of motion about the equilibrium position in Lagrange's form are

$$\frac{d}{dt}\left(\frac{\partial T}{\partial q_s'}\right) = \frac{\partial U}{\partial q_s} \quad (s = 1, \ldots, k),$$

or

$$\sum_j (a_{sj}q_j'' - b_{sj}q_j) = 0. \tag{7.31}$$

The coordinates $x_\nu = \sum_s \alpha_{\nu s}q_s$ in which the kinetic energy and the force function have the form $2T = \sum_\nu x_\nu'^2$ and $2U = \sum \lambda_\nu x_\nu^2$, respectively, are said to be *normal*. The equations of motion in normal coordinates x_ν have a simple form

$$x_\nu'' = \lambda_\nu x_\nu \quad (\nu = 1, \ldots, k). \tag{7.32}$$

Let us find λ_ν. Each of the equations in (7.32) is equivalent to a linear combination of those in (7.31). We multiply the latter by h_s and sum up the results:

$$\sum_{s,j} h_s a_{sj}q_j'' = \sum_{s,j} h_s b_{sj}q_j.$$

Let these relations be equivalent to the equation

$$x_\nu'' = \lambda_\nu x_\nu,$$

or

$$\sum_j \alpha_{\nu j}q_j'' = \lambda_\nu \sum \alpha_{\nu j}q_j.$$

From this we obtain

$$\alpha_{\nu j} = \sum_s h_s a_{sj} = \frac{1}{\lambda_\nu}\sum_s h_s b_{sj},$$

or (omitting the subscript of λ_ν)

$$\sum_s (\lambda a_{sj} - b_{sj})h_s = 0 \quad (j = 1, \ldots, k). \tag{7.33}$$

To find the values of λ_ν, we get an equation

$$\Delta(\lambda) = \|\lambda a_{sj} - b_{sj}\| = 0.$$

All the roots of this equation are real (according to Sylvester's law of inertia). If λ is a root, then there are nonzero constants h_s which satisfy (7.33):

$$\lambda \sum_s a_{sj} h_s = \sum b_{sj} h_s.$$

We multiply these equations by $\bar{h}_j$ and add the results:

$$\lambda = \frac{\sum b_{sj} h_s \bar{h}_j}{\sum a_{sj} h_s \bar{h}_j}.$$

If λ is complex, then its conjugate $\bar{\lambda}$ is $(a_{sj} = a_{js}, \; b_{sj} = b_{js})$

$$\bar{\lambda} = \frac{\sum b_{sj} \bar{h}_s h_j}{\sum a_{sj} \bar{h}_s h_j} = \lambda.$$

We have proved that λ_ν is real.

We distinguish between two cases.

1°. All λ_s are negative. Then the solution of (7.32) for $\nu_s = \sqrt{-\lambda_s}$ is

$$x_s = A_s \cos(\nu_s t - \varphi_s) \quad (s = 1, \ldots, k),$$

where A_s is the amplitude, ν_s is the natural frequency, φ_s is the phase of the sth fundamental, or natural, vibrations of the system. For small initial deviations, the system only differs a little from the equilibrium position. The equilibrium is stable.

2°. Some of the roots λ_s are positive. The normal coordinates corresponding to these roots and their velocities may increase without limit with time. The equilibrium is unstable. If $\lambda_1 > 0$, then

$$x_1 = A e^{t\sqrt{\lambda_1}} + B e^{-t\sqrt{\lambda_1}}.$$

The number of positive λ_s is called the *degree of instability of the system*.

205. At a stable equilibrium all λ_s are negative and $2U$ is a negative definite quadratic form. We can easily see in the normal coordinates that the greatest λ_s (the lowest in magnitude) is the maximum of the function $2U$ when the variables are constrained by the condition $\sum x_\nu^2 = 1$ (or by the condition $2T = 1$ when x_ν' in T are replaced by x_ν).

When the inertia increases, the fundamental tone (the lowest frequency) drops, or at least does not become higher. In this case the increase in the inertia is interpreted as a transition to a system with a kinetic energy T' such that $T' - T$ is never negative, the force function remaining unchanged.

When the system becomes more rigid, the fundamental tone becomes higher or, at least, does not drop. The increase in the rigidity is interpreted as a transition to a system whose kinetic energy is the same but whose force function differs from the previous one by an extra negative quadratic form of constant sign.

206. Assume that a new constraint is imposed on the mechanical system in the vicinity of the equilibrium position $x_\nu = 0$ and that in the first approximation, for small x_ν, the constraint has the form

$$\sum_\nu A_\nu x_\nu = 0 \quad \text{or} \quad \sum_\nu A_\nu \delta x_\nu = 0 \quad (A_\nu \text{ are constants}).$$

From Hamilton's principle

$$\int_{t_0}^{t_1} \sum (x_\nu'' - \lambda_\nu x_\nu) \delta x_\nu dt = 0,$$

by the Lagrange method of undetermined multipliers, we get the following equations of motion:

$$x_\nu'' - \lambda_\nu x_\nu - \mu A_\nu = 0 \quad (\nu = 1, \ldots, k).$$

Let the normal vibration of the altered system be

$$x_\nu = \alpha_\nu e^{\sqrt{\lambda}t}, \quad \mu = m e^{\sqrt{\lambda}t}.$$

From this we have

$$\alpha_\nu(\lambda - \lambda_\nu) - mA_\nu = 0.$$

From equations of constraints we have $\sum A_\nu \alpha_\nu = 0$, or

$$\sum_\nu \frac{A_\nu^2}{\lambda - \lambda_\nu} = 0.$$

This equation defines $k - 1$ values of λ which lie between $\lambda_1, \ldots, \lambda_k$.

207. Assume that a system is subjected to two kinds of forces, those defined by the force function U and dissipative forces whose virtual work is defined by the formula

$$-\sum_\nu \frac{\partial f}{\partial x_\nu'} \delta x_\nu,$$

where $2f = \sum c_{\mu\nu} x_\mu' x_\nu'$ is a positive quadratic form of x_ν', of constant sign, and $c_{\mu\nu} = c_{\nu\mu}$ are constants.

Very often dissipative forces are due to viscosity. Lagrange's equations in normal coordinates x_ν assume the form

$$x_\nu'' = \lambda_\nu x_\nu - \frac{\partial f}{\partial x_\nu'} \quad (\nu = 1, \ldots, k). \tag{7.34}$$

We multiply the equations of motion (7.34) by x_ν' and add the results. We get

$$\frac{d(T - U)}{dt} = -2f.$$

Consequently, if the equilibrium position $x_\nu = 0$ is stable without the dissipative forces, it remains stable when the dissipative forces act. When the dissipative forces act, the total energy $H = T - U$ is dissipated with the velocity $2f$, where f is Rayleigh's dissipative function. If f is explicitly dependent on all x_ν', then the dissipation is complete, otherwise it is incomplete.

208. Assume that the conservative and dissipative forces are complemented by other forces x, i.e.

$$x_\nu'' = \lambda_\nu x_\nu - \frac{\partial f}{\partial x_\nu'} + X_\nu \quad (\nu = 1, \ldots, k).$$

The forces X_ν are *gyroscopic* if the work they do over a true displacement is zero, i.e. $\Sigma X_\nu dx_\nu = 0$. Hence

$$\frac{d(T - U)}{dt} = -2f.$$

If the equilibrium was stable when conservative forces alone acted, then it will be stable when dissipative and gyroscopic forces are added.

209. Under certain conditions, gyroscopic forces may transform an unstable equilibrium into a stable one, i.e. gyroscopic stabilization occurs:

$$x_\nu'' = \lambda_\nu x_\nu + \Sigma g_{\nu\mu} x_\nu' \quad (g_{\mu\nu} = -g_{\nu\mu}).$$

The characteristic roots x are defined by the equation

$$F(x) = \det \| \delta_{\mu\nu}(x^2 - \lambda_\nu) + x g_{\mu\nu} \| = 0.$$

Hence $F(+\infty) > 0$ and $F(0) = (-\lambda_1) \ldots (-\lambda_k)$. If the initial instability was odd in degree and $|\lambda_\nu| > 0$, then $F(0) < 0$ and, consequently, there will be at least one positive root x. The equilibrium will remain unstable. The initial instability must be even in degree for gyroscopic stabilization to be possible.

Consider the following example to see how gyroscopic stabilization occurs:

$$x'' = \alpha x + gy', \quad y'' = \beta y - gx' \quad (\alpha > 0, \beta > 0),$$
$$(g^2 - \alpha - \beta)^2 - 4\alpha\beta > 0, \quad g^2 - \alpha - \beta > 0.$$

210. *Lyapunov's characteristic number* for function z is a real number λ such that the function $ze^{(\lambda + \varepsilon)t}$ is unbounded for every positive constant ε and vanishing for every negative constant ε.

The characteristic number of the sum of two functions is equal to the smallest of their characteristic numbers. The characteristic number of the product of two functions is not smaller than the sum of their characteristic numbers.

211. For the unperturbed motion Poincaré's equations in variations have coefficients dependent on time. The smallest of the characteristic numbers of the functions which form a particular solution ξ_s, η_s of Poincaré's equations in variations is the *characteristic number* of this solution. Let it be $\varkappa$ and assume that $\varkappa'$ is the characteristic number of another solution ξ_s', η_s' for which the invariant is nonzero, i.e.

$$\Sigma\,(\xi_s\eta_s' - \xi_s'\eta_s) = \text{const} \gtreqless 0.$$

From this it follows that $\varkappa + \varkappa' \leqslant 0$.

If all $\varkappa$ are positive, then the solutions of the equations in variations yield stability, and if at least one of the characteristic numbers is negative then they yield instability. It follows from the last inequality that for the unperturbed motion to be stable according to Poincaré's equations in variations, it is necessary that all characteristic numbers $\varkappa$ be zero. For the case of reducible Poincaré's equations in variations, this proposition means that perturbed motions are vibrations in the vicinity of a stable unperturbed motion. What the frequency of the normal vibrations is has not yet been found.

212. For a holonomic conservative system, perturbed motions which are close to an unperturbed one are defined by different values of the constants α_s, β_s of Jacobi's complete integral $V(t, q_s, \alpha_s)$.

Let $T = \dfrac{1}{2} \sum g_{ij}p_ip_j$ and assume that perturbations are produced by the variations of the constant β_s alone. We have

$$\eta_i = \sum_j \frac{\partial^2 V}{\partial q_i \partial q_j}\,\xi_j.$$

The first group of Poincaré's equations in variations immediately yields equations

$$\frac{d\xi_i}{dt} = \sum_{s,j} \xi_s \frac{\partial}{\partial q_s}\left(g_{ij}\frac{\partial V}{\partial q_j}\right) \quad (i = 1, \ldots, k).$$

These equations depend only on ξ_s (for the general form of T a similar equation can be easily derived). If the last equations are regular[*] and if

---

[*] A system of linear differential equations $\dfrac{dx_s}{dt} = \sum p_{sr}x_r$ is *regular* if there is a relation $s + \mu = 0$ for it, where s is the sum of the characteristic numbers of the solutions of the normal system, μ is the characteristic number of the function $\exp -\int \sum p_{ss}dt$.

the motion is stable when only the constants β_s are perturbed, then the characteristic number of the expression

$$\exp \int \sum_{ij} \frac{\partial}{\partial q_i}\left(g_{ij}\frac{\partial V}{\partial q_j}\right) dt$$

must be zero in accordance with Abel's theorem.

Vibrations necessary for the analogy between the optics and mechanics occur when stable equilibria or motions are perturbed. Further study of this topic is beyond the scope of this book.

7.9 Stability

213. Mechanical systems oscillate about a stable equilibrium position and in the vicinity of stable motions. Stability has the following fundamental significance.

When we describe a real motion of a mechanical system using the principle of analytical dynamics we have studied, we must solve problems concerning the constraints imposed on the system and the active forces. Of all the possible hypotheses concerning the constraints imposed on the system and the active forces we adopt one which states that the equations of mechanics should yield theoretical values f_s for F_s, which deviate from the empirical values φ_s by less than the possible error ε of the experiment:

$$|f_s - \varphi_s| < \varepsilon. \tag{7.35}$$

The requirement that the deviations of the theory from the experiment be small is fundamental for the whole system of scientific knowledge.

214. The requirement of small deviations of the theory from the experiment makes it possible to establish a general practical proposition known as the principle of stability.

Imagine a holonomic system acted on by conservative forces. Let q_i be its Lagrange coordinates. Assume that the constraint hypothesis, which yields an expression $T(t, q_i, q_i)$ for the kinetic energy, and the hypothesis concerning the force function $U(q_i)$ of the active forces provide a good explanation of specific motions of the system. If we introduce the impulses $p_i = \partial T/\partial q_i'$, then we can accept Hamilton's equations

$$\frac{dq_i}{dt} = \frac{\partial H}{\partial p_i}, \quad \frac{dp_i}{dt} = -\frac{\partial H}{\partial q_i} \tag{7.36}$$

with the Hamiltonian function $H = \Sigma\, p_i q_i' - T - U$ as equations of motion.

In a specific experiment the mechanical system moves in a field of real forces which in theory (as a function of U) can be fully accounted for with an accuracy to within insignificant perturbing forces or, perhaps, with only slightly different initial data which are due to the inaccuracy of the experimental readings. The small deviations which really exist affect the specific motion of the mechanical system. If the real value of the Hamiltonian function is H^*, then the equations of the real motion have the form

$$\frac{dq_i}{dt} = \frac{\partial H^*}{\partial p_i}, \quad \frac{dp_i}{dt} = -\frac{\partial H^*}{\partial q_i}. \tag{7.37}$$

The theoretical values f_s of the quantities $F_s(t, q_i, p_i)$ which are observed result if we replace q_i and p_i in the function $F_s(t, q_i, p_i)$, by their values obtained from the equations of the theoretical, or unperturbed, motions (7.36). The experimental values φ_s of F_s result when we replace q_i and p_i by the solutions of the real, or perturbed, equations (7.37). It follows that the requirement of small deviations of the theory from the experiment (7.35)

$$|f_s - \varphi_s| < \varepsilon$$

is the requirement for the stability of the theoretical motion (H) for the perturbing forces ($H^* - H$) with respect to the observed F_s.

215. It is very difficult to study stability when we have perturbing forces since by definition the expressions for the perturbing forces $H^* - H$ are unknown, and we only know their nature and their order of smallness. Lyapunov's stability can be used to find the necessary conditions for stability under perturbing forces on the basis of the following theorem.

If, for the initial data $t = t_0$, $q_i = q_i^0$, $p_i = p_i^0$, the theoretical motion H is stable with respect to the quantities F_s for the perturbing forces $H^* - H$ and for the perturbation of the initial data, then the real motion in the field H^* for the initial data $t = t_0$, $q_i = q_i^0$, $p_i = p_i^0$ is stable when only the initial data are perturbed.

For instance, assume that f_s is the value of F_s for the theoretical unperturbed motion, φ_s is the value of F_s for the perturbed motion, and φ_s^0 is the value of F_s for the motion with the Hamiltonian function H^* when $t = t_0$, $q_i = q_i^0$, $p_i = p_i^0$. From the inequalities

$$|f_s - \varphi_s| < \varepsilon \text{ and } |f_s - \varphi_s^0| < \varepsilon$$

it follows that

$$|\varphi_s - \varphi_s^0| < 2\varepsilon.$$

216. Lyapunov's direct method of studying the stability of a perturbed motion

$$\frac{dx_s}{dt} = X_s(t, x_1, \ldots, x_n), \quad X_s(t, 0, \ldots, 0) = 0 \quad (s = 1, \ldots, n)$$

is to find some one-valued continuous bounded function $V(t, x_1, \ldots, x_n)$ whose complete derivative

$$V' = \frac{dV}{dt} = \frac{\partial V}{\partial t} + \sum \frac{\partial V}{\partial x_s} X_s$$

possesses certain properties.

Assume that the function V vanishes when all x_s are zero. If the function V is independent of t and vanishes in a sufficiently small domain $\sum x_s^2 \leqslant \varepsilon$ about the origin only when all x_s are zero, then V has a *definite sign*. If the partial derivatives $\partial V/\partial x_s$ of the function V of a definite sign are continuous, then the surfaces $V = c$ are closed in a sufficiently small domain $\sum x_s^2 \leqslant \varepsilon$ if $|c|$ is smaller than the minimum $|V|$ on the sphere $\sum x_s^2 = \varepsilon$. Indeed, in this case there will be no way from the origin $x_s = 0$ to any point on the sphere $\sum x_s^2 = \varepsilon$ on which there is no point where $V = c$.

The function $V(t, x_1, \ldots, x_n)$ is of definite sign only if there is a positive definite function $W(x_1, \ldots, x_n)$ for it, which is independent of t and for which one of the two expressions, $V - W$ and $-V - W$, represents a positive function.

In a sufficiently small domain $\sum x_s^2 \leqslant \varepsilon$, where $W = c$ is a surface enclosing the origin $x_s = 0$, the surfaces $V = c' < c$ are also closed for every value of t if $V - W$ is positive, since there is no way from the origin $x_s = 0$ to a point of the surface $W = c$ on which there is no point where $V = c'$.

217. Lyapunov's fundamental theorem on stability states that *if the differential equations of a perturbed motion are such that there is a function V of definite sign whose derivative V', by virtue of these equations, either has constant sign opposite to that of V or is identically zero, then the perturbed motion is stable.*

Proof. Assume that V is positive definite. Then, in accordance with the theorem, the inequalities

$$V' \leqslant 0 \quad \text{and} \quad V \geqslant W,$$

where W is a positive definite function of the variables x_s, independent of t, are satisfied in a small domain $\sum x_s^2 \leqslant \varepsilon$.

Assume that l is the greatest lower bound of the function W on the sphere $\sum x_s^2 = \varepsilon$. Let the domain $\sum x_s^2 < \eta$ be common for the domains $V < l$ for t varying from t_0 to ∞.

Evidently, a motion which begins with the initial perturbations of $x_s = \xi_s$, which lie within $\sum \xi_s^2 < \eta$, will not go beyond the domain $\sum x_s^2 \leqslant \varepsilon$. Indeed, from the relation

$$V - V_0 = \int_{t_0}^{t} V' \, dt \leqslant 0$$

we have

$$l > V_0 \geqslant V \geqslant W.$$

218. The general theorem on instability.[*] For simplicity we assume that the interval of variation of time (t_0, ∞) is closed. We agree to say that the domain $V > 0$ exists within $\Sigma x_s^2 \leqslant \varepsilon$ if for every t on the interval (t_0, ∞) there are x_s satisfying the inequality $V > 0$.

The instability theorem states: *if the differential equations of a perturbed motion are such that (1) for a function V, admitting of an infinitely small upper limit*[**]*, there is a domain VV' > 0 and (2) for certain arbitrarily small values of the quantities x_s, we can single out in this domain (VV' > 0) a subdomain in which a function W > 0 on whose boundary (W = 0) the complete time derivative W' has the same definite sign, then the unperturbed motion is unstable.*

Proof. Let $V > 0$ in a domain satisfying item (1) and assume that $W' > 0$ on the boundary $W = 0$. We choose an initial value of the perturbation $x_s = \xi_s$ within the domain $W > 0$ indicated in item (2). Let the initial value of V be V_0. For the function V which admits of an infinitesimal upper limit we can find λ that is smaller than the values of Σx_s^2 possible for the inequality $V \geqslant V_0$. Assume that l is the minimum of V' for our choice of ξ_s for t varying on (t_0, ∞) and under the condition

$$\lambda \leqslant \Sigma x_s^2 \leqslant \varepsilon.$$

The equation

$$V - V_0 = \int_{t_0}^{t} V' dt$$

yields an inequality

$$V > V_0 + l(t - T). \tag{7.38}$$

But the function V is bounded in the domain $\Sigma x_s^2 \leqslant \varepsilon$, i.e. $V < L$ and the last inequality (7.38) states that for the values of t greater than the value $t_0 + \dfrac{L - V_0}{l}$ the motion must leave the domain $W > 0$, which lies within $VV' > 0$, via the sphere $\Sigma x_s^2 = \varepsilon$.

Remark. In the formulation of the theorem, the condition that W' has a definite sign on $W = 0$ means that the derivative W' does not assume

---

[*] N.G. Chetaev, *Dokl. Akad. Nauk SSSR* **1**, 9: 529-531 (1934).

[**] The bounded function V admits of an infinitesimal upper limit if for every $l > 0$, however small, there is a number $\lambda \neq 0$ such that $|V| < l$ for $t \geqslant t_0$ and $\sum_s x_s^2 \leqslant \lambda$. — *Ed.*

different signs on $W = 0$, having either any one sign or being zero. In the last case, $W' = 0$, it is necessary that in the corresponding place the surface $W = 0$ be simple (not multiple).

219. The principle of stability was used by Lagrange, Laplace, Poisson, Poincaré, and Lyapunov in their work on cosmogony. It was extensively used when Lagrange's theorem on the stability of an equilibrium with the existence of a force function was employed to describe the development of equilibria of slowly varying mechanical systems. The fundamental laws of physics, namely, Hooke's law, the laws concerning entropy and the Lorentz's force, Newton's law of gravitation, satisfy the necessary conditions of the principle of stability.

CHAPTER 8

THE THEORY OF ATTRACTION

220. The *law of gravitation* states that two particles of masses m and m' a distance r apart are attracted with the force

$$F = -f\frac{mm'}{r^2}.$$

The constant f is equal to the force of attraction between two particles of unit mass a unit distance apart. Using appropriate units of measurement, we can make f equal to unity.

221. Assume that we have n fixed particles m_i ($i = 1, \ldots, n$) and one moving particle of mass μ at a distance r_i from the particles m_i. The resultant force of attraction acting on the mass μ is the geometric sum of the separate attraction forces

$$F_i = -f\frac{\mu m_i}{r_i^2}.$$

During an elementary displacement of the mass μ the work done by the resultant attraction force is

$$-\sum f\frac{\mu m_i}{r_i^2}\, dr_i = df\mu \sum \frac{m_i}{r_i}.$$

Let

$$U = \sum \frac{m_i}{r_i}.$$

Consequently, the attraction forces have a force function $f\mu U$.

The set of forces acting on a unit mass $\mu = 1$ in its various positions P is known as a *field of forces*, the direction and magnitude of such a force at P is called the *direction* and *intensity of the field*. The force function of the forces of gravitation $f\mu U$ is known as *Newton's potential* and the function U is called a *potential*.

222. Consider an element $d\sigma$ of a surface. We draw a normal n to it in some direction. Let α, β, γ be the direction cosines of the normal, $\mathbf{F}$ the force of attraction of a particle P of the element $d\sigma$, $\mathbf{F}_n$ a component of the force $\mathbf{F}$ along the normal, and θ the angle between the normal and the force (Fig. 148). The expression

$$F_n d\sigma = F \cos\theta d\sigma = f\left(\frac{\partial U}{\partial x}\alpha + \frac{\partial U}{\partial y}\beta + \frac{\partial U}{\partial z}\gamma\right)d\sigma = f\frac{\partial U}{\partial n}d\sigma$$

is known as the *flux of the force* through the element of the surface in the direction n. The flux of force through the surface S in the direction of the chosen normal n is the sum of the elementary fluxes through different elements $d\sigma$ of this surface, i.e.

$$\iint_S F_n d\sigma = f\iint_S \frac{\partial U}{\partial n}d\sigma.$$

223. When we have a single attracting particle m, it is easy to find the flux of a force through a closed surface. Let us find the flux of force through the element $d\sigma$. We draw rays through $d\sigma$ from the attracting particle m. The resulting cone cuts spheres of radii r and 1 along the elements $d\omega'$ and $d\omega$ respectively (Fig. 149):

$$d\omega' = d\sigma \cos\theta \text{ and } d\omega = r^2 d\omega.$$

The flux of force through the element $d\sigma$ is

$$-f\frac{m}{r^2}\cos\theta\, d\sigma = -fm\, d\omega.$$

If we reverse the direction of the normal, the expression for the flux of force will change sign. The quantity $d\omega$ is a solid angle at which we see the element $d\sigma$ from m. If we see the negative side of the element $d\sigma$, we take the minus sign in the formula and if we see the positive side of the element $d\sigma$, we take the plus sign. Hence, the outward flux of force through a closed surface (we shall take n to be the direction of the exterior normal) is $-4\pi fm$ if the particle m is inside the surface, it is $-2\pi fm$ if the particle

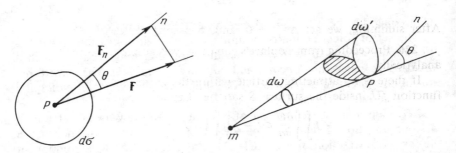

Figure 148   Figure 149

lies on the surface where the surface has a definite tangent plane, and it is zero if the particle m lies outside the surface.

224. Let $m_1, \ldots, m_n$ be attracting particles and assume that some of them are inside and some outside a closed surface.

Gauss' theorem states that the outward flux of force through the surface is equatl to $-4\pi f$ multiplied by the sum of the masses of the particles within the surface.

Indeed,

$$f\iint_S \frac{\partial U}{\partial n}\, d\sigma = -f\sum \iint_S \frac{m_i}{r_1^2} \cos(n, r_i) d\sigma,$$

where the integrals on the right-hand side are zeros for the particles m_i lying outside the surface S and are equal to $4\pi m_i$ for the particles m_i lying within the surface S.

225. The potential satisfies Laplace's equation. Indeed,

$$\Delta U = \Delta \sum \frac{m_i}{r_i} = \sum m_i \Delta \frac{1}{r_i}.$$

We designate the coordinates of the particle m_i as x_i, y_i, z_i and the coordinates of the particle P as x, y, z. Then we have

$$r_i = \sqrt{(x - x_i)^2 + (y - y_i)^2 + (z - z_i)^2},$$

and, hence

$$\frac{\partial^2 \frac{1}{r_i}}{\partial x^2} = 3\frac{(x - x_i)^2}{r_i^5} - \frac{1}{r_i^3}, \quad \frac{\partial^2 \frac{1}{r_i}}{\partial y^2} = 3\frac{(y - y_i)^2}{r_i^5} - \frac{1}{r_i^3},$$

$$\frac{\partial^2 \frac{1}{r_i}}{\partial z^2} = 3\frac{(z - z_i)^2}{r_i^5} - \frac{1}{r_i^3}.$$

After summing, we get $\Delta \frac{1}{r_i} = 0$, and, consequently, $\Delta U = 0$.

226. Proceeding from Laplace's equation, we can prove Gauss' theorem analytically.

If there is no attracting particle within the closed surface S, then the function fU inside the surface S satisfies Laplace's equation, i.e.

$$f\iint_S \frac{\partial U}{\partial n}\, d\sigma = f\iiint_V \Delta U\, d\tau = 0.$$

Assume that the closed surface S contains $m_1, \ldots, m_p$ of the masses m_i.

The outward flux of forces through the surface S is

$$f \iint_S \frac{\partial U}{\partial n}\, d\sigma,$$

where n is the exterior normal. We enclose the particles $m_1, \ldots, m_p$ in spheres $S_1, \ldots, S_p$ of an arbitrarily small radius (Fig. 150). Let us consider a volume V bounded by the surface S and the spheres $S_1, \ldots, S_p$. For the particles within the volume V the function fU is continuous together with its first-order derivatives, i.e.,

$$f \iint_{S + S_1 + \ldots + S_p} \frac{\partial U}{\partial n}\, d\sigma = f \iiint_V \Delta U\, d\tau = 0.$$

Hence

$$f \iint_S \frac{\partial U}{\partial n}\, d\sigma + f \iint_{S_1} \frac{\partial U}{\partial n}\, d\sigma + \ldots + f \iint_{S_p} \frac{\partial U}{\partial n}\, d\sigma = 0.$$

We calculate the integral $I_1 = \iint_{S_1} \frac{\partial U}{\partial n}\, d\sigma$. We have $U = \dfrac{m_1}{r_1} + U_1$, where $U_1 = \displaystyle\sum_{i=2}^{n} \frac{m_i}{r_i}$ is a continuous function satisfying Laplace's equation within the sphere S_1. From this we have

$$I_1 = \iint_{S_1} \frac{\partial \dfrac{m_1}{r_1}}{\partial n}\, d\sigma + \iint_{S_1} \frac{\partial U_1}{\partial n}\, d\sigma.$$

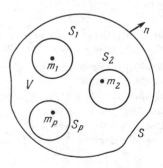

Figure 150

According to Green's formula the second integral is evidently zero. The normal n, exterior with respect to the volume V, is interior with respect to the sphere S_1, $dn = -dr$:

$$\frac{\partial \dfrac{m_1}{r_1}}{\partial n} = -\frac{\partial \dfrac{m_1}{r_1}}{\partial r_1} = \frac{m_1}{r_1^2}.$$

Consequently,

$$I_1 = \int\!\!\int_{S_1} \frac{m_1}{r_1^2}\, d\sigma = \frac{m_1}{r_1^2} \int\!\!\int_{S_1} d\sigma = \frac{m_1}{r_1^2}\, 4\pi r_1^2 = 4\pi m_1,$$

and this means that

$$f\int\!\!\int_{S} \frac{\partial U}{\partial n}\, d\sigma = -4\pi f(m_1 + \ldots + m_p) = -4\pi f M',$$

where $M' = m_1 + \ldots + m_p$ is the sum of the masses inside S.

8.1 Continuous Masses

227. If we consider a continuous mass composed of a large number of infinitesimal masses dm, which are at the distance r from a point P, we get a formula

$$U = \int \frac{dm}{r}$$

for the potential of the continuous mass. For an arbitrary point P, lying outside the attracting masses dm, r is never zero, and hence

$$\Delta U = \int \Delta \frac{1}{r}\, dm = 0 \qquad \left(\Delta U = \frac{\partial^2 U}{\partial x^2} + \frac{\partial^2 U}{\partial y^2} + \frac{\partial^2 U}{\partial z^2} \right).$$

The potential U of the continuous mass satisfies Laplace's equation for the particles outside the attracting mass.

A continuous mass can lie along a curve, on a surface, or in a volume. The last two cases are more frequent.

228. The potential of a simple layer. Let the attracting masses lie on a surface S and assume that $d\sigma$ is an element of the surface and dm is the mass of the attracting matter on this element. The quantity $dm/d\sigma = \varrho$ is the surface *density of the simple layer.* The potential U is

$$U = \int \frac{dm}{r} = \int\!\!\int_{S} \frac{\varrho\, d\sigma}{r}.$$

229. Consider the potential of a simple homogeneous spherical layer. Assume that R is the radius of the sphere and ϱ is the constant surface density.

1°. A point P is inside the sphere. For every element $d\sigma$ there is an element $d\sigma'$ on the sphere on the other side of the point P, $d\sigma'$ is cut by a cone of rays originating at $d\sigma$ and passing through P (Fig. 151). The forces of attraction of these elements towards the point P are $f\varrho d\sigma/r^2$ and $f\varrho d\sigma'/r'^2$. But

$$\frac{d\sigma}{r^2} = \frac{d\sigma'}{r'^2},$$

and, consequently, the attraction forces of the elements $d\sigma$ and $d\sigma'$ at P are equal and opposite. We can partition the whole surface of the sphere into pairs of elements of this kind. Consequently, the total force of attraction of a simple homogeneous spherical layer towards an interior point is zero. This means that the potential U is constant inside the spherical layer. Choosing a point at the centre of the sphere, we find the potential for the interior point:

$$U = \frac{1}{R} \int dm = \frac{M}{R}.$$

2°. The point P is outside of the sphere (Fig. 152). Let P' be a point which satisfies the relation $OP' \cdot OP = R^2$. The triangles OAP and $OP'A$ are similar and therefore the angle O is bounded by the proportional sides. Hence

$$\frac{AP}{AP'} = \frac{r}{r'} = \frac{OP}{R},$$

and therefore,

$$U_P = \int \frac{dm}{r} = \frac{R}{OP} \int \frac{dm}{r'} = \frac{R}{OP} U_{P'} = \frac{R}{OP} \frac{M}{R} = \frac{M}{OP}.$$

The potential of the spherical layer at the exterior point is the same as it would be if the total mass of the layer were concentrated at its centre.

230. The potential of a simple homogeneous spherical layer is a continuous function of the coordinates of a point P. The force of attraction of the simple layer suffers a discontinuity when passing through the layer. Indeed, inside the layer there is no force of attraction. In accordance with the formula we have derived, the attraction force at the exterior point P is directed towards the centre of the layer and is

$$f\frac{dU_e}{dOP} = -f\frac{M}{OP^2} = -f\frac{4\pi R^2 \varrho}{OP^2}.$$

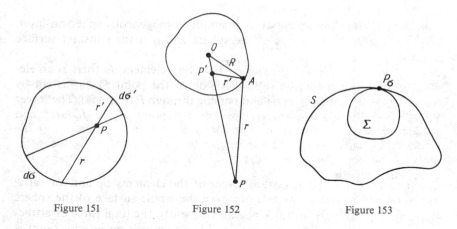

Figure 151   Figure 152   Figure 153

In the limit, as $OP \to R$, we have

$$f \frac{\partial U_e}{\partial n} = -4\pi f \varrho$$

if n is an exterior normal, or better still,

$$\frac{\partial U_i}{\partial n} - \frac{\partial U_e}{\partial n} = 4\pi \varrho.$$

231. Assume that a surface S is filled with an attracting mass with a variable surface density ϱ. The potential

$$U = \iint_S \frac{\varrho \, d\sigma}{r}$$

is a continuous function and its normal derivatives suffer a discontinuity when passing through the surface of the layer.

The surface S and the osculating sphere Σ have a common infinitesimal element σ at a point P (Fig. 153). Assume that at P the density is ϱ' and n is an exterior normal:

$$U = \iint_S \frac{\varrho \, d\sigma}{r} = \iint_{S-\sigma} \frac{\varrho \, d\sigma}{r} - \iint_{\Sigma-\sigma} \frac{\varrho' \, d\sigma}{r} + \iint_\Sigma \frac{\varrho' \, d\sigma}{r}.$$

The derivatives of the first two integrals on the right-hand side with respect to the normal n at P are continuous since P is an exterior point for them. We have explained the discontinuity of the normal derivatives of the spherical layer Σ (the third integral). Hence

$$\frac{\partial U_i}{\partial n} - \frac{\partial U_e}{\partial n} = 4\pi\varrho'.$$

You can find a rigorous proof of this theorem in more specialized courses. We can take any sphere which has a tangent plane in common with the surface S at the point P.

232. A double layer. Consider the potential of an elementary infinitesimal magnet. Let ε be the length of the magnet and m the magnetic mass of the poles, positive at the North Pole and negative at the South Pole. The potential is

$$U = \frac{m}{r'} - \frac{m}{r} = \frac{m(r - r')}{rr'}.$$

But $r - r' = \varepsilon \cos\theta$ (Fig. 154) and hence $U = m\varepsilon \cos\theta/r^2$. We can assign another form to the potential, namely,

$$U = m\varepsilon \frac{(1/r') - (1/r)}{\varepsilon} = m\varepsilon \frac{\partial(1/r)}{\partial n},$$

if we assume n to be directed along the magnet from the South Pole to the North Pole. Designating the direction cosines of n as α, β, γ, we have

$$U = m\varepsilon \left(\alpha \frac{\partial(1/r)}{\partial x} + \beta \frac{\partial(1/r)}{\partial y} + \gamma \frac{\partial(1/r)}{\partial z} \right),$$

whence, we immediately derive Laplace's equation $\Delta U = 0$ for points P outside of the magnet since α, β, γ are constants, x, y, z being the coordinates of the particles of the magnet and ΔU referring to the coordinates of the running point P.

233. Consider a surface S. Assume that elementary magnets have been constructed on the exterior normal n at every point of the surface S. Let the magnetic mass of the element $d\sigma$ of the surface S be (Fig. 155)

$$\mp m = \mp \varrho \, d\sigma$$

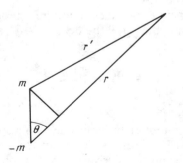

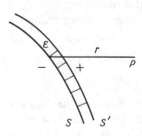

Figure 154   Figure 155

and $\varrho\varepsilon = \mu$. The potential of this double layer is

$$U = \int\int_S \frac{\mu\, d\sigma \cos\theta}{r^2} = \int\int_S \mu\, \frac{\partial(1/r)}{\partial n}\, d\sigma.$$

Outside the layer this function satisfies Laplace's equations.

234. Let us now consider the case when the attracting masses dm fill up the volume V. Assume that $d\tau(a, b, c)$ is an element of the volume and ϱ is the volume density $dm = \varrho\, d\tau$.

The components X, Y, Z of the attraction forces at $P(m = 1$, let its coordinates be $x, y, z)$ which is at the distance r from the element $d\tau$ are

$$X = f\int\int\int_V \frac{a - x}{r^3}\, \varrho\, d\tau, \quad Y = f\int\int\int_V \frac{b - y}{r^3}\, \varrho\, d\tau,$$

$$Z = f\int\int\int_V \frac{c - z}{r^3}\, \varrho\, d\tau \quad (d\tau = da\, db\, dc).$$

The potential

$$U = \int\int\int_V \frac{\varrho\, d\tau}{r}.$$

If P lies outside V, then the functions X, Y, Z, U have derivatives of different orders with respect to x, y, z, and

$$X = f\frac{\partial U}{\partial x}, \quad Y = f\frac{\partial U}{\partial y}, \quad Z = f\frac{\partial U}{\partial z}.$$

If the point P lies inside the attracting masses V, these propositions must be proved.

235. We shall prove that U, X, Y, Z are definite finite functions whatever the position of the point P. We assume that P is inside the attracting masses. We introduce a system of polar coordinates with origin at P, i.e.

$$a - x = r\sin\theta\cos\varphi, \quad b - y = r\sin\theta\sin\varphi, \quad c - z = r\cos\theta.$$

From this we have $d\tau = r^2\sin\theta\, dr\, d\theta\, d\varphi$ and, consequently,

$$X = f\int\int\int_V \varrho\sin^2\theta\cos\varphi\, dr\, d\theta\, d\varphi,$$

$$Y = f\int\int\int_V \varrho\sin^2\theta\sin\varphi\, dr\, d\theta\, d\varphi,$$

$$Z = f\int\int\int_V \varrho\sin\theta\cos\theta\, dr\, d\theta\, d\varphi,$$

$$U = \int\int\int_V \varrho r\sin\theta\, dr\, d\theta\, d\varphi.$$

The integrands of all these functions are finite for a finite volume and, hence, U, X, Y, Z are finite and definite.

236. The relation $X = f \dfrac{\partial U}{\partial x}$ is valid when the point P lies inside V. Let V_1 be a small volume around P and V_2 a volume resulting from V after the subtraction of V_1,

$$U = U_1 + U_2, \quad X = X_1 + X_2, \quad \dots .$$

Assume that P' is a point with coordinates $x + \Delta x$, y, z inside V_1, and $U' = U_1' + U_2'$ is the potential at the point P',

$$\frac{U' - U}{\Delta x} = \frac{U_1' - U_1}{\Delta x} + \frac{U_2' - U_2}{\Delta x}.$$

Since P' is a point exterior with respect to V_2, the limit of the second term exists as $P' \to P$ and is $\partial U_2 / \partial x$.

Let us study

$$\frac{U_1' - U_1}{\Delta x} = \iiint\limits_{V_1} \frac{\varrho \, d\tau}{\Delta x} \left(\frac{1}{r'} - \frac{1}{r} \right).$$

From the triangle $PP' \, d\tau$ we have $\Delta x > |r - r'|$ and, hence,

$$\left| \frac{1}{\Delta x} \left(\frac{1}{r'} - \frac{1}{r} \right) \right| = \left| \frac{r - r'}{\Delta x} \right| \frac{1}{rr'} < \frac{1}{rr'}.$$

From the inequality $\left(\dfrac{1}{r'} - \dfrac{1}{r} \right)^2 > 0$ we obtain $\dfrac{1}{2} \left(\dfrac{1}{r'^2} + \dfrac{1}{r^2} \right) > \dfrac{1}{rr'}$. Let ϱ_1 be the maximum of ϱ for V_1. We have

$$\left| \frac{U_1' - U_1}{\Delta x} \right| < \frac{1}{2} \iiint\limits_{V_1} \frac{\varrho_1 d\tau}{r^2} + \frac{1}{2} \iiint\limits_{V_1} \frac{\varrho_1 d\tau}{r'^2}.$$

In polar coordinates

$$\iiint\limits_{V_1} \frac{\varrho_1 d\tau}{r^2} = \iiint\limits_{V_1} \varrho_1 \sin \theta \, dr \, d\theta \, d\varphi.$$

Let C be the maximum diameter of the volume V_1, then we have

$$\iiint\limits_{V_1} \frac{\varrho_1 d\tau}{r^2} < C \varrho_1 \int\limits_0^{2\pi} d\varphi \int\limits_0^{\pi} \sin \theta \, d\theta = 4\pi \varrho_1 C.$$

In the same way we can prove that

$$\iiint\limits_{V_1} \frac{\varrho_1 d\tau}{r'^2} < 4\pi\varrho_1 C,$$

and, consequently,

$$\left| \frac{U_1' - U}{\Delta x} \right| < 4\pi\varrho_1 C.$$

We can take arbitrarily small volume V_1, and then C will prove to be smaller than any given number and X_2 will differ arbitrarily from X. In the limit we obtain

$$\frac{\partial U}{\partial x} = \frac{X}{f}.$$

Thus the potential U everywhere has first derivatives with respect to the coordinates.

237. The second derivatives of the potential evidently exist for the exterior point P, and then $\Delta U = 0$.

Let now P be an interior point. The second derivatives of the potential cannot now be obtained by a simple differentiation inside the integral since inside the integral $1/r$ turns into infinity. As we have shown earlier, we have

$$\frac{X}{f} = \frac{\partial U}{\partial x} = \iiint\limits_{V} \varrho \frac{a - x}{r^3} d\tau.$$

But

$$\frac{a - x}{r^3} \varrho = -\varrho \frac{\partial(1/r)}{\partial a} = -\frac{\partial(\varrho/r)}{\partial a} + \frac{1}{r} \frac{\partial \varrho}{\partial a},$$

and this means that

$$\frac{X}{f} = \frac{\partial U}{\partial x} = -\iiint\limits_{V} \frac{\partial(\varrho/r)}{\partial a} d\tau + \iiint\limits_{V} \frac{1}{r} \frac{\partial \varrho}{\partial a} d\tau.$$

Let us enclose the point P in a sphere s of a very small radius ε (Fig. 156). Green's formula yields

$$\iiint\limits_{V-(s)} \frac{\partial(\varrho/r)}{\partial a} d\tau = \iint\limits_{S} \frac{\varrho}{r} \alpha\, d\sigma + \iint\limits_{S} \frac{\varrho}{r} \alpha\, d\sigma,$$

the last integral being equal to zero. Consequently,

$$\frac{X}{f} = \frac{\partial U}{\partial x} = -\iint\limits_{S} \frac{\varrho}{r} \alpha\, d\sigma + \iiint\limits_{V} \frac{1}{r} \frac{\partial \varrho}{\partial a} d\tau.$$

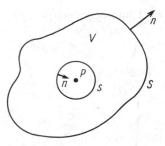

Figure 156

The first integral is the potential of the simple layer with surface density $\varrho\alpha$, and the second integral is the volume potential with density $\partial\varrho/\partial a$. These potentials have derivatives. Thus we have

$$\frac{\partial^2 U}{\partial x^2} = -\iint_S \frac{a-x}{r^3}\,\varrho\alpha\,d\sigma + \iiint_V \frac{a-x}{r^3}\frac{\partial\varrho}{\partial a}\,d\tau.$$

In a similar way we find $\partial^2 U/\partial y^2$, $\partial^2 U/\partial z^2$. The second derivatives of the potential U are continuous inside the volume, they suffer a discontinuity only when passing through the surface S.

238. After we have derived the equality $X = f\dfrac{\partial U}{\partial x}$, ..., we see that the outward flux of the attraction force through the surface Σ satisfies Gauss' theorem

$$\iint_\Sigma F_n\,d\sigma = f\iint_\Sigma \frac{\partial U}{\partial n}\,d\sigma = -4\pi M'f,$$

where M' is the mass within Σ. According to Green's theorem and the relation

$$M' = \iiint_W \varrho\,d\tau$$

it follows from Gauss' theorem that

$$\iiint_W (\Delta U + 4\pi\varrho)d\tau = 0,$$

where W is the volume within the surface Σ. Since Σ is arbitrary, we find that

$$\Delta U = -4\pi\varrho,$$

i.e. the potential U within the volume satisfies Poisson's equation.

239. The potential of a homogeneous solid sphere of radius R. 1°. A point P lies outside the sphere. We partition the sphere into infinitesimal layers, and then we have

$$U_e = \int_0^R \frac{4\pi r^2 \varrho}{OP}\, dr = \frac{4}{3}\,\pi\varrho\,\frac{R^3}{OP} = \frac{M}{OP},$$

where ϱ is the density.

2°. The point P is inside the solid sphere $OP < R$. A sphere passing through the point P of radius OP divides each spherical layer into two parts, the layers of one part are of radius $r \leqslant OP$ (for them the point P is exterior) and the layers of the other part are of radius $r \geqslant OP$ (for them the point P is interior):

$$U_i = \int_0^{OP} \frac{4\pi r^2 \varrho\, dr}{OP} + \int_{OP}^R 4\pi\varrho r\, dr$$

$$= \frac{4}{3}\,\pi OP^2 \varrho + 2\pi\varrho R^2 - 2\pi\varrho OP^2 = 2\pi\varrho R^2 - \frac{2}{3}\,\pi\varrho OP^2.$$

240. The potential of a bounded body at a very distant point. Assume that P is a very distant point being attracted (Fig. 157), m is a particle of the attracting mass with coordinates a, b, c, and $mP = r$. We assume O to be the origin and set $mO = \delta$, $\angle mOP = \gamma$, $OP = R$. From the triangle mOP we find that $r^2 = R^2 + \delta^2 - 2R\delta \cos\gamma$. Hence

$$\frac{1}{r} = \frac{1}{R}\,\frac{1}{\sqrt{1 - \dfrac{2\delta \cos\gamma}{R} + \dfrac{\delta^2}{R^2}}}.$$

Expanding the last expression in powers of $1/R$ and truncating to the first three terms, we obtain

$$\frac{1}{r} = \frac{1}{R} + \frac{\delta}{R^2}\cos\gamma + \frac{\delta^2}{2R^3}(3\cos^2\gamma - 1) + \frac{\varepsilon}{R^4}.$$

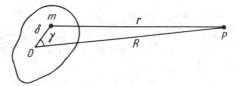

Figure 157

If OP is the x-axis (Fig. 157), then $a = \delta \cos \gamma$. Hence

$$U = \iiint_V \frac{\varrho \, d\tau}{r} = \frac{1}{R} \iiint_V \varrho \, d\tau + \frac{1}{R^2} \iiint_V a\varrho \, d\tau$$

$$+ \frac{1}{2R^3} \iiint_V \varrho(3a^2 - \delta^2) \, d\tau + \frac{k}{R^4}.$$

But

$$\iiint_V \varrho \, d\tau = M, \quad \iiint_V a\varrho \, d\tau = M\xi,$$

where ξ is the x-coordinate of the centre of gravity of the attracting body. Since $\quad 3a^2 - \delta^2 = 3a^2 - (a^2 + b^2 + c^2) = 2(a^2 + b^2 + c^2) - 3(b^2 + c^2)$, the third integral is

$$\iiint_V \varrho(3a^2 - \delta^2) d\tau$$

$$= 2\iiint_V \varrho(a^2 + b^2 + c^2) d\tau - 3\iiint_V \varrho(b^2 + c^2) d\tau = 2\mathfrak{M}_0 - 3I,$$

where $\mathfrak{M}_0$ is the moment of inertia of the body about O, I is the moment of inertia of the body about the axis OP, and

$$U = \frac{M}{R} + \frac{M\xi}{R^2} + \frac{2\mathfrak{M}_0 - 3I}{R^3} + \frac{k}{R^4}.$$

As $R \to \infty$, we have $U \to M/R$.

241. Gauss' formula. Assume that a volume V bounded by a surface S is filled up by a homogeneous mass of density ϱ. Assume furthermore that $d\sigma$ is an element of the surface S, α, β, γ are the direction cosines of the exterior normal to the surface S, a, b, c are the coordinates of a particle of the element $d\tau$ of the volume, x, y, z are the coordinates of a running point P,

$$r = \sqrt{(a - x)^2 + (b - y)^2 + (c - z)^2}.$$

Consider a vector with projections

$$A = \frac{1}{2} \frac{a - x}{r}, \quad B = \frac{1}{2} \frac{b - y}{r}, \quad C = \frac{1}{2} \frac{c - z}{r}.$$

Hence

$$\frac{\partial A}{\partial a} + \frac{\partial B}{\partial b} + \frac{\partial C}{\partial c} = \frac{1}{r}.$$

The potential U is

$$U = \iiint_V \frac{\varrho\, d\tau}{r} = \varrho \iiint_V \left(\frac{\partial A}{\partial a} + \frac{\partial B}{\partial b} + \frac{\partial C}{\partial c} \right) d\tau$$

$$= \varrho \iint_S (A\alpha + B\beta + C\gamma)d\sigma$$

$$= \frac{\varrho}{2} \iint_S \frac{(a - x)\alpha + (b - y)\beta + (c - z)\gamma}{r}\, d\sigma$$

$$= \frac{\varrho}{2} \iint_S \cos(r, n)d\sigma.$$

The expression $(a - x)\alpha + (b - y)\beta + (c - z)\gamma = p$ is the algebraic distance of the point P from the plane tangent to the surface S at the point (a, b, c). This means that

$$U = \frac{\varrho}{2} \iint_S \frac{p}{r}\, d\sigma.$$

Consequently, the potential of the mass enclosed in the volume U is equal to the potential of a simple layer with the surface density $\frac{1}{2}\varrho p$. This fictitious density varies with the position of P. It is easy to find the components of the attraction:

$$\frac{1}{f} X = -\frac{\varrho}{2} \iint_S \frac{\alpha}{r}\, d\sigma + \frac{\varrho}{2} \iint_S \frac{p(a - x)}{r^3}\, d\sigma,$$

since $\dfrac{\partial p}{\partial x} = -\alpha$ and $\dfrac{\partial(1/r)}{\partial x} = \dfrac{a - x}{r^3}$. But

$$\frac{1}{f} X = \varrho \iiint_V \frac{a - x}{r^3}\, d\tau = -\varrho \iiint_V \frac{\partial(1/r)}{\partial a}\, d\tau = -\varrho \iint_S \frac{\alpha}{r}\, d\sigma,$$

and this means that

$$X = f\varrho \iint_S \frac{P(a - x)}{r^3}\, d\sigma.$$

8.2 Potential of a Homogeneous Ellipsoid

242. Assume that ϱ is the density and x', y', z' are the coordinates of the attracting particles. We have to calculate

$$U = \varrho \iiint \frac{d\tau}{r}, \quad r = \sqrt{(x - x')^2 + (y - y')^2 + (z - z')^2},$$

where the integration is over the volume of the ellipsoid

$$\frac{x'^2}{a^2} + \frac{y'^2}{b^2} + \frac{z'^2}{c^2} \leqslant 1.$$

It is difficult to calculate the potential U since the limits of integration are variable. Dirichlet proposed a method of avoiding this difficulty by using a discontinuity factor:

$$\frac{2}{\pi} \int\limits_0^\infty \frac{\sin \varphi}{\varphi} \cos (f\varphi)d\varphi = \begin{cases} 1, & \text{if } f < 1, \\ 1/2, & \text{if } f = 1, \\ 0, & \text{if } f > 1. \end{cases}$$

For the ellipsoid in question the discontinuity factor has the form

$$\frac{2}{\pi} \int\limits_0^\infty \frac{\sin \varphi}{\varphi} \cos \left[\left(\frac{x'^2}{a^2} + \frac{y'^2}{b^2} + \frac{z'^2}{c^2} \right) \varphi \right] d\varphi.$$

Using the discontinuity factor, we have

$$U = \frac{2\varrho}{\pi} \int\limits_0^\infty \frac{\sin \varphi}{\varphi} d\varphi \iiint\limits_{-\infty}^{+\infty} \cos \left[\left(\frac{x'^2}{a^2} + \frac{y'^2}{b^2} + \frac{z'^2}{c^2} \right) \varphi \right] \frac{dx'dy'dz'}{r},$$

or

$$U = \text{Re} \frac{2\varrho}{\pi} \int\limits_0^\infty \frac{\sin \varphi}{\varphi} d\varphi \iiint\limits_{-\infty}^{+\infty} \exp \left[\left(\frac{x'^2}{a^2} + \frac{y'^2}{b^2} + \frac{z'^2}{c^2} \right) \varphi i \right] \frac{dx'dy'dz'}{r}.$$

From the formula known from analysis

$$\int\limits_0^\infty \frac{e^{r^2\psi i}}{\sqrt{\psi}} d\psi = \frac{\sqrt{\pi}}{r} e^{\frac{\pi}{4}i}$$

we have

$$U = \text{Re} \frac{2}{\pi} \frac{\varrho}{\sqrt{\pi}} e^{-\frac{\pi}{4}i} \int\limits_0^\infty d\psi \int\limits_0^\infty \frac{\sin \varphi}{\varphi \sqrt{\psi}} d\varphi$$

$$\times \iiint\limits_{-\infty}^{+\infty} \exp\left\{\left[\left(\frac{x'^2}{a^2} + \frac{y'^2}{b^2} + \frac{z'^2}{c^2}\right)\varphi + r^2\psi\right]i\right\} dx'\,dy'\,dz'.$$

But

$$r^2 = (x - x')^2 + (y - y')^2 + (z - z')^2$$
$$= x^2 + y^2 + z^2 - 2xx' - 2yy' - 2zz' + x'^2 + y'^2 + z'^2.$$

We introduce the integrals

$$Q_x = \int\limits_{-\infty}^{+\infty} \exp\left\{\left[\left(\psi + \frac{\varphi}{a^2}\right)x'^2 - 2xx'\psi\right]i\right\} dx',\ Q_y = \ldots,\ Q_z = \ldots$$

and use them to write the potential as

$$U = \operatorname{Re} \frac{2\varrho}{\pi\sqrt{\pi}} e^{-\frac{\pi}{4}i} \int\limits_0^\infty \int\limits_0^\infty \frac{\sin\varphi}{\varphi\sqrt{\psi}} \exp\left\{(x^2 + y^2 + z^2)\psi i\right\} Q_x Q_y Q_z d\varphi\,d\psi.$$

243. From analysis we know the formula

$$\int\limits_{-\infty}^{+\infty} \exp\left\{(hu^2 - 2ku)i\right\} du = \sqrt{\frac{\pi}{h}} \exp\left\{\left(\frac{\pi}{4} - \frac{k^2}{h}\right)i\right\},$$

whence we have

$$Q_x = \sqrt{\frac{\pi}{\psi + \varphi/a^2}} \exp\left\{\left(\frac{\pi}{4} - \frac{x^2\psi^2}{\psi + \varphi/a^2}\right)i\right\}.$$

The expressions for Q_y and Q_z are similar. Consequently,

$$Q_x Q_y Q_z = \pi^{3/2} e^{\frac{3\pi}{4}i} \frac{1}{\sqrt{(\psi + \varphi/a^2)(\psi + \varphi/b^2)(\psi + \varphi/c^2)}}$$
$$\times \exp\left\{-\psi^2\left(\frac{x^2}{\psi + \varphi/a^2} + \frac{y^2}{\psi + \varphi/b^2} + \frac{z^2}{\psi + \varphi/c^2}\right)i\right\}.$$

We set

$$S\varphi = (x^2 + y^2 + z^2)\psi - \psi^2\left(\frac{x^2}{\psi + \varphi/a^2} + \frac{y^2}{\psi + \varphi/b^2} + \frac{z^2}{\psi + \varphi/c^2}\right)$$
$$= \psi\left[x^2\left(1 - \frac{\psi}{\psi + \varphi/a^2}\right) + y^2\left(1 - \frac{\psi}{\psi + \varphi/b^2}\right) + z^2\left(1 - \frac{\psi}{\psi + \varphi/c^2}\right)\right]$$
$$= \psi\left[x^2 \frac{\varphi/a^2}{\psi + \varphi/a^2} + y^2 \frac{\varphi/b^2}{\psi + \varphi/b^2} + z^2 \frac{\varphi/c^2}{\psi + \varphi/c^2}\right]$$

$$= \varphi\psi\left(\frac{x^2}{\varphi + a^2\psi} + \frac{y^2}{\varphi + b^2\psi} + \frac{z^2}{\varphi + c^2\psi}\right)$$

$$= \varphi\left(\frac{x^2}{a^2 + \varphi/\psi} + \frac{y^2}{b^2 + \varphi/\psi} + \frac{z^2}{c^2 + \varphi/\psi}\right).$$

From this we have

$$U = \mathrm{Re}\, 2\varrho e^{\frac{\pi}{2}i} \int_0^\infty \int_0^\infty \frac{\sin\varphi}{\varphi\sqrt{\psi}}\, \frac{e^{S\varphi i}}{\sqrt{(\psi + \varphi/a^2)(\psi + \varphi/b^2)(\psi + \varphi/c^2)}}\, d\varphi\, d\psi.$$

We introduce a new variable instead of ψ, namely, $s = \dfrac{\varphi}{\psi}$, $ds = -\dfrac{\varphi}{\psi^2}\, d\psi$. We have

$$U = \mathrm{Re}\, 2\varrho e^{\frac{\pi}{2}i} \int_0^\infty \frac{ds}{\sqrt{(1 + s/a^2)(1 + s/b^2)(1 + s/c^2)}} \int_0^\infty \frac{\sin\varphi}{\varphi^2} e^{S\varphi i}\, d\varphi,$$

if we take into account that when ψ varies over the interval $(0, \infty)$, the variable s runs over the interval $(\infty, 0)$;

$$S = \frac{x^2}{a^2 + s} + \frac{y^2}{b^2 + s} + \frac{z^2}{c^2 + s}.$$

Now let us pass to the real function

$$U = -2\varrho \int_0^\infty \frac{ds}{\sqrt{(1 + s/a^2)(1 + s/b^2)(1 + s/c^2)}} \int_0^\infty \frac{\sin\varphi}{\varphi^2} \sin S\varphi\, d\varphi.$$

244. We integrate with respect to φ as follows:

$$\frac{\partial U}{\partial x} = -4\varrho x \int_0^\infty \frac{ds}{(a^2 + s)D} \int_0^\infty \frac{\sin\varphi}{\varphi} \cos S\varphi\, d\varphi,$$

$$D = \sqrt{(1 + s/a^2)(1 + s/b^2)(1 + s/c^2)}.$$

But

$$\int_0^\infty \frac{\sin\varphi}{\varphi} \cos S\varphi\, d\varphi = \begin{cases} \dfrac{\pi}{2}, & \text{if } S < 1, \\[2mm] 0, & \text{if } S > 1. \end{cases}$$

For points inside the ellipsoid $\dfrac{x^2}{a^2} + \dfrac{y^2}{b^2} + \dfrac{z^2}{c^2} < 1$ we have s varying in the interval $(0, \infty)$ and

$$S = \frac{x^2}{a^2 + s} + \frac{y^2}{b^2 + s} + \frac{z^2}{c^2 + s} < 1.$$

Consequently, for an interior point we have

$$\frac{\partial U}{\partial x} = -2\pi\varrho x \int_0^\infty \frac{ds}{(a^2 + s)D}, \quad \frac{\partial U}{\partial y} = -2\pi\varrho y \int_0^\infty \frac{ds}{(b^2 + s)D},$$

$$\frac{\partial U}{\partial z} = -2\pi\varrho z \int_0^\infty \frac{ds}{(c^2 + s)D}.$$

Let a point (x, y, z) be exterior. An ellipsoid cofocal with the given ellipsoid passes through this point:

$$\frac{x^2}{a^2 + \sigma} + \frac{y^2}{b^2 + \sigma} + \frac{z^2}{c^2 + \sigma} = 1.$$

From this, we have $S > 1$ for $s < \sigma$, and this means that

$$\int_0^\infty \frac{\sin \varphi}{\varphi} \cos S\varphi \, d\varphi = 0,$$

and we have $S < 1$ for $s > \sigma$, and this means that

$$\int_0^\infty \frac{\sin \varphi}{\varphi} \cos S\varphi \, d\varphi = \frac{\pi}{2}.$$

Consequently, for the exterior point we have

$$\frac{\partial U}{\partial x} = -2\pi\varrho x \int_\sigma^\infty \frac{ds}{(a^2 + s)D}, \quad \frac{\partial U}{\partial y} = -2\pi\varrho y \int_\sigma^\infty \frac{ds}{(b^2 + s)D},$$

$$\frac{\partial U}{\partial z} = -2\pi\varrho z \int_\sigma^\infty \frac{ds}{(c^2 + s)D}.$$

But $dU = \frac{\partial U}{\partial x} dx + \frac{\partial U}{\partial y} dy + \frac{\partial U}{\partial z} dz$, and hence

$$U = \pi\varrho \int_0^\infty \left(C - \frac{x^2}{a^2 + s} - \frac{y^2}{b^2 + s} - \frac{z^2}{c^2 + s} \right) \frac{ds}{D} \quad \text{for the interior point,}$$

$$U = \pi \varrho \int\limits_{\sigma}^{\infty} \left(C - \frac{x^2}{a^2 + s} - \frac{y^2}{b^2 + s} - \frac{z^2}{c^2 + s} \right) \frac{ds}{D} \text{ for the exterior point.}$$

The constants in these expressions are equal since we must have the same expression for the potential U for any point on the ellipsoid. For an infinitely distant point $\sigma = \infty$ we find from the second expression the value $C = 1$.

245. Rodrigues suggested another method for determining the potential of a homogeneous ellipsoid. Assume that ϱ is the density, ξ, η, ζ are the coordinates of the attracting particle, x, y, z are the coordinates of the particle being attracted, a, b, c are the semi-major, semi-mean, and semi-minor axes of the ellipsoid respectively. We consider an ellipsoid similar to the given one and cofocal with it, with the semi-major, semi-mean, and semi-minor axes a', b', c':

$$a'^2 = a^2 + \delta\lambda, \quad b'^2 = b^2 + \delta\lambda, \quad c'^2 = c^2 + \delta\lambda. \tag{8.1}$$

Let us determine the variation of the potential

$$U = \varrho \iiint \frac{d\xi \, d\eta \, d\zeta}{r}$$

when passing from the given ellipsoid to the cofocal one. We change the variables

$$\xi = aR \cos\theta, \quad \eta = bR \sin\theta \cos\psi, \quad \zeta = cR \sin\theta \sin\psi. \tag{8.2}$$

The new variables vary within the limits $0 \leqslant R \leqslant 1$, $0 \leqslant \theta \leqslant \pi$, $0 \leqslant \psi \leqslant 2\pi$:

$$d\xi \, d\eta \, d\zeta = abcR^2 \sin\theta \, dR \, d\theta \, d\psi,$$

and this means that

$$U = \varrho abc \int\limits_0^1 \int\limits_0^\pi \int\limits_0^{2\pi} \frac{R^2 \sin^2\theta \, dR \, d\theta \, d\psi}{r};$$

from which we have

$$\delta\left(\frac{U}{abc}\right) = \varrho \int\limits_0^1 \int\limits_0^\pi \int\limits_0^{2\pi} \delta\left(\frac{1}{2}\right) R^2 \sin\theta \, dR \, d\theta \, d\psi.$$

In accordance with (8.2) and (8.1) we have $\delta a^2 = \delta b^2 = \delta c^2 = \delta\lambda$,

$$\frac{\delta\xi}{\xi} = \frac{\delta a}{a} = \frac{1}{2}\frac{\delta\lambda}{a^2}, \quad \frac{\delta\eta}{\eta} = \frac{\delta b}{b} = \frac{1}{2}\frac{\delta\lambda}{b^2}, \quad \frac{\delta\zeta}{\zeta} = \frac{\delta c}{c} = \frac{1}{2}\frac{\delta\lambda}{c^2}.$$

Consequently,

$$\delta\left(\frac{1}{r}\right) = -\frac{(\xi - x)\delta\xi + (\eta - y)\delta\eta + (\zeta - z)\delta\zeta}{r^3}$$

$$= -\frac{\delta\lambda}{2r^3}\left[\frac{(\xi - x)\xi}{a^2} + \frac{(\eta - y)\eta}{b^2} + \frac{(\zeta - z)\zeta}{c^2}\right] = -\frac{\delta\lambda}{2r^3}\,N,$$

and

$$\delta\left(\frac{U}{abc}\right) = -\frac{\varrho\delta\lambda}{2}\int_0^1 \int_0^\pi \int_0^{2\pi} \frac{N}{r^3}\,R^2\sin\theta\,dR\,d\theta\,d\psi, \qquad (8.3)$$

where $N = \dfrac{(\xi - x)\xi}{a^2} + \dfrac{(\eta - y)\eta}{b^2} + \dfrac{(\zeta - z)\zeta}{c^2}$.

246. We set R constant and obtain, from (8.2), an ellipsoid similar to the given one:

$$\frac{\xi^2}{a^2} + \frac{\eta^2}{b^2} + \frac{\zeta^2}{c^2} = R^2.$$

At the same time we consider an ellipsoid $R + dR$. Designating the points of the ellipsoid R as (ξ, η, ζ) and the segment of the normal (α, β, γ) to the ellipsoid R between the ellipsoids R and $R + dR$ as ε, we get an equation of the ellipsoid $R + dR$ in the form

$$\frac{(\xi + \alpha\varepsilon)^2}{a^2} + \frac{(\eta + \beta\varepsilon)^2}{b^2} + \frac{(\zeta + \gamma\varepsilon)^2}{c^2} = (R + dR)^2,$$

or, with an accuracy to within the first-order quantities with respect to ε (Fig. 158),

$$\varepsilon\left(\frac{\alpha\xi}{a^2} + \frac{\beta\eta}{b^2} + \frac{\gamma\zeta}{c^2}\right) = R\,dR.$$

But

$$\alpha = \frac{\xi/a^2}{\sqrt{\dfrac{\xi^2}{a^4} + \dfrac{\eta^2}{b^4} + \dfrac{\zeta^2}{c^4}}}, \qquad \beta = \frac{\eta/b^2}{\sqrt{\dfrac{\xi^2}{a^4} + \dfrac{\eta^2}{b^4} + \dfrac{\zeta^2}{c^4}}},$$

$$\gamma = \frac{\zeta/c^2}{\sqrt{\dfrac{\xi^2}{a^4} + \dfrac{\eta^2}{b^4} + \dfrac{\zeta^2}{c^4}}},$$

and hence (after multiplying these relations by α, β, γ and adding) we obtain from the preceding formula a relation

$$\varepsilon\sqrt{\frac{\xi^2}{a^4} + \frac{\eta^2}{b^4} + \frac{\zeta^2}{c^4}} = R\,dR.$$

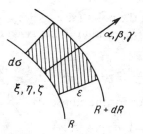

Figure 158

Hence

$$\alpha R \, dR = \frac{\varepsilon \xi}{a^2}, \quad \beta R \, dR = \frac{\varepsilon \eta}{b^2}, \quad \gamma R \, dR = \frac{\varepsilon \zeta}{c^2}.$$

From the figure we immediately find that an element of the volume is

$$abcR^2 \sin \theta \, dR \, d\theta \, d\psi = \varepsilon d\sigma,$$

where $d\sigma$ is an element of the surface of the ellipsoid R, and this means that

$$\frac{N}{r^3} R^2 \sin \theta \, dR \, d\theta \, d\psi$$

$$= \frac{N}{r^3} \frac{\varepsilon d\sigma}{abc} = \frac{\varepsilon d\sigma}{r^3 abc} \left[\frac{(\xi - x)\xi}{a^2} + \frac{(\eta - y)\eta}{b^2} + \frac{(\zeta - z)\zeta}{c^2} \right]$$

$$= \frac{d\sigma}{r^2 abc} \left[\frac{\xi - x}{r} \frac{\varepsilon \xi}{a^2} + \frac{\eta - y}{r} \frac{\varepsilon \eta}{b^2} + \frac{\zeta - z}{r} \frac{\varepsilon \zeta}{c^2} \right] = \frac{d\sigma}{abc} \frac{\cos(r, n)}{r^2} R \, dR.$$

Consequently (see (8.3)),

$$\delta \left(\frac{U}{abc} \right) = - \frac{\varrho \delta \lambda}{2abc} \int_0^1 \left[\int \frac{\cos(r, n) \, d\sigma}{r^2} \right] R \, dR. \qquad (8.4)$$

247. Assume that the point (x, y, z) being attracted lies outside the ellipsoid $R = 1$. It also lies outside any similar ellipsoid $0 \leqslant R \leqslant 1$. According to Gauss' theorem, we have the following relation for R on the interval $(0, 1)$:

$$\int \frac{\cos(r, n) \, d\sigma}{r^2} = 0.$$

Formula (8.4) yields Maclaurin's theorem, namely,

$$\delta \left(\frac{U}{abc} \right) = 0, \quad \frac{U}{abc} = \text{const.}$$

248. Assume now that the point (x, y, z) lies within the ellipsoid $R = 1$. There is one ellipsoid $R = R_1$, similar to the given one and passing through the point (x, y, z):

$$\frac{x^2}{a^2} + \frac{y^2}{b^2} + \frac{z^2}{c^2} = R_1^2 < 1.$$

The point (x, y, z) is interior for the ellipsoids $R > R_1$. According to Gauss' theorem,

$$\int \frac{d\sigma \cos(r, n)}{r^2} = 4\pi.$$

For the ellipsoids $R < R_1$ the point (x, y, z) is exterior and, consequently,

$$\int_0^{R_1} \left[\int \frac{d\sigma \cos(r, n)}{r^2} \right] R \, dR = 0.$$

This means that

$$\delta\left(\frac{U}{abc}\right) = -\frac{2\pi\varrho\delta\lambda}{abc} \int_{R_1}^1 R \, dR = \frac{\pi\varrho\delta\lambda}{abc}(1 - R_1^2)$$

$$= -\frac{\pi\varrho\delta\lambda}{abc}\left(1 - \frac{x^2}{a^2} - \frac{y^2}{b^2} - \frac{z^2}{c^2}\right).$$

249. Having prepared the formulas we need, we begin calculating the potential U. Assume that we have to find the potential of the ellipsoid with the semi-major, semi-mean, and semi-minor axes α, β, γ respectively. The whole space, which is exterior with respect to the ellipsoid, can be covered with cofocal ellipsoids whose semi-major, semi-mean, and semi-minor axes are $\sqrt{\alpha^2 + \lambda}$, $\sqrt{\beta^2 + \lambda}$, $\sqrt{\gamma^2 + \lambda}$ respectively. The value $\lambda = 0$ corresponds to the given ellipsoid, while $\lambda = \infty$ corresponds to an infinitely large ellipsoid. We know how the potential changes from the given ellipsoid to a cofocal one. Consequently, it is sufficient to know the value of U for any one of the cofocal ellipsoids to determine the potential of any ellipsoid between $0 \leqslant \lambda \leqslant \infty$. The expression $\lambda = \infty$ represents the whole space filled with the mass ϱ. When we pass from the ellipsoid λ to the ellipsoid $\lambda + d\lambda$, we use the preceding formulas. Assume that the point (x, y, z) lies within the ellipsoid $\lambda = 0$. It lies within any ellipsoid $0 \leqslant \lambda \leqslant \infty$. We set $a^2 = \alpha^2 + \lambda$, $b^2 = \beta^2 + \lambda$, $c^2 = \gamma^2 + \lambda$ and have

$$d\frac{U}{\sqrt{(\alpha^2 + \lambda)(\beta^2 + \lambda)(\gamma^2 + \lambda)}} = -\frac{\pi\varrho \, d\lambda}{\sqrt{(\alpha^2 + \lambda)(\beta^2 + \lambda)(\gamma^2 + \lambda)}}$$

$$\times \left(1 - \frac{x^2}{\alpha^2 + \lambda} - \frac{y^2}{\beta^2 + \lambda} - \frac{z^2}{\gamma^2 + \lambda}\right),$$

whence, if we set $\Delta = \sqrt{(\alpha^2 + \lambda)(\beta^2 + \lambda)(\gamma^2 + \lambda)}$, we get

$$\left[\frac{U}{\Delta}\right]_\infty^0 = -\pi\varrho \int\limits_\infty^0 \left(1 - \frac{x^2}{\alpha^2 + \lambda} - \frac{y^2}{\beta^2 + \lambda} - \frac{z^2}{\gamma^2 + \lambda}\right)\frac{d\lambda}{\Delta}.$$

But $(U/\Delta)_{\lambda = \infty} = 0$, and we have

$$U = \pi\varrho\alpha\beta\gamma \int\limits_0^\infty \left(1 - \frac{x^2}{\alpha^2 + \lambda} - \frac{y^2}{\beta^2 + \lambda} - \frac{z^2}{\gamma^2 + \lambda}\right)\frac{d\lambda}{\Delta}.$$

Assume now that the point (x, y, z) lies outside the ellipsoid $\lambda = 0$. Then there will be one cofocal ellipsoid $\lambda = \lambda_0$, which passes through the point (x, y, z). For the ellipsoids $0 \leqslant \lambda \leqslant \lambda_0$ the point (x, y, z) is exterior and $\delta(U/\Delta) = 0$. For the ellipsoids $\lambda_0 < \lambda \leqslant \infty$ the point (x, y, z) is interior and, hence, for $\lambda_0 < \lambda \leqslant \infty$ we have

$$d\left(\frac{U}{\Delta}\right) = -\frac{\pi\varrho d\lambda}{\Delta}\left(1 - \frac{x^2}{\alpha^2 + \beta} - \frac{y^2}{\beta^2 + \lambda} - \frac{z^2}{\gamma^2 + \lambda}\right).$$

Integrating this formula from λ_0 to ∞ and using Maclaurin's theorem

$$\left[\frac{U}{\Delta}\right]_{\lambda_0}^0 = 0$$

we immediately obtain

$$U = \pi\varrho\alpha\beta\gamma \int\limits_{\lambda_0}^\infty \left(1 - \frac{x^2}{a^2 + \lambda} - \frac{y^2}{b^2 + \lambda} - \frac{z^2}{c^2 + \lambda}\right)d\lambda.$$

8.3 General Properties of Harmonic Functions

250. Functions which satisfy Laplace's equation play a very significant part in mechanics and physics. We shall restrict our discussion to continuous functions which are single-valued together with their partial derivatives up to the second order inclusive.

From Green's formula

$$\iiint\limits_V \Delta U d\tau = \iint\limits_S \frac{\partial U}{\partial n}\, d\sigma,$$

for an infinitesimal volume $d\tau$ containing a point (x, y, z) we obtain

$$\Delta U = \frac{1}{d\tau} \iint\limits_S \frac{\partial U}{dn}\, d\sigma,$$

and this proves that ΔU is invariant (independent of the choice of the orthogonal axes).

If the function U is harmonic in the volume V and finite together with its partial derivatives of the first and the second order, then from Green's formula we get

$$\iint_S \frac{\partial U}{\partial n}\, d\sigma = 0.$$

This relation yields a number of corollaries.

If the harmonic function U is finite together with its derivatives up to the second order in the volume V, then it has neither an isolated maximum nor an isolated minimum in this volume.

Indeed, let U be a maximum at a point P. We encircle P by a sphere S of a very small radius. In this case

$$\iint_{S'} \frac{\partial U}{\partial n}\, d\sigma$$

is negative (n is an exterior normal), and this contradicts the property we have proved. Consequently, U does not have an isolated maximum (or minimum) inside V. Hence, if U is the potential of a Newtonian attraction force, then it follows that U does not have a maximum or minimum outside the attracting masses, and according to the converse of Lagrange's theorem there is no stable equilibrium in the Newtonian force field outside the attracting masses.

In just the same way we can prove that a harmonic function cannot have a constant value larger or smaller than all the adjacent (close) values either on a curve, or on a surface, or in a volume. It follows that if the harmonic function U is zero on the surface S, then it is zero in the volume V inside S if in the volume V the function U is harmonic everywhere and finite together with its partial derivatives.

Corollary. If two functions U_1 and U_2 are finite together with their derivatives of the first and the second order in the volume V and assume the same values on a bounding surface, then they are identical within the volume V. A harmonic function, which is single-valued, finite, and definite together with its derivatives up to the second order inclusive at all finite as well as infinite points of space, is constant.

251. Dirichlet's principle. There is a single-valued, harmonic, and finite function with the same partial derivatives up to the second order inclusive in a domain V which assumes preassigned continuous values on a bounded surface S.

Riemann proved the existence of such a function by an empirical physical experiment.

Dirichlet proved it as follows. Assume that φ is a function of x, y, z which is finite together with its partial derivatives of the first and the second order in the domain V and assumes the preassigned values on S and let

$$I_\varphi = \iiint_V \left[\left(\frac{\partial\varphi}{\partial x}\right)^2 + \left(\frac{\partial\varphi}{\partial y}\right)^2 + \left(\frac{\partial\varphi}{\partial z}\right)^2 \right] d\tau.$$

There will be one function (let it be φ) for which the integral I is at a minimum. Other functions of this kind have the form

$$\Phi = \varphi + \varepsilon\psi,$$

where ψ is a function which is arbitrary in the domain V and assumes zero values on S. We have

$$I_\Phi = \iiint_V \left[\frac{\partial(\varphi + \varepsilon\psi)^2}{\partial x} + \ldots \right] d\tau = I_\psi + 2\varepsilon H + \varepsilon^2 I_\psi,$$

where
$$H = \iiint_V \left(\frac{\partial\varphi}{\partial x}\frac{\partial\psi}{\partial x} + \frac{\partial\varphi}{\partial y}\frac{\partial\psi}{\partial y} + \frac{\partial\varphi}{\partial z}\frac{\partial\psi}{\partial z} \right) d\tau.$$

From the condition that I_φ is a minimum it follows that $H = 0$ for an arbitrary function ψ. Green's theorem yields

$$H = \iint_S \psi\,\frac{\partial\varphi}{\partial n}\,d\sigma - \iiint_V \psi\Delta\varphi d\tau.$$

On the surface S the function ψ is zero and the first integral is eliminated. From the condition

$$H = -\iiint_V \psi\Delta\varphi d\tau = 0$$

for an arbitrary ψ we get

$$\Delta\varphi = 0,$$

i.e. the function φ for which I_φ is at a minimum is harmonic. This is precisely Dirichlet's proof. Weierstrass demonstrated that this proof of the existence of a function φ for which I_φ is at a minimum is not rigorous.

252. Neumann's problem. Assume that $\partial U/\partial n$ is zero on the surface S and the function U is harmonic inside S. From Green's formula

$$\iiint_V U\Delta U d\tau + \iiint_V \left[\left(\frac{\partial U}{\partial x}\right)^2 + \left(\frac{\partial U}{\partial y}\right)^2 + \left(\frac{\partial U}{\partial z}\right)^2 \right] d\tau = \iint_S U\,\frac{\partial U}{\partial n}\,d\sigma$$

we find that $\dfrac{\partial U}{\partial x} = \dfrac{\partial U}{\partial y} = \dfrac{\partial U}{\partial z} = 0$ inside S, i.e., the function U is constant.

Corollary. Assume that in the volume V two functions U_1 and U_2 are finite together with their derivatives of the first two orders and let $\dfrac{\partial U_1}{\partial n} = \dfrac{\partial U_2}{\partial n}$ on the bounding surface S. Then the difference $U_1 - U_2$ inside V is constant. To put it another way, the specification of $\partial U/\partial n$ on S fully defines U inside V with an accuracy of an additive constant.

253. Determining a harmonic function U from the given U and $\partial U/\partial n$ on a surface. Let a, b, c be the coordinates of a point A inside V (Fig. 159). We encircle A in a sphere σ of a very small radius. Green's formula yields

$$\iint\limits_{S} \left(\frac{1}{r} \frac{\partial U}{\partial n} - U \frac{\partial(1/r)}{\partial n} \right) d\sigma + \iint\limits_{\sigma} \left(\frac{1}{r} \frac{\partial U}{\partial n} - U \frac{\partial(1/r)}{\partial n} \right) d\sigma = 0.$$

For the sphere σ we have

$$\iint\limits_{\sigma} \frac{1}{r} \frac{\partial U}{\partial n} \, d\sigma = \frac{1}{r} \iint\limits_{\sigma} \frac{\partial U}{\partial n} \, d\sigma = -\frac{1}{r} \iiint \Delta U d\tau = 0$$

and

$$\iint\limits_{\sigma} U \frac{\partial(1/r)}{\partial n} \, d\sigma = -\frac{U}{r^2} \iint\limits_{\sigma} d\sigma = -4\pi U(a, b, c).$$

Consequently,

$$U(a, b, c) = \frac{1}{4\pi} \iint\limits_{S} \left(\frac{1}{r} \frac{\partial U}{\partial n} - U \frac{\partial(1/r)}{\partial n} \right) d\sigma.$$

254. The mean value of a harmonic function on a spherical surface is equal to the value of that function at the centre of the sphere. Indeed, assume that S is a sphere of radius r and a, b, c are the coordinates of the centre of S. The preceding formula yields

$$U(a, b, c) = \frac{1}{4\pi r} \iint\limits_{S} \frac{\partial U}{\partial n} \, d\sigma + \frac{1}{4\pi r^2} \iint\limits_{S} U d\sigma = \frac{1}{4\pi r^2} \iint\limits_{S} U d\sigma.$$

255. Green's function. The function $G(x, y, z, a, b, c)$ is called a *Green function* if (1) it is harmonic within the volume V bounded by S, (2) it is finite together with its derivatives of the first two orders within V, (3) it assumes the value $1/r$ on the surface S. Using the relation

$$\iint \left(G \frac{\partial U}{\partial n} - U \frac{\partial G}{\partial n} \right) d\sigma = 0,$$

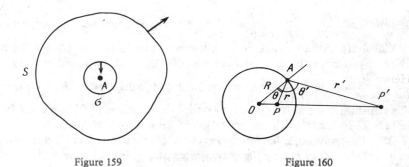

Figure 159   Figure 160

we obtain

$$U(a, b, c) = \frac{1}{4\pi} \int\int_S U \left(\frac{\partial G}{\partial n} - \frac{\partial (1/r)}{\partial n} \right) d\sigma.$$

Using Green's function, it is easy to solve Dirichlet's problem.

Example. Assume that R is the radius of the sphere S, O is its centre, a, b, c are the coordinates of a point P within the sphere, l is the distance OP, $l' = OP'$, where P' is a conjugate of P, which is harmonic with respect to the endpoints of the diameter OP (Fig. 160),

$$\frac{1}{r} = \frac{R}{OP} \frac{1}{r'}.$$

From this we can find Green's function for the sphere. It is

$$G = \frac{R}{OP} \frac{1}{r'}.$$

Hence the solution of Dirichlet's problem is

$$U(a, b, c) = -\frac{1}{4\pi} \int\int U \left(\frac{R}{l} \frac{\partial (1/r')}{\partial n} - \frac{\partial (1/r)}{\partial n} \right) d\sigma,$$

but

$$\frac{\partial (1/r')}{\partial n} = \frac{1}{r'^2} \cos \theta', \quad \frac{\partial (1/r)}{\partial n} = \frac{1}{r^2} \cos \theta,$$

and

$$l^2 = R^2 + r^2 - 2rR \cos \theta,$$
$$R^2 = l^2 + r^2 - 2rl \cos \theta', \quad l^2 - R^2 = r(-R \cos \theta + l \cos \theta').$$

We have

$$\frac{R \cos \theta'}{r'^2} - \frac{\cos \theta}{r^2} = - \frac{R^2 - l^2}{Rr^3}.$$

Hence

$$U(a, b, c) = \frac{1}{4\pi R} \iint\limits_{S} \frac{R^2 - l^2}{r^3} \, U d\sigma.$$

8.4 Total Energy

256. The entire force function of a system of n particles with masses m_i and with the distance r_{ik} between the particles m_i and m_k is

$$W = \sum_{k=1}^{n} m_k \left(\sum_{i=k+1}^{n} \frac{m_i}{r_{ik}} \right) = \frac{1}{2} \sum_{1}^{n} m_k \left(\sum \frac{m_i}{r_{ik}} \right) = \frac{1}{2} \sum_{1}^{n} m_k U_k.$$

From this we find that for a continuous body of density ϱ the total energy is

$$W = \int \varrho \, \frac{U}{2} \, d\tau.$$

The work done by the force of mutual attraction in deforming a body is expressed by the formula

$$\delta W = \int \frac{\delta \varrho U + \varrho \delta U}{2} \, d\tau.$$

257. Poincaré's formula. If U and V are two functions which are continuous within the volume T bounded by a surface S, then

$$\iint\limits_{S} U \frac{\partial V}{\partial n_e} \, d\sigma \equiv \iiint\limits_{T} U \Delta V d\tau + \iiint\limits_{T} \left(\frac{\partial U}{\partial x} \frac{\partial V}{\partial x} + \frac{\partial U}{\partial y} \frac{\partial V}{\partial y} + \frac{\partial U}{\partial z} \frac{\partial V}{\partial z} \right) d\tau.$$

If U and V are the potentials of bounded masses, then at the infinity $(R \to \infty)$ U and V have order $1/R$ and their first derivatives are of order $1/R^2$. Then Green's formula can be used in the region T' between the surface S and a sphere of an infinitely large radius R. The integral $\iint\limits_{S} U \frac{\partial V}{\partial n} \, d\sigma$ along a sphere of an infinitely large radius in the limit of $R \to \infty$ is zero and therefore

$$\iint\limits_{S} U \frac{\partial V}{\partial n_i} \, d\sigma = \iiint\limits_{T'} U \Delta V d\tau + \iiint\limits_{T'} \left(\frac{\partial U}{\partial x} \frac{\partial V}{\partial x} + \frac{\partial U}{\partial y} \frac{\partial V}{\partial y} + \frac{\partial U}{\partial z} \frac{\partial V}{\partial z} \right) d\tau.$$

Adding the two formulas together, we have

$$\iiint U\Delta V d\tau + \iiint \left(\frac{\partial U}{\partial x}\frac{\partial V}{\partial x} + \frac{\partial U}{\partial y}\frac{\partial V}{\partial y} + \frac{\partial U}{\partial z}\frac{\partial V}{\partial z}\right) d\tau = 0,$$

where the integration is over the whole space. Exchanging U and V, we obtain

$$\iiint (U\Delta V - V\Delta U) d\tau = 0,$$

or, if we take into account Poisson's equations $\Delta U = -4\pi\varrho$ and $\Delta V = -4\pi\varrho'$, then

$$\iiint (\varrho V - \varrho' U) d\tau = 0,$$

i.e. the integral is over the whole space.

Let U be the potential of a slightly deformable body K for which

$$V = U + \delta U, \quad \varrho' = \varrho + \delta\varrho.$$

The last formula yields

$$\iiint (\varrho\delta U - U\delta\varrho) d\tau = 0.$$

Problem 1. Show that

$$W = \frac{1}{8\pi} \iiint \left[\left(\frac{\partial U}{\partial x}\right)^2 + \left(\frac{\partial U}{\partial y}\right)^2 + \left(\frac{\partial U}{\partial z}\right)^2\right] d\tau.$$

Problem 2. Show that

$$\delta W = \frac{1}{2} \iiint U\delta\varrho \, d\tau = \iiint \frac{1}{2}\varrho\delta U \, d\tau.$$

Problem 3. Assume that a homogeneous incompressible body of density ϱ is deformed and ζ is a segment along an exterior normal to S between the nondeformable surface S and the surface S' which bounds the body which can be infinitesimally deformed. Show that $\delta W = \iint \varrho U \zeta \, d\sigma$.

Problem 4. Show that a body on whose surface the potential U is constant has an extremum in W, provided that the medium is incompressible and homogeneous.

Problem 5. Show that the total energy W of a finite, homogeneous, incompressible body is bounded from above. Show that when we have Steiner's symmetry, the total energy increases and W is at a maximum for a sphere.

APPENDIX

A.1 Analogy Between Optics and Mechanics

Most of the results obtained in dynamics after Lagrange are associated with the analogy between mechanics and optics, which Hamilton discovered in 1824*.

1. The following principle lies at the basis of Huygens' wave theory. If Σ is the front of a light wave at an instant t (Fig. 161), then every point P on this front must be regarded as a source of a new, secondary wave and the front of the wave Σ', by the moment t' ($t' > t$) is an envelope of these secondary waves.

Hamilton developed an analytical notation for Huygens' principle.

The time needed for light to pass between the points $P(x, y, z)$ and $P'(x', y', z')$ in an isotropic medium depends only on their coordinates, i.e.

$$t' - t = V(x, y, z, x', y', z').$$

Hamilton called V a *characteristic function**.

Assume that Σ is the front of the wave at time t, Σ' is the front of the wave at some subsequent moment t', α, β, γ are the cosines of the normal to the wave Σ at the point P, and α', β', γ' are the cosines of the normal to Σ' at the point P'.

In accordance with Huygens' principle, the front Σ' is an envelope of the secondary waves emanating from various points P of the front Σ at the moment t. The equation of a secondary wave, which emanates at time t from the point (x, y, z) by the moment t' becomes

$$t' - t = V(x, y, z, x', y', z'),$$

where x', y', z' are running coordinates. To find the envelope of the secondary waves, we must exclude the parameters x, y, z from their equations. Then we get a relation

$$\frac{\partial V}{\partial x} \delta x + \frac{\partial V}{\partial y} \delta y + \frac{\partial V}{\partial z} \delta z = 0$$

---

* W.R. Hamilton, *Transac. Royal Irish Acad.*,**15**: 69 (1828); **16**: 4, 93 (1833); **17**: 1 (1837).

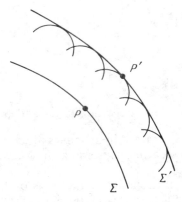

Figure 161

for any displacement δx, δy, δz along the front Σ, which are naturally orthogonal to their normal:

$$\alpha\delta x + \beta\delta y + \gamma\delta z = 0.$$

It follows from these two relations that

$$\frac{\partial V/\partial x}{\alpha} = \frac{\partial V/\partial y}{\beta} = \frac{\partial V/\partial z}{\gamma} = \varkappa \quad (\alpha^2 + \beta^2 + \gamma^2 = 1), \tag{1}$$

with the ratio $\varkappa$ depending only on x, y, z. We put a point (x, y, z) on the secondary wave σ into correspondence with a point (x', y', z') which belongs to the envelope Σ' of the secondary waves, i.e. the point P' at which the normal to the secondary wave σ coincides with the normal to the front Σ'. For these normals to be parallel

$$\frac{\partial V/\partial x'}{\alpha'} = \frac{\partial V/\partial y'}{\beta'} = \frac{\partial V/\partial z'}{\gamma'} = \lambda \quad (\alpha'^2 + \beta'^2 + \gamma'^2 = 1), \tag{2}$$

the ratio λ depending only on x', y', z'. The sequence of points (x', y', z') constructed in this way for different times t' is known as a *light ray*.

That relations (1) and (2) have the same form proves the principle of recursion, i.e. that the reverse motion of a wave Σ' with respect to Σ can also be obtained according to Huygens' principle and that in the case of the symmetry of the characteristic function $V(x, y, z, x', y', z') = V(x', y', z', x, y, z)$ the reverse motion can also be obtained according to Huygens' principle as a straight wave emanating from a point of the front Σ' but going in the reverse direction.

The total differential of the characteristic function V is

$$dV = \varkappa(\alpha dx + \beta dy + \gamma dz) + \lambda(\alpha' dx' + \beta' dy' + \gamma' dz'). \tag{3}$$

To find the meaning of the ratios $\varkappa$ and λ, we designate the quantity which is the inverse of the speed of light v at the point (x, y, z) as μ and a similar quantity at the point (x', y', z') as μ'. When the point P is displaced (dx, dy, dz) along a ray of light in the positive direction by ds and the point P' is displaced (dx', dy', dz') along a ray of light ds', we have

$$dV = dt' - dt = \frac{ds'}{v'} - \frac{ds}{v} = \mu' ds' - \mu ds$$
$$= \mu'(\alpha'^2 + \beta'^2 + \gamma'^2)ds' - \mu(\alpha^2 + \beta^2 + \gamma^2)ds$$
$$= \mu'(\alpha' dx' + \beta' dy' + \gamma' dz') - \mu(\alpha dx + \beta dy + \gamma dz).$$

Consequently, a comparison with formula (3) yields

$$\lambda = \mu', \quad \varkappa = -\mu.$$

If, instead of the cosines $\alpha, \beta, \gamma, \alpha', \beta', \gamma'$, we introduce new variables

$$\xi = \mu\alpha, \quad \eta = \mu\beta, \quad \zeta = \mu\gamma, \quad \xi' = \mu'\alpha', \quad \eta' = \mu'\beta', \quad \zeta' = \mu'\gamma',$$

then, substituting λ and $\varkappa$ into (3), we find that the variables $x, y, z, \xi,$ η, ζ and $x', y', z', \xi', \eta', \zeta'$, which are related by (1) and (2), satisfy the condition of a canonical, or contact, transformation* valid for arbitrary $dx, dy, dz, dx', dy', dz'$.

$$dV = \xi' dx' + \eta' dy' + \zeta' dz' - (\xi dx + \eta dy + \zeta dz).$$

2. An analogy between the two phenomena arises when the groups of transformations corresponding to them coincide or differential equations which govern the phenomena are similar.

For a mechanical system subjected to holonomic constraints and acted by forces which have a force function U the action

$$V(t, q_1, \ldots, q_k, q_1^0, \ldots, q_k^0) = \int_{t_0}^{t} (T + U)dt$$

satisfies the relation

$$\delta V = \sum p_i \delta q_i - \sum p_i^0 \delta q_i^0. \tag{4}$$

---

* Chetaev used the terms "canonical" and "contact" transformations as synonyms. — *Ed.*

This relation makes us infer that the transformation involving q_i and p_i and their initial values q_i^0, p_i^0 constitutes a group of contact transformations. The action here is a characteristic function.

We can choose the moments t_0 and t arbitrarily, and therefore we can regard the motion of a holonomic, conservative mechanical system as a chain of contact transformations.

Consequently, the group of motions of the holonomic system is the same as the group of the propagation of light in an isotropic medium according to Huygens' wave theory. This is the analogy between optics and mechanics Hamilton discovered.

We can note that (4) is significant in principle. Assume that the moment t is infinitely close to the moment t_0, i.e. $t = t_0 + dt$. Then the action $V = (T + U)dt$ and, with an accuracy to within the second-order quantities with respect to dt, we have

$$q_i^0 = q_i - \frac{dq_i}{dt} dt + \ldots, \quad p_i^0 = p_i - \frac{dp_i}{dt} dt + \ldots.$$

With the same degree of accuracy relation (4) yields

$$\delta(T + U)dt = \sum \left(\frac{dp_i}{dt} \delta q_i + p_i \delta \frac{dq_i}{dt} \right) dt + \ldots.$$

Cancelling dt from the relation and taking the limit as $dt \to 0$, we get

$$\sum \left(\frac{\partial H}{\partial q_i} + \frac{dp_i}{dt} \right) \delta q_i + \sum \left(\frac{\partial H}{\partial p_i} - \frac{dq_i}{dt} \right) \delta p_i = 0, \tag{5}$$

where $H = \Sigma p_i q_i' - T - U$. Relation (5) must hold true for arbitrary δq_i, δp_i and therefore (5) yields the canonical equations of motion

$$\frac{dq_i}{dt} = \frac{\partial H}{\partial p_i}, \quad \frac{dp_i}{dt} = -\frac{\partial H}{\partial q_i}.$$

That we can derive the equations of motion from (4) makes us infer that the analogy between mechanics and optics not only results from the special properties of the motion of mechanical systems, but is also an independent principle of dynamics which completely governs the motions of a holonomic mechanical system acted on by forces with a force function.

Hamilton's principle (4), which contains the geometric construction of a trajectory in accordance with Huygens' wave theory, enables us, as does other principle of dynamics, to cover in detail the methods of integrating the differential equations of motion.

The action is known for infinitesimal interval of time dt. The construction of secondary waves which refers to this action only defined the way the front Σ varies.

If the action V is known for the whole time interval, then we can construct the secondary waves for the whole time interval and thus easily determine the trajectory. This follows from principle (4). Writing the expression for δV explicitly and comparing the terms with δq_i and δq_i^0, we have

$$p_i = \frac{\partial V}{\partial q_i}, \quad -p_i^0 = \frac{\partial V}{\partial q_i^0}.$$ (6)

This system of equations defines the trajectory.

3. Canonical, or contact, transformations possess a rather useful property which makes the integration of the equations of dynamics easier. Instead of the old variables q_i and p_i, we introduce new variables α_i and β_i which satisfy the condition of the contact transformation

$$\sum p_i \delta q_i - \sum \beta_i \delta \alpha_i = \delta W,$$

or

$$p_i = \frac{\partial W}{\partial q_i}, \quad -\beta_i = \frac{\partial W}{\partial \alpha_i},$$ (7)

where the characteristic function $W(t, q_1, \ldots, q_n, \alpha_1, \ldots, \alpha_n)$ is assumed to be a single-valued function of time t and the variables $q_1, \ldots, q_n$ and $\alpha_1, \ldots, \alpha_n$.

In the new variables α_i and β_i Hamilton's principle

$$\delta \int_{t_0}^{t_1} \left(\sum p_s dq_s - H dt \right) = 0$$

assumes the form

$$\delta \int_{t_0}^{t_1} \left(\sum \beta_s d\alpha_s - H^* dt \right) = 0,$$

where the new function $H^* = \frac{\partial W}{\partial t} + H$ is assumed to be a function of t and the new variables $\alpha_1, \ldots, \alpha_n$ and $\beta_1, \ldots, \beta_n$ since

$$\sum p_s dq_s = \sum \beta_s d\alpha_s + dW - \frac{\partial W}{\partial t} dt$$

and for a fixed position of the system at t_0 and t_1 (for the fixed endpoints in Hamilton's principle of least action) we have

$$\delta \int_{t_0}^{t_1} dW = \delta(W_1 - W_0) = 0.$$

In the new variables the differential equations of motion of the mechanical system have the canonical form since the principle of least action has the same form as in the initial variables, i.e.

$$\frac{d\alpha_i}{dt} = \frac{\partial H^*}{\partial \beta_i}, \quad \frac{d\beta_i}{dt} = -\frac{\partial H^*}{\partial \alpha_i} \quad (i = 1, \ldots, n). \tag{8}$$

These equations acquire the simplest form when the function H^* is zero. The characteristic function for the transformation W must then satisfy for relations (7) the equality

$$\frac{\partial W}{\partial t} + H = 0$$

or the first-order partial differential equation

$$\frac{\partial W}{\partial t} + H\left(t, q_1, \ldots, q_n, \frac{\partial W}{\partial q_1}, \ldots, \frac{\partial W}{\partial q_n}\right) = 0. \tag{9}$$

But W must depend on $q_1, \ldots, q_n, \alpha_1, \ldots, \alpha_n$. Therefore W must be a complete integral of equation (9). All that we have said above leads to the theorem* derived by Jacobi, which states that *if $W(t, q_1, \ldots, q_n, \alpha_1, \ldots, \alpha_n)$ is a complete integral of the first-order partial differential equation* (9), *then the solution of the dynamics problem is defined by* (7) *in which, according to* (8), *the quantities $\alpha_1, \ldots, \alpha_n, \beta_1, \ldots, \beta_n$ are constants.*

4. In many problems of dynamics the search for the complete integral of the Hamilton-Jacobi equation (9) can be simplified by Imshenetsky's substitution**.

5. For problem solving, Lagrange developed a general approximate method in dynamics based on the variation of arbitrary constants***.

The complications arising in the integration of the equations of motion often make it necessary to use approximations.

Every approximation presupposes an exact solution of a problem when some quantities are assumed to be small enough to be discarded while the remaining principal forces acting on the system are retained. This approximation is corrected by gradually accounting for the disregarded quantities.

Assume that a mechanical system is acted by smooth holonomic constraints and by forces with a force function. Assume that q_s and p_s are its coordinates and impulses, T is its kinetic energy, H_0 is the Hamiltonian function when the principal forces have a force function U, and W is the force function of the perturbing or discarded forces in the approximation.

---

* K.G.J. Jacobi, *Vorlesungen über Dynamik*, Zweite, revidirte Ausgabe (Druck and Verlag von. G. Reimer, Berlin: 1884).
** Chap. 7, p. 244.
*** See Dynamique, Sec. 5 in Lagrange's *Méchanique Analitique* (Paris: 1788).

We assume the approximate solution to be defined in terms of the formulas for the *contact* transformation

$$\sum p_s \delta q_s - \sum \beta_s \delta \alpha_s = \delta V_s \tag{10}$$

whose characteristic function $V(t, q_1, \ldots, q_n, \alpha_1, \ldots, \alpha_n)$ is the complete integral of the first-order partial differential equation

$$\frac{\partial V}{\partial t} + H_0 \left(t, q_1, \ldots, q_n, \frac{\partial V}{\partial q_1}, \ldots, \frac{\partial V}{\partial q_n} \right) = 0. \tag{11}$$

We can now see whether α_j and β_j (the constants in the approximate motion defined by H_0) can be replaced by some time functions so that when substituting these functions into the solution for the unperturbed motion

$$p_s = \frac{\partial V}{\partial q_s}, \quad -\beta_s = \frac{\partial V}{\partial \alpha_s} \tag{12}$$

we would have solutions for the perturbed motion.

This problem was solved by Lagrange. When perturbing forces with a force function W act on a system, the differential equations of motion of the mechanical system are defined by the principle of least action

$$\delta \int_{t_0}^{t_1} \left[\sum p_s dq_s - (H_0 - W)dt \right] = 0.$$

We pass to new variables α_j and β_j according to the formulas for the contact transformation (10)

$$\delta \int_{t_0}^{t_1} \left[\sum \beta_j d\alpha_j - \left(\frac{\partial V}{\partial t} + H_0 - W \right) dt \right] = 0.$$

According to (11) we have

$$\delta \int_{t_0}^{t_1} \left[\sum \beta_j d\alpha_j + W dt \right] = 0,$$

or

$$\frac{d\alpha_j}{dt} = - \frac{\partial W}{\partial \beta_j}, \quad \frac{d\beta_j}{dt} = \frac{\partial W}{\partial \alpha_j} \quad (j = 1, \ldots, n) \tag{13}$$

where the variables q_s and p_s in the function W are replaced by their values depending on the new variables α_j, β_j according to formulas (12).

Equations (13) are Lagrange's fundamental equations in the perturbation theory.

Questions. When can we restrict our calculations to the first approximation without exceeding a permissible error? When will variations α_j, β_j be stable?

6. For a holonomic mechanical system acted on by forces with a force function, Poincaré* derived equations for the perturbed motion when the perturbations are induced by small deviations of the initial values of coordinates q_s and impulses p_s.

Assume that for such a system H is the Hamiltonian function. We isolate a motion from the motions of this system which we accept as the unperturbed motion and which is expressed by the particular solution

$$q_s = q_s(t), \quad p_s = p_s(t)$$

of the equations of motion

$$\frac{dq_i}{dt} = \frac{\partial H}{\partial p_i}, \quad \frac{dp_i}{dt} = - \frac{\partial H}{\partial q_i}. \tag{14}$$

Let us consider a motion which is close to the unperturbed motion. We agree to write its coordinates and impulses in the form

$$q_s = q_s(t) + \xi_s, \quad p_s = p_s(t) + \eta_s,$$

where ξ_s and η_s are the deviations or variations of the coordinates and impulses. A perturbed motion is a motion of the mechanical system which begins with other initial values of q_s and p_s and continues due to the action of the same forces. Therefore its coordinates and impulses must satisfy equations (14), i.e.

$$\frac{d[q_i(t) + \xi_i]}{dt} = \frac{\partial H(t, \, q_s(t) + \xi_s, \, p_s(t) + \eta_s)}{\partial p_i},$$

$$\frac{d[p_i(t) + \eta_i]}{dt} = - \frac{\partial H(t, \, q_s(t) + \xi_s, \, p_s(t) + \eta_s)}{\partial q_i}.$$

After expanding the right-hand sides in Taylor's series with respect to the small values of ξ_s, η_s and using (14), we get, in the first approximation, Poincaré's equations in variations

$$\frac{d\xi_i}{dt} = \sum_s \left(\frac{\partial^2 H}{\partial p_i \partial q_s} \xi_s + \frac{\partial^2 H}{\partial p_i \partial p_s} \eta_s \right),$$

$$\frac{d\eta_i}{dt} = - \sum_s \left(\frac{\partial^2 H}{\partial q_i \partial q_s} \xi_s + \frac{\partial^2 H}{\partial q_i \partial p_s} \eta_s \right). \tag{15}$$

---

* H. Poincaré, "Sur la problème de trois corps", *Acta Math.* **13**: 1-270 (1890).

These equations possess a number of properties closely related to Hamilton's optico-mechanical analogy.

If ξ_s, η_s and ξ_s', η_s' are two solutions of the equations in variations (15), then

$$\sum(\xi_s\eta_s' - \xi_s'\eta_s) = \text{const.} \tag{16}$$

This proposition can be proved by direct differentiation with respect to t:

$$\frac{d}{dt} \sum (\xi_s\eta_s' - \xi_s'\eta_s) = \ldots = 0.$$

This proposition is attributed to Lagrange*.

We can note that equations (15) can also be obtained using Lagrange's arguments. Indeed, if we substitute the general solution of (14), which depends on $2n$ constants, into (14), we get an identity which is independent of the numerical values of these constants. This identity can be differentiated with respect to the constants. Differentiating once and denoting the differentials of the coordinates by ξ_s and the differentials of the impulses by η_s, we get equations in variations (15).

(We find the connection between the invariant (16) and the analogy between optics and mechanics if we note that the left-hand side of (16) is another expression for the bilinear covariant of Hamilton's principle of analogy

$$\sum(\Delta p_s\delta q_s - \Delta q_s\delta p_s).$$

But we shall discuss this later on.)

Invariant (16) shows that the solutions of equations in variations are interrelated. This relation yields new integrals or solutions in some cases.

By direct differentiation we can prove that if equations (15) have a linear integral

$$\sum(A_s\xi_s + B_s\eta_s) = \text{const},$$

they have a particular solution

$$\xi_s' = -B_s, \quad \eta_s' = A_s.$$

If $\varphi(t, q_s, p_s) = \text{const}$ is an integral of the canonical equations (14), then the expression

$$\sum \left(\frac{\partial\varphi}{\partial q_s} \xi_s + \frac{\partial\varphi}{\partial p_s} \eta_s \right) = \text{const}$$

---

* See Lagrange's *Méchanique analitique* (Paris: 1788).

is a linear integral of equations (15), and

$$\xi_s' = - \frac{\partial \varphi}{\partial p_s}, \quad \eta_s' = \frac{\partial \varphi}{\partial q_s}$$

is a particular solution of (15).

Hence, if $\varphi(t, q_s, p_s) = a$ and $\psi(t, q_s, p_s) = b$ are two integrals of the canonical equations of motion (14), then the equations in variations have particular solutions

$$\xi_s = - \frac{\partial \varphi}{\partial p_s}, \quad \eta_s = \frac{\partial \varphi}{\partial p_s}, \quad \xi_s' = - \frac{\partial \psi}{\partial p_s}, \quad \eta_s' = \frac{\partial \psi}{\partial q_s}.$$

Substituting these particular solutions into (16), we find that the expression

$$(\varphi, \psi) = \sum \left(\frac{\partial \psi}{\partial q_s} \frac{\partial \varphi}{\partial p_s} - \frac{\partial \psi}{\partial p_s} \frac{\partial \varphi}{\partial q_s} \right) = \text{const}$$

is an integral of the canonical equations. This theorem was first derived by Poisson.

In his *Méchanique analytique* Lagrange said that the most significant application of invariant (16) is in approximate integration when there are perturbing forces with a force function W.

Assume that the perturbed motion

$$q_s = q_s(t) + \xi_s, \quad p_s = p_s(t) + \eta_s$$

is the motion of the system when there are perturbing forces with a force functions W, i.e.

$$\frac{d[q_s(t) + \xi_s]}{dt} = \frac{\partial \{H(t, q_s(t) + \xi_s, p_s(t) + \eta_s) - W(q_s(t) + \xi_s)\}}{\partial p_s},$$

$$\frac{d[p_s(t) + \eta_s]}{dt} = - \frac{\partial \{H(t, q_s(t) + \xi_s, p_s(t) + \eta_s) - W(q_s(t) + \xi_s)\}}{\partial q_s}.$$

Expanding these relations in Taylor's series and retaining only the linear terms, we have

$$\frac{d\xi_s}{dt} = \sum_j \left(\frac{\partial^2 H}{\partial p_s \partial q_j} \xi_j + \frac{\partial^2 H}{\partial p_s \partial p_j} \eta_j \right),$$

$$\frac{d\eta_s}{dt} = - \sum_j \left(\frac{\partial^2 H}{\partial q_s \partial q_j} \xi_j + \frac{\partial^2 H}{\partial q_s \partial p_j} \eta_j \right) + \sum_j \frac{\partial^2 W}{\partial q_s \partial q_j} \xi_j + \frac{\partial W(q_s(t))}{\partial q_s}. \tag{17}$$

Let ξ_s, η_s and ξ_s', η_s' be two particular solutions of (17). We have

$$\frac{d}{dt} \sum (\xi_s \eta_s' - \xi_s' \eta_s) = \sum \left\{ \eta_s' \frac{\partial^2 H}{\partial p_s \partial q_j} \xi_j + \eta_s' \frac{\partial^2 H}{\partial p_s \partial p_j} \eta_j - \xi_s \frac{\partial^2 H}{\partial q_s \partial q_j} \xi_j' \right.$$

$$- \xi_s \frac{\partial^2 H}{\partial q_s \partial p_j} \eta_j' + \xi_s \frac{\partial^2 W}{\partial q_s \partial q_j} \xi_j' - \eta_s \frac{\partial^2 H}{\partial p_s \partial q_j} \xi_j'$$

$$- \eta_s \frac{\partial^2 H}{\partial p_s \partial p_j} \eta_j' + \xi_s' \frac{\partial^2 H}{\partial q_s \partial q_j} \xi_j + \xi_s' \frac{\partial^2 H}{\partial q_s \partial p_j} \eta_j - \xi_s' \frac{\partial^2 W}{\partial q_s \partial q_j} \xi_j$$

$$+ \sum (\xi_s - \xi_s') \frac{\partial W(q_s(t))}{\partial q_s},$$

or

$$\frac{d}{dt} \sum (\xi_s \eta_s' - \eta_s \xi_s') = \sum (\xi_s - \xi_s') \frac{\partial W}{\partial q_s}.$$

7. Let us see how the action can be determined for Hamilton's equation

$$\frac{\partial V}{\partial t} + H \left(q, \frac{\partial V}{\partial q}, t \right) = 0$$

with the use of a complete integral (in Lagrange's sense).

Let $f(q_1, \ldots, q_n, t, a_1, \ldots, a_n)$ be a complete integral. We introduce the designation

$$S = f(q, t, a) - f(q^0, t_0, a),$$

where q^0 and t_0 are the initial values of the variables and time. Let

$$\frac{\partial S}{\partial a_r} = \frac{\partial f}{\partial a_r} - \frac{\partial f_0}{\partial a_r} = 0 \quad (r = 1, \ldots, n).$$

These equations make it possible to eliminate $a_r = a_r (q, t, q^0, t_0)$ in S. We have

$$\frac{\partial S}{\partial t} = \frac{\partial f}{\partial t} + \sum \frac{\partial S}{\partial a_m} \frac{\partial a_m}{\partial t} = \frac{\partial f}{\partial t}, \quad \frac{\partial S}{\partial q_r} = \frac{\partial f}{\partial q_r} + \sum \frac{\partial S}{\partial a_m} \frac{\partial a_m}{\partial q_r} = \frac{\partial f}{\partial q_r},$$

and, consequently, S as a function of q, t, q^0, t_0 satisfies the equation

$$\frac{\partial S}{\partial t} + H \left(q, \frac{\partial S}{\partial q}, t \right) = 0.$$

8. Let us consider a system

$$\frac{\partial f}{\partial a_r} = \frac{\partial f_0}{\partial a_r} \qquad (18)$$

assuming a_r to be an arbitrary constant $\Big($from Jacobi's theorem we have

$\dfrac{\partial f}{\partial a_r} = \beta_r$ and the sense of the constant $\dfrac{\partial f_0}{\partial a_r}$ is the same$\Big)$; $f(q,\ t,\ a)$ is a complete integral and therefore

$$\left\| \frac{\partial^2 f}{\partial a_r \partial q_s} \right\|$$

is nonzero. This means that we can solve system (18) and obtain

$$q_r = q_r(t,\ t_0,\ q^0,\ a). \tag{19}$$

It follows from (19) that

$$q_r^0 = q_r(t_0,\ t_0,\ q^0,\ a);$$

$p_r = \dfrac{\partial f}{\partial q_r}$ and hence

$$p_r = p_r(t,\ t_0,\ q^0,\ a). \tag{20}$$

The functions defined in (19) and (20) comprise the general solution of a mechanics problem for arbitrary a_r. From (18) we get

$$\sum \frac{\partial^2 f}{\partial a_r \partial q_s} \frac{dq_s}{dt} + \frac{\partial^2 f}{\partial a_r \partial t} = 0, \tag{21}$$

but

$$\frac{\partial f}{\partial t} + H(q,\ p,\ t) = 0, \quad p = \frac{\partial f}{\partial q} \tag{22}$$

and, hence, differentiating (22) with respect to a_r, we obtain

$$\frac{\partial^2 f}{\partial a_r \partial t} + \sum \frac{\partial H}{\partial p_s} \frac{\partial^2 f}{\partial a_r \partial q_s} = 0,$$

whence we get

$$\frac{dq_s}{dt} = \frac{\partial H}{\partial p_s}.$$

It follows from (22) that

$$\frac{dp_r}{dt} = \frac{\partial^2 f}{\partial q_r \partial t} + \sum \frac{\partial^2 f}{\partial q_r \partial q_s} \frac{dq_s}{dt} = \frac{\partial^2 f}{\partial q_r \partial t} + \sum \frac{\partial^2 f}{\partial q_r \partial q_s} \frac{\partial H}{\partial p_s}.$$

Differentiating (22) with respect to q_r, we get

$$\frac{\partial^2 f}{\partial q_r \partial t} + \sum \frac{\partial H}{\partial p_s} \frac{\partial^2 f}{\partial q_r \partial q_s} + \frac{\partial H}{\partial q_r} = 0,$$

and, hence,

$$\frac{dp_r}{dt} = - \frac{\partial H}{\partial q_r}.$$

By definition an action is

$$V = \int_{t_0}^{t} \left(\sum p_r \frac{\partial H}{\partial p_r} - H \right) dt = \int_{t_0}^{t} \left(\sum \frac{\partial f}{\partial q_r} \frac{dq_r}{dt} + \frac{\partial f}{\partial t} \right) dt$$

$$= f(q, t, a) - f(q^0, t_0, a); \qquad \frac{\partial V}{\partial a} = 0.$$

9. Geometric interpretation. $F = f(q, t, a)$ is a complete integral. In the (F, t, q) space the integral yields a system of surfaces dependent on $n + 1$ parameters $a_1, \ldots, a_n, a_{n+1}$. We seek a family of curves passing through the point $(0, t_0, q^0)$:

$$0 = f_0 + a_{n+1}.$$

From this we have

$$F = f(q, t, a) - f(q^0, t_0, a).$$

Then we seek an *envelope* of the family:

$$\frac{\partial f}{\partial a_r} - \frac{\partial f_0}{\partial a_r} = 0.$$

Excluding a, we have Hamilton's principal function (action integral)*

$$F = S(q, t, t_0, q^0).$$

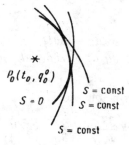

Figure 162

---

* See the footnote on p. 303.

Assume that

$$S = f - f_0 = \text{const}$$

for a definite t (Fig. 162). The system of surfaces depends on the parameters $a_1, \ldots, a_n$. We seek their envelope

$$\frac{\partial f}{\partial a_r} - \frac{\partial f_0}{\partial a_r} = 0.$$

The envelope defines the principal function

$$S = S(t, q, t_0, q^0) = \text{const}.$$

A.2 Poincaré's Equations

1. Holonomic coordinates. Assume that for every t the position of the system is defined by the variables $q_1, \ldots, q_n$, or, to put it otherwise,

$$
\begin{aligned}
x_\nu &= x_\nu(t, q_1, \ldots, q_n), \\
y_\nu &= y_\nu(t, q_1, \ldots, q_n), \\
z_\nu &= z_\nu(t, q_1, \ldots, q_n).
\end{aligned}
\tag{1}
$$

These coordinates were called *defining* by Chaplygin and *holonomic* by Appell.

Let T be the kinetic energy calculated from (1):

$$
T = \sum \frac{m_\nu}{2} \left\{ \left(\frac{\partial x_\nu}{\partial t} + \sum \frac{\partial x_\nu}{\partial q_s} q_s' \right)^2 + \left(\frac{\partial y_\nu}{\partial t} + \sum \frac{\partial y_\nu}{\partial q_s} q_s' \right)^2 \right.
$$
$$
\left. + \left(\frac{\partial z_\nu}{\partial t} + \sum \frac{\partial z_\nu}{\partial q_s} q_s' \right)^2 \right\}.
\tag{2}
$$

It follows from (1) that

$$
\frac{\partial x_\nu'}{\partial q_s'} = \frac{\partial x_\nu}{\partial q_s}, \quad \frac{\partial x_\nu'}{\partial q_s} = \frac{d}{dt} \frac{\partial x_\nu}{\partial q_s}.
\tag{3}
$$

Similar formulas are valid for y and z.

The Euler-Lagrange principle yields

$$
\sum \left[\frac{d}{dt} \left(\frac{\partial T}{\partial q_s'} \right) - \frac{\partial T}{\partial q_s} - Q_s \right] \delta q_s = 0,
\tag{4}
$$

where

$$
Q_s = \sum \left(X_\nu \frac{\partial x_\nu}{\partial q_s} + Y_\nu \frac{\partial y_\nu}{\partial q_s} + Z_\nu \frac{\partial z_\nu}{\partial q_s} \right),
$$

i.e. $Q_s \, \delta q_s$ is the work done by the forces over the virtual displacement defined by the variation of one variable q_s by the value δq_s.

If the virtual displacements of the system include one defined by the relations

$$\delta q_1 \neq 0, \quad \delta q_2 = 0, \quad \ldots, \quad \delta q_n = 0, \tag{5}$$

then from (4) we have

$$\frac{d}{dt} \left(\frac{\partial T}{\partial q_1'} \right) - \frac{\partial T}{\partial q_1} = Q_1. \tag{6}$$

If, moreover, the work of the given forces over displacement (5) is zero ($Q_1 = 0$) and if the kinetic energy T does not change when q_1 $\left(\dfrac{\partial T}{\partial q_1} = 0 \right)$ changes, then equation (6) yields the first integral

$$\frac{\partial T}{\partial q_1'} = \text{const.}$$

A coordinate q_1 with these properties is called *cyclic*. The proposition we have proved is a generalization of the theorem on the motion of the centre of gravity and the theorem on the angular momentum since

1°. if q_1 defines a translation along the x-axis, then

$$\frac{\partial x_\nu}{\partial q_1} = 1, \quad \frac{\partial y_\nu}{\partial q_1} = 0, \quad \frac{\partial z_\nu}{\partial q_1} = 0;$$

2°. if q_1 denotes a rotation about the z-axis as a rigid body, then

$$\frac{\partial x_\nu}{\partial q_1} = -y_\nu, \quad \frac{\partial y_\nu}{\partial q_1} = x_\nu, \quad \frac{\partial z_\nu}{\partial q_1} = 0.$$

Warning! It is very important that T be determined from formula (2) with the use of the defining relations (1). We cannot take any expression of the type

$$T = T(t, q_1, \ldots, q_n, q_1', \ldots, q_n')$$

as T. The coordinates $q_1, \ldots, q_n$ are not yet constrained by the condition of independence. The condition requires that at least one of the derivatives

$$\frac{\partial x_\nu}{\partial q_s}, \quad \frac{\partial y_\nu}{\partial q_s}, \quad \frac{\partial z_\nu}{\partial q_s} \quad (\nu = 1, \ldots, m)$$

does not vanish for the defining coordinate q_s. Therefore it would be erroneous to define the position of a particle with mass m in a plane by the polar coordinates $q_1 = r$, $q_2 = \theta$ and the sector area $q_s = \sigma$, and take the kinetic energy of this particle to be

$$T = \frac{m}{2}\left(q_1'^2 + \frac{1}{2}q_1^2 q_2'^2 + \frac{1}{2}\frac{q_3'^2}{q_1^2}\right)$$

and write relation (4).

2. Independent defining coordinates, or Lagrange's coordinates. If $q_1, \ldots, q_n$ are independent defining coordinates, then from formula (4) on p. 226 follows a complete system of Lagrange's equations

$$\frac{d}{dt}\left(\frac{\partial T}{\partial q_s'}\right) - \frac{\partial T}{\partial q_s} = Q_s \quad (s = 1, \ldots, n).$$

The energy integral. The kinetic energy has the form

$$T = T_2 + T_1 + T_0.$$

Assume that the forces have a force function U, i.e.

$$Q_s = \frac{\partial U}{\partial q_s}.$$

We multiply Lagrange's equations by q_s' and add the results. If T does not depend explicitly on time $\left(\dfrac{\partial T}{\partial t} = 0\right)$, then

$$\frac{d}{dt}\left(\sum \frac{\partial T}{\partial q_s'}q_s'\right) - \left(\sum \frac{\partial T}{\partial q_s'}q_s'' + \frac{\partial T}{\partial q_s}q_s'\right) = \sum \frac{\partial U}{\partial q_s}q_s',$$

or

$$\frac{d}{dt}(2T_2 + T_1) - \frac{d}{dt}(T_2 + T_1 + T_0) = \frac{dU}{dt}.$$

Integrating, we get the kinetic energy integral

$$T_2 - T_0 = U + h$$

in a more general form than the one we proved earlier. This integral was found by Jacobi.

3. Cyclic coordinates. Let q_α be cyclic coordinates

$$(L = T + U), \quad \frac{\partial L}{\partial q_\alpha} = 0 \quad (\alpha = s + 1, \ldots, k).$$

From this we get the cyclic integrals

$$\frac{\partial L}{\partial q_\alpha'} = \beta_\alpha.$$

Routh introduced a function $R = L - \sum q'_\alpha \beta_\alpha$, which, after the elimination of the cyclic velocities with the use of the cyclic integrals, has the form

$$R(t, q_1, \ldots, q_s, q'_1, \ldots, q'_s, \beta_{s+1}, \ldots, \beta_k).$$

Varying both expressions for the Routh function

$$\delta R = \sum \frac{\partial R}{\partial q_j} \delta q_j + \sum \frac{\partial R}{\partial q'_j} \delta q_j' + \sum \frac{\partial R}{\partial \beta_\alpha} \delta \beta_\alpha,$$

$$\delta R = \sum \frac{\partial L}{\partial q_j} \delta q_j + \sum \frac{\partial L}{\partial q'_j} \partial q_j' - \sum q'_\alpha \delta \beta_\alpha,$$

we obtain

$$\frac{\partial L}{\partial q_j} = \frac{\partial R}{\partial q_j}, \quad \frac{\partial L}{\partial q'_j} = \frac{\partial R}{\partial q_j'}, \quad -q'_\alpha = \frac{\partial R}{\partial \beta_\alpha}.$$

The problem reduces to the integration of the equations

$$\frac{d}{dt}\left(\frac{\partial R}{\partial q'_j}\right) - \frac{\partial R}{\partial q_j} = 0 \quad (j = 1, \ldots, s),$$

and then

$$q_\alpha = -\int \frac{\partial R}{\partial \beta_\alpha} \, dt.$$

4. Poincaré parameters. We denote the defining coordinates by $x_1, \ldots, x_n$. Assume that for every t they define the position of the mechanical system under holonomic constraints

$$\omega_j = \sum a_{js} \delta x_s = 0 \quad (j = k + 1, \ldots, n). \tag{7}$$

The matrix $\|a_{js}\|$ has at least one nonzero determinant of order $n - k$.

If we have a form

$$\omega_0 = a_1 \delta x_1 + \ldots + a_n \delta x_n,$$

we can derive a new form from it, i.e.

$$\delta \omega_{\delta'} - \delta' \omega_\delta = \sum a_i (\delta \delta' x_i - \delta' \delta x_i) + \sum (\delta a_i \delta' x_i - \delta' a_i \delta x_i).$$

Assume that the two differentiation symbols δ, δ' are commutative, i.e.

$$\delta \delta' x_i = \delta' \delta x_i.$$

The right-hand side of this formula is known as the *bilinear covariant* of the form ω_δ, or the exterior derivative ω'_δ:

$$\omega'_\delta = \sum \begin{vmatrix} \delta a_i & \delta x_i \\ \delta' a_i & \delta' x_i \end{vmatrix} = \sum [\delta a_i, \delta x_i] = \sum \left(\frac{\partial a_j}{\partial x_i} - \frac{\partial a_i}{\partial x_j}\right)(\delta x_i, \delta x_j).$$

In the last sum the indices are not repeated. If $\omega_\delta' = 0$, then ω_δ is a total differential. The derivative of the derivative ω_δ' of some form ω_δ is identically zero.

Let us consider the term $a\delta x_1$ in ω. In ω' it is associated with a term $[\delta a, \delta x_1]$. If a depends only on x_1, then this term is zero, and if a depends on something else in addition to x_1, then we can take a as x_2. The derivative of $(\delta x_2, \delta x_1)$ is zero since the coefficient is 1.

By definition, system (7) must be integrable for holonomic constraints. For the Pfaffian system (7) to be completely integrable, it is necessary that all the derivatives ω_j' be eliminated by virtue of equations (7).

Indeed, if system (7) can be reduced to the form

$$\delta y_{k+1} = 0, \ \ldots, \ \delta y_n = 0,$$

then the forms ω_j are linear with respect to $\delta y_{k+1}, \ldots, \delta y_n$, i.e.

$$\omega_j = A_{j,k+1}\delta y_{k+1} + \ldots + A_{j,n}\delta y_n, \quad \det \|A_{ji}\| \neq 0.$$

Hence the derivative

$$\omega_j' = \sum_{i=k+1}^{n} [\delta A_{ji}, \delta y_i]$$

is zero for $\delta y_{k+1} = 0, \ldots, \delta y_n = 0$.

To prove the contrary, we write (7) in another form, viz.,

$$\bar{\omega}_{k+1} \equiv \beta_{k+1,\ k+1}\omega_{k+1} + \ldots + \beta_{k+1,\ n}\omega_n = 0,$$
$$\cdots\cdots\cdots\cdots\cdots\cdots\cdots\cdots\cdots\cdots\cdots\cdots\cdots\cdots\cdots$$
$$\bar{\omega}_n \equiv \beta_{n,k+1}\omega_{k+1} + \ldots + \beta_{n,n}\omega_n = 0,$$

$$\det \|\beta_{r,\ s}\| \neq 0.$$

From this we have

$$\bar{\omega}_\alpha' = \beta_{\alpha,k+1}\omega_{k+1}' + \ldots + \beta_{\alpha,n}\omega_n' + [\delta\beta_{\alpha,k+1},\ \omega_{k+1}] + \ldots + [\delta\beta_{\alpha,\ n}\omega_n],$$

and, consequently, if ω_α' are eliminated by virtue of $\omega_\alpha = 0$ ($\alpha = k+1, \ldots, n$), then $\bar{\omega}$ are eliminated by virtue of $\bar{\omega}_\alpha = 0$. The property is independent of the method of notation.

We assume that the converse proposition has been proved for $n-1$ variables and we shall prove it for n variables. By hypothesis, ω_i', which are eliminated by virtue of (7), will be eliminated if we assume that $\delta x_n = 0$, or, in other words, if we regard x_n as a fixed parameter. The hypothesis reduces system (7) to the form

$$\delta y_{k+1} = 0, \ \ldots, \ \delta y_n = 0,$$

where $y_{k+1}, \ldots, y_n$ are understood as the functions of $x_1, \ldots, x_{n-1}$. If we do not regard x_n as a constant, then (7) evidently reduces to the form

$$\tilde{\omega}_{k+1} \equiv \delta y_{k+1} + b_{k+1}\delta x_n = 0,$$
$$\ldots\ldots\ldots\ldots\ldots\ldots\ldots\ldots\ldots \qquad (8)$$
$$\tilde{\omega}_n \equiv \delta y_n + b_n\delta x_n = 0;$$

where b_α are functions of $y_{k+1}, \ldots, y_n$ and, say, $x_1, x_2, \ldots, x_k$. Hence

$$\tilde{\omega}'_{k+1} = [\delta b_{k+1}, \delta x_n], \ldots, \tilde{\omega}'_n = [\delta b_n, \delta x_n].$$

By virtue of (8) we have

$$\tilde{\omega}'_\alpha = \frac{\partial b_\alpha}{\partial x_2} [\delta x_2, \delta x_n] + \ldots + \frac{\partial b_\alpha}{\partial x_k} [\delta x_k, \delta x_n].$$

This means that for $\tilde{\omega}_\alpha$ to be eliminated by virtue of (8), we must have

$$\frac{\partial b_\alpha}{\partial x_2} = 0, \ldots, \frac{\partial b_\alpha}{\partial x_k} = 0,$$

i.e. b_α depend only on $y_{k+1}, \ldots, y_n, x_n$. But then system (8) can be reduced to the form

$$\delta y_{k+1} = 0, \ldots, \delta y_n = 0.$$

We have obtained **Frobenius' theorem**, which states that the *equalities* $\delta y_{k+1} = 0, \ldots, \delta y_n = 0$ *are the necessary and sufficient condition for the integrability of system* (7).

Thus, if the constraints are expressed by the integrable equations (7), then, according to Frobenius' theorem, we have

$$[\omega_{k+1}, \ldots, \omega_n, \omega'_\alpha] = 0 \quad (\alpha = k+1, \ldots, n). \qquad (9)$$

The equation

$$[F, f_1, f_2, \ldots, f_n] = 0,$$

where F is an exterior form, yields a condition necessary and sufficient for F to be eliminated when the variables satisfy the relations

$$f_1 = 0, \quad f_2 = 0, \ldots, f_n = 0.$$

5. We choose k linear differential forms

$$\omega_i \equiv \sum a_{is}\delta x_s \quad (i = 1, \ldots, k)$$

which are independent both of one another and of the forms $\omega_{k+1}, \ldots, \omega_n$. This means that

$$\det \|a_{ij}\| \neq 0 \quad (i, j = 1, \ldots, n).$$

We can therefore solve the system

$$\omega_r = \sum a_{rs}\delta x_s \quad (r = 1, \ldots, n)$$

for δx_s and get

$$\delta x_s = \sum \xi_s^{(r)}\omega_r.$$

Hence

$$\delta f = \sum \frac{\partial f}{\partial x_s}\delta x_s = \sum \xi_s^{(r)}\frac{\partial f}{\partial x_s}\omega_r = \sum_{r=1}^{n}\omega_r X_r f,$$

where

$$X_r f = \sum \xi_s^{(r)}\frac{\partial f}{\partial x_s}.$$

Let δ be a virtual displacement when the conditions of constraints

$$\omega_{k+1} = 0, \ldots, \omega_n = 0 \tag{10}$$

are fulfilled. For this understanding of δ

$$\delta f = \sum_{\alpha=1}^{k}\omega_\alpha X_\alpha f.$$

For holonomic constraints the first integrals in (10) $y_{k+1}, \ldots, y_n$ satisfy the conditions $\delta y_{k+1} = 0, \ldots, \delta y_n = 0$ for any $\omega_1, \ldots, \omega_k$. This yields a system of k linear partial differential equations for $y_{k+1}, \ldots, y_n$, i.e.,

$$X_\alpha f = 0 \quad (\alpha = 1, \ldots, k). \tag{11}$$

This system has $n - k$ independent solutions $y_{k+1}, \ldots, y_n$. System (11) is said to be complete if it has $n - k$ independent solutions. For every holonomic (integrable) system (10), system (11) is always complete.

Since

$$\delta(X_j, f) = \sum \omega_i X_i X_j f,$$

we have

$$0 = \sum \omega_j X_j f = \sum [\delta(X_j f), \omega_j] = \sum \omega_j X_j f + \sum X_i X_j f[\omega_i, \omega_j]$$

from the formula

$$\delta f = \sum_{j=1}^{n}\omega_j X_j f.$$

But we can always express n covariants of ω_j' $(j = 1, \ldots, n)$ as $\omega_j' = -\sum_{r,s}{}^{*}c_{rsj}(\omega_r, \omega_s)$ $(\sum^{*}$ is a sum without repetitions). It follows that

$$0 = -\sum{}^* c_{rsj}[\omega_r, \omega_s] X_j f + \sum{}^* (X_i, X_j) f [\omega_i, \omega_j]$$
$$= \sum{}^* [\omega_r, \omega_s][(X_r, X_s) f - \sum c_{rsj} X_j f],$$

where
$$(X_i, X_j) f = X_i(X_j f) - X_j(X_i f)$$

is a *commutator*. It follows that

$$(X_r, X_s) f = \sum c_{rsj} X_j f.$$

If the constraints are holonomic, then the system $\omega_j = 0$ $(j = k + 1, \ldots, n)$ is integrable. According to Frobenius' theorem ω_j' $(j = k + 1, \ldots, n)$ must be eliminated at the same time as ω_j. This means that the expression for ω_j' $(j = k + 1, \ldots, n)$ must not contain terms without ω_j. Hence

$$c_{rsj} = 0 \quad (j = k + 1, \ldots, n; \ r, s = 1, \ldots, n),$$

and this means that X_α $(\alpha = 1, \ldots, k)$ is a closed system in the sense that

$$(X_\alpha X_\beta) f = \sum c_{\alpha\beta\gamma} X_\gamma f \quad (\alpha, \beta, \gamma = 1, \ldots, k).$$

6. Properties of a commutator. Let a and b be constants.

I. $(aX_\alpha + bX_\beta, X_\gamma) f = (aX_\alpha + bX_\beta) X_\gamma f - X_\gamma (aX_\alpha + bX_\beta) f = aX_\alpha X_\gamma f + bX_\beta X_\gamma f - aX_\gamma X_\alpha f - bX_\gamma X_\beta f = a(X_\alpha, X_\gamma) f + b(X_\beta, X_\gamma) f,$

II. $(X_\alpha, X_\beta) = -(X_\beta, X_\alpha),$

III. $(X_\alpha, (X_\beta, X_\gamma)) + (X_\beta, (X_\gamma, X_\alpha)) + (X_\gamma, (X_\alpha, X_\beta)) = 0.$

The system X_α $(\alpha = 1, \ldots, k)$ is a closed system of infinitesimal linear operators, or, as it was customary to say, an infinitesimal group whose commutator satisfies properties I, II, III. It remains for us to choose auxiliary forms $\omega_1, \ldots, \omega_k$ which will help us to simplify the system and get a Lie infinitesimal group*.

Assume that in the system defining the constraints

$$\omega_j \equiv \sum_{i=1}^{n} a_{ij} \delta x_i = 0 \quad (j = k + 1, \ldots, n),$$

$$\begin{vmatrix} a_{k+1, \, k+1} & \cdots & a_{k+1, \, n} \\ \cdots\cdots\cdots\cdots\cdots\cdots \\ a_{n, \, k+1} & \cdots & a_{n, \, n} \end{vmatrix} \neq 0;$$

then $\delta x_j = \sum_{\alpha = 1}^{k} b_{j\alpha} \delta x_\alpha.$

_____

* Lie algebra in modern terminology. — *Ed.*

If we assume that $\omega_\alpha \equiv \delta x_\alpha$ $(\alpha = 1, \ldots, k)$, then

$$\delta f = \sum \frac{\partial f}{\partial x_i} \, \delta x_i = \sum_{\alpha = 1}^{k} \left(\frac{\partial f}{\partial x_\alpha} + \sum_{j = k + 1}^{n} \frac{\partial f}{\partial x_j} \, b_{j\alpha} \right) \omega_\alpha ,$$

and, hence,

$$X_\alpha f = \frac{\partial f}{\partial x_\alpha} + \sum_{j = k + 1}^{n} b_{j\alpha} \frac{\partial f}{\partial x_j} \quad (\alpha = 1, \ldots, k).$$

It follows that

$$(X_\alpha, X_\beta) f \equiv 0,$$

and, hence, for all α, β, i we have

$$c_{\alpha\beta i} = 0 \quad (\alpha, \beta, i = 1, \ldots, k).$$

The system of operators $X_\alpha f$ $(\alpha = 1, \ldots, k)$ forms an infinitesimal subgroup of the Lie group. For brevity we shall call it a Lie group X_α $(\alpha = 1, \ldots, k)$.

Assumption. We shall assume that the infinitesimal group of virtual displacements $X_\alpha f$ $(\alpha = 1, \ldots, k)$ is an infinitesimal subgroup of a Lie group and is defined by its structural constants $c_{\alpha\beta i}$:

$$(X_\alpha, X_\beta) f = \sum c_{\alpha\beta i} X_i f \quad (\alpha, \beta, i = 1, \ldots, k).$$

From this we can find the variation of f for a virtual displacement, in terms of the parameters ω_α $(\alpha = 1, \ldots, k)$, from the formula

$$\delta f = \sum_{\alpha = 1}^{k} \omega_\alpha X_\alpha f.$$

7. True Displacements. The true displacements of a system subjected to holonomic constraints are constrained by Pfaffian integrable equations

$$\pi_j \equiv \sum_{s = 1}^{n} a_{js} dx_s + a_j dt = 0 \quad (j = k + 1, \ldots, n).$$

We supplement this system by the forms

$$\pi_\alpha = \sum a_{\alpha s} dx_s \quad (\alpha = 1, \ldots, k)$$

(by the previous ω_α $(\alpha = 1, \ldots, k)$ with a substitution of d for δ) and $\pi_0 = dt$. Then, in accordance with what we have proved for virtual displacements, we can prove for true displacements that

$1°.$ $df = dtX_0f + \sum \pi_\alpha X_\alpha f$

(where X_α ($\alpha = 1, \ldots, k$) are the virtual displacement operators),

$2°.$ the set

$$X_0f, \ X_1f, \ \ldots, \ X_kf$$

is closed in the sense that

$$(X_i, \ X_j)f = \sum c_{ijs} X_s f \quad (i, \ j, \ s = 0, \ 1, \ \ldots, \ k),$$

i.e. they constitute an infinitesimal group,

$3°.$ since $X_1f, \ \ldots, \ X_kf$ form a Lie group of virtual displacements and do not contain the derivative $\partial f/\partial t$, and

$$X_0f = \frac{\partial f}{\partial t} + \sum \beta_i \frac{\partial f}{\partial x_i},$$

we infer that $(X_0, \ X_\alpha)f$ does not include $\partial f/\partial t$, and, consequently,

$$c_{0\alpha 0} = 0 \quad (\alpha = 1, \ \ldots, \ k),$$

$4°.$ $X_1, \ \ldots, \ X_k$ is a Lie subgroup of the infinitesimal group X_0, $X_1, \ \ldots, \ X_k$ of true displacements.

We use the notation $\dfrac{\pi_\alpha}{dt} = \eta_\alpha$ which yields

$$\frac{df}{dt} = X_0f + \sum_{\alpha = 1}^{k} \eta_\alpha X_\alpha f.$$

Assumption. The infinitesimal group, $X_0, \ X_1, \ \ldots, \ X_k$ is assumed to be an infinitesimal subgroup of a Lie group for which all the $c_{\alpha\beta\gamma}$ are constants ($\alpha, \ \beta, \ \gamma = 0, \ 1, \ \ldots, \ k$).

Example. Assume that

$$\begin{vmatrix} a_{k+1, \ k+1} & \cdots & a_{k+1, \ n} \\ \cdots\cdots\cdots\cdots\cdots\cdots \\ a_{n, \ k+1} & \cdots & a_{n, \ n} \end{vmatrix} \neq 0$$

in the system

$$\pi_j \equiv \sum a_{ji} dx_i + a_j dt = 0 \quad (j = k+1, \ \ldots, \ n)$$

and then

$$dx_j = \sum_{\alpha = 1}^{k} b_{j\alpha} dx_\alpha + b_j dt \quad (j = k+1, \ \ldots, \ n).$$

If we accept that $\pi_0 = dt$, $\pi_\alpha = dx_\alpha$, then

$$df = \sum \frac{\partial f}{\partial x_i} dx_i + \frac{\partial f}{\partial t} dt$$

$$= \pi_0 \left(\frac{\partial f}{\partial t} + \sum_{j=k+1}^{n} b_j \frac{\partial f}{\partial x_j} \right) + \sum_{\alpha=1}^{n} \pi_\alpha \left(\frac{\partial f}{\partial x_\alpha} + \sum b_{j\alpha} \frac{\partial f}{\partial x_j} \right),$$

and this means that

$$X_0 = \frac{\partial f}{\partial t} + \sum_{j=k+1}^{n} b_j \frac{\partial f}{\partial x_j}, \quad X_\alpha = \frac{\partial f}{\partial x_\alpha} + \sum_{j=k+1}^{n} b_{j\alpha} \frac{\partial f}{\partial x_j}$$

$$(\alpha = 1, \ldots, k).$$

Hence all $c_{\alpha\beta\gamma} = 0$ ($\alpha, \beta, \gamma = 0, 1, \ldots, k$). If the constraints are holonomic and the operators d and δ are commutable, then $\delta df = d\delta f$. We immediately get

$$0 = d \left(\sum \omega_\alpha X_\alpha f \right) - \delta \left(X_0 f + \sum \eta_\alpha X_\alpha f \right) dt$$

$$= \sum (d\omega_\alpha - \delta\eta_\alpha dt) X_\alpha f + \sum \omega_\alpha (dt X_0 X_\alpha f + \sum \eta_\beta X_\beta X_\alpha f dt)$$

$$- \sum \omega_\alpha X_\alpha X_0 f dt - \sum \eta_\alpha \sum \omega_\beta X_\beta X_\alpha f dt$$

$$= \sum (d\omega_\alpha - \delta\eta_\alpha dt) X_\alpha f + \sum \omega_\alpha dt (X_0 X_\alpha - X_\alpha X_0) f$$

$$+ \sum \omega_\alpha \eta_\beta (X_\beta X_\alpha - X_\alpha X_\beta) = \sum (d\omega_\alpha - \delta y_\alpha dt) X_\alpha f + \sum \omega_\alpha \eta_\beta c_{\beta\alpha i} X_i f dt$$

$$+ \sum \omega_\alpha dt c_{0\alpha\beta} X_\beta f = dt \left\{ \sum X_i f \left[\frac{d\omega_i}{dt} - \delta\eta_i + \sum c_{\beta\alpha i} \eta_\beta \omega_\alpha + \sum c_{0\alpha i} \omega_\alpha \right] \right\}.$$

Since f is arbitrary, we have

$$\frac{d\omega_i}{dt} = \delta\eta_i - \sum c_{\beta\alpha i} \eta_\beta \omega_\alpha - \sum c_{0\alpha i} \omega_\alpha.$$

8. Equations of motion. The rectangular coordinates of a particle of a mechanical system are $u(t, x_1, \ldots, x_n)$, $v(t, x_1, \ldots, x_n)$, $w(t, x_1, \ldots, x_n)$ and m is its mass. Assume for simplicity that the forces have a force function. The Euler-Lagrange principle is

$$\sum m \left(\frac{d^2 u}{dt^2} \delta u + \frac{d^2 v}{dt^2} \delta v + \frac{d^2 w}{dt^2} \delta w \right) = \delta U,$$

but

$$\delta f = \sum \omega_\alpha X_\alpha f.$$

Hence the principle assumes the form

$$\sum \omega_\alpha \left[\sum m \left(\frac{d^2 u}{dt^2} X_\alpha u + \frac{d^2 v}{dt^2} X_\alpha v + \frac{d^2 w}{dt^2} X_\alpha w \right) - X_\alpha U \right] = 0.$$

This must occur for arbitrary ω_α. It follows that

$$\sum m \left(\frac{d^2 u}{dt^2} X_\alpha u + \frac{d^2 v}{dt^2} X_\alpha v + \frac{d^2 w}{dt^2} X_\alpha w \right) = X_\alpha U,$$

or

$$\frac{d}{dt} \sum m \left(\frac{du}{dt} X_\alpha u + \frac{dv}{dt} X_\alpha v + \frac{dw}{dt} X_\alpha w \right)$$

$$- \sum m \left(\frac{du}{dt} \frac{dX_\alpha u}{dt} + \frac{dv}{dt} \frac{dX_\alpha v}{dt} + \frac{dw}{dt} \frac{dX_\alpha w}{dt} \right) = X_\alpha U, \quad (12)$$

but

$$\frac{du}{dt} = u' = X_0 U + \sum \eta_\alpha X_\alpha u.$$

Hence $\dfrac{\partial u'}{\partial \eta_\alpha} = X_\alpha u$,

$$X_i(u') = X_i \left(X_0 u + \sum \eta_\alpha X_\alpha u \right) = X_i(X_0 u) + \sum \eta_\alpha X_i X_\alpha u$$

$$= X_0(X_i u) + \sum c_{i0\beta} X_\beta u + \sum \eta_\alpha X_\alpha X_i u + \sum \eta_\alpha c_{i\alpha\beta} X_\beta u$$

$$= \frac{dX_i(u)}{dt} + \sum c_{i\alpha\beta} \eta_\alpha X_\beta u + \sum c_{i0\beta} X_\beta u.$$

Similar relations exist for v and w and therefore we can transform (12) as

$$\frac{d}{dt} \sum m \left(u' \frac{\partial u'}{\partial \eta_\alpha} + v' \frac{\partial v'}{\partial \eta_\alpha} + w' \frac{\partial w'}{\partial \eta_\alpha} \right)$$

$$- \sum m \left[u' \left(X_\alpha u' - \sum c_{\alpha i\beta} \eta_i X_\beta u \frac{\partial u'}{\partial \eta_\beta} - \sum c_{0i\beta} \frac{\partial u'}{\partial \eta_\beta} \right) \right.$$

$$+ v' \left(X_\alpha v' - \sum c_{\alpha i\beta} \eta_i X_\beta v \frac{\partial v'}{\partial \eta_\beta} - \sum c_{0i\beta} \frac{\partial v'}{\partial \eta_\beta} \right)$$

$$\left. + w' \left(X_\alpha w' - \sum c_{\alpha i\beta} \eta_i X_\beta w \frac{\partial w'}{\partial \eta_\beta} - \sum c_{0i\beta} \frac{\partial w'}{\partial \eta_\beta} \right) \right] = X_\alpha U.$$

We introduce the kinetic energy of the system

$$T = \sum \frac{m}{2} (u'^2 + v'^2 + w'^2).$$

Then the preceding relation becomes

$$\frac{d}{dt} \left(\frac{\partial T}{\partial \eta_\alpha} \right) - X_\alpha T + \sum c_{\alpha i \beta} \eta_i \frac{\partial T}{\partial \eta_\beta} + \sum c_{0i\beta} \frac{\partial T}{\partial \eta_\beta} = X_\alpha U,$$

or

$$\frac{d}{dt} \left(\frac{\partial T}{\partial \eta_i} \right) = \sum c_{\alpha i \beta} \eta_\alpha \frac{\partial T}{\partial \eta_\beta} + \sum c_{0i\beta} \frac{\partial T}{\partial \eta_\beta} + X_i(T + U) \quad (i = 1, \ldots, k).$$

These equations are known as Poincaré's equations. They were derived for the case $X_0 = \frac{\partial f}{\partial t}$ and $c_{0\alpha\beta} = 0$ $(\alpha, \beta = 1, \ldots, k)$*.

Poincaré's equations have sense for the infinitesimal group X_α ($\alpha = 1, \ldots, k$) for which $c_{\alpha i \beta}$ may vary. This remark was made by P. A. Shirokov in 1928. We assume X_α to be an infinitesimal subgroup of a Lie group for which $c_{\alpha i \beta}$ are structural constants.

It should be pointed out that Poincaré's equations have been derived without the use of relations following from the assumption that d and δ are commutable. (There was a mistake in Poincaré's paper in the indices of $c_{\alpha i \beta}$.) Lagrange's second-order equations are a particular case of Poincaré's equations.

Example 1. We shall derive, from Poincaré's equations, Euler's equations for the motion of a rigid body with one fixed point. Assume that x, y, z are fixed coordinate axes and $\omega_1, \omega_2, \omega_3$ are the parameters of virtual displacements. From Euler's formulas

$$\begin{pmatrix} \delta x \\ \delta y \\ \delta z \end{pmatrix} = \begin{vmatrix} \omega_1 & \omega_2 & \omega_3 \\ x & y & z \end{vmatrix},$$

$$\delta x = \omega_2 z - \omega_3 y, \quad \delta y = \omega_3 x - \omega_1 z, \quad \delta z = \omega_1 y - \omega_2 x,$$

$$\delta f = \sum \frac{\partial f}{\partial x} (\omega_2 z - \omega_3 y) + \sum \frac{\partial f}{\partial y} (\omega_3 x - \omega_1 z) + \sum \frac{\partial f}{\partial z} (\omega_1 y - \omega_2 x)$$

---

* H. Poincaré, *Comp. Rend.* **132**: 369 (1901).
H. Poincaré, "Sur la precession der corps deformables", *Bull. astronomique* (1910).
N. Četaev, *Comp. Rend.* **185**: 1577 (1927).

$$= \omega_1 \sum \left(y \frac{\partial f}{\partial z} - z \frac{\partial f}{\partial y} \right) + \omega_2 \sum \left(z \frac{\partial f}{\partial x} - x \frac{\partial f}{\partial z} \right) + \omega_3 \sum \left(x \frac{\partial f}{\partial y} - y \frac{\partial f}{\partial x} \right),$$

$$Xf = \sum \left(y \frac{\partial f}{\partial z} - z \frac{\partial f}{\partial y} \right),$$

$$Yf = \sum \left(z \frac{\partial f}{\partial x} - x \frac{\partial f}{\partial z} \right),$$

$$Zf = \sum \left(x \frac{\partial f}{\partial y} - y \frac{\partial f}{\partial x} \right),$$

$$(X, Y)f = \sum \left(y \frac{\partial f}{\partial z} - z \frac{\partial f}{\partial y} \right) \sum \left(z \frac{\partial f}{\partial x} - x \frac{\partial f}{\partial z} \right)$$

$$- \sum \left(z \frac{\partial f}{\partial x} - x \frac{\partial f}{\partial z} \right) \sum \left(y \frac{\partial f}{\partial z} - z \frac{\partial f}{\partial y} \right)$$

$$= \sum \left(y \frac{\partial f}{\partial x} - x \frac{\partial f}{\partial y} \right) = -Zf.$$

Similarly,

$$(Y, Z) = -X, \quad (Z, X) = -Y,$$

$c_{123} = -1$, $c_{231} = -1$, $c_{312} = -1$, $c_{213} = 1$, $c_{321} = 1$, $c_{132} = 1$, p, q, r are the projections of the absolute angular velocity of a right body on the fixed axes x, y, z. It follows from Euler's theorem and the choice of ω_1, ω_2, ω_3 that

$$\eta_1 = p, \quad \eta_2 = q, \quad \eta_3 = r.$$

Poincaré's equation for $i = 1$ is

$$\frac{d}{dt} \frac{\partial T}{\partial p} = c_{213} \eta_2 \frac{\partial T}{\partial \eta_3} + c_{312} \eta_3 \frac{\partial T}{\partial \eta_2} + X(T + U),$$

or

$$\frac{d}{dt} \frac{\partial T}{\partial p} = q \frac{\partial T}{\partial r} - r \frac{\partial T}{\partial q} + X(T + U).$$

For the general case

$$T = \frac{1}{2} (Ap^2 + Bq^2 + Cr^2 - 2Dqr - 2Erp - 2Fpq),$$

$$Xf = \sum \left(y \frac{\partial f}{\partial z} - z \frac{\partial f}{\partial y} \right),$$

$$XA = X\sum m(y^2 + z^2) = 2\sum m(yz - zy) = 0,$$

$$XB = X\sum m(z^2 + x^2) = 2\sum myz = 2D,$$

$$XC = X\sum m(x^2 + y^2) = -2\sum mzy = -2D,$$

$$XD = X\sum myz = \sum m(y^2 - z^2) = C - B,$$

$$XE = X\sum mzx = \sum myx = F,$$

$$XF = X\sum mxy = -\sum mzx = -E,$$

$$q \frac{\partial T}{\partial r} - r \frac{\partial T}{\partial q} + X(T) = q(Cr - Dq - Ep) - r(Bq - Dr - Fp)$$

$$+ Dq^2 - Dr^2 - (C - B)qr - Frp + Epq = 0.$$

Consequently, Poincaré's equation reduces to

$$\frac{d}{dt} \frac{\partial T}{\partial p} = XU.$$

There are similar equations for the other axes. This is an expression for the theorem on the angular momentum

$$\sigma_x = \frac{\partial T}{\partial p}, \quad \sigma_y = \frac{\partial T}{\partial q}, \quad \sigma_z = \frac{\partial T}{\partial r},$$

since

$$XU, \ YU, \ ZU$$

are the moments of the given forces about the axes.

   Example 2. Assume that x, y, z are moving axes associated with a rigid body and originating at a fixed point (Fig. 163), p, q, r are the projections of the absolute angular velocity of the rigid body on the moving axes. The coordinates x, y, z of the points of the rigid body do not vary and do not define the position of the rigid body in the fixed space. The position of the rigid body is defined by the cosines of the angles β_i^k between the fixed axis $\bar{x}_i$ and the moving axis x_k. We have used here the ordinary notation $x_1 = x$, $x_2 = y$, $x_3 = z$. The axes $\bar{x}$, $\bar{y}$, $\bar{z}$ are fixed. We have

$$\sum \beta_\alpha^k \beta_\alpha^s = \delta_{ks}, \quad \sum \beta_k^\alpha \beta_s^\alpha = \delta_{ks}. \tag{13}$$

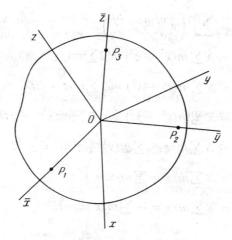

Figure 163

Let

$$p_{ks} = \sum \beta_\alpha^k \frac{d\beta_\alpha^s}{dt},$$

and then

$$p_{ks} = -p_{sk},$$

since we find from (12) that

$$\sum \beta_\alpha^k \frac{d\beta_\alpha^s}{dt} + \sum \beta_\alpha^s \frac{d\beta_\alpha^k}{dt} = 0, \quad p_{kk} = 0.$$

From this we have

$$\sum_k \beta_r^k p_{ks} = \sum_\alpha \left(\sum_k \beta_r^k \beta_\alpha^k \right) \frac{d\beta_\alpha^s}{dt} = \sum_\alpha \delta_{r\alpha} \frac{d\beta_\alpha^s}{dt} = \frac{d\beta_r^s}{dt}.$$

And the final result is

$$\frac{d\beta_r^s}{dt} = \sum \beta_r^k p_{ks}.$$

We can regard these to be constraint relations

$$\frac{d\beta_r^1}{dt} = r\beta_r^2 - q\beta_r^3, \quad \frac{d\beta_r^2}{dt} = p\beta_r^3 - r\beta_r^1, \quad \frac{d\beta_r^3}{dt} = q\beta_r^1 - p\beta_r^2.$$

These are Poisson's equations and they can be written with a direct use of the fact that the points P_1, P_2, P_3 are fixed and lie on the fixed axes,

$$P_1(\bar{x} = 1, \bar{y} = 0, \bar{z} = 0), \quad P_2(\bar{x} = 0, \bar{y} = 1, \bar{z} = 0),$$
$$P_3(\bar{x} = 0, \bar{y} = 0, \bar{z} = 1).$$

Poincaré's equations state that the absolute velocity of the point P_r (β_r^1, β_r^2, β_r^3) is zero. The absolute velocity of the point P_r consists of the relative velocity

$$\frac{d\beta_r^1}{dt}, \quad \frac{d\beta_r^2}{dt}, \quad \frac{d\beta_r^3}{dt}$$

and the transportation velocity (according to Euler's theorem)

$$\left\| \begin{matrix} p & q & r \\ \beta_r^1 & \beta_r^2 & \beta_r^3 \end{matrix} \right\|.$$

From Euler's theorem it follows that

$$p = -p_{23} = p_{32}, \quad q = -p_{31} = p_{13}, \quad r = -p_{12} = p_{21}.$$

Let us consider now a function of the position of the rigid body $f(t, \beta_r^s)$. For this function we have

$$\frac{df}{dt} = \sum \frac{\partial f}{\partial \beta_r^s} \frac{d\beta_r^s}{dt} + \frac{\partial f}{\partial t} = \sum \frac{\partial f}{\partial \beta_r^s} \beta_r^k p_{ks} + \frac{\partial f}{\partial t}$$

$$= p \sum_\alpha \left(\frac{\partial f}{\partial \beta_\alpha^2} \beta_\alpha^3 - \frac{\partial f}{\partial \beta_\alpha^3} \beta_\alpha^2 \right)$$

$$+ q \sum_\alpha \left(\frac{\partial f}{\partial \beta_\alpha^3} \beta_\alpha^1 - \frac{\partial f}{\partial \beta_\alpha^1} \beta_\alpha^3 \right) + r \sum_\alpha \left(\frac{\partial f}{\partial \beta_\alpha^1} \beta_\alpha^2 - \frac{\partial f}{\partial \beta_\alpha^2} \beta_\alpha^1 \right) + \frac{\partial f}{\partial t}$$

and, consequently,

$$X_0 = \frac{\partial f}{\partial t}, \quad X = \sum \left(\frac{\partial f}{\partial \beta_\alpha^2} \beta_\alpha^3 - \frac{\partial f}{\partial \beta_\alpha^3} \beta_\alpha^2 \right),$$

$$Y = \sum \left(\frac{\partial f}{\partial \beta_\alpha^3} \beta_\alpha^1 - \frac{\partial f}{\partial \beta_\alpha^1} \beta_\alpha^3 \right), \quad Z = \sum \left(\frac{\partial f}{\partial \beta_\alpha^1} \beta_\alpha^2 - \frac{\partial f}{\partial \beta_\alpha^2} \beta_\alpha^1 \right),$$

$$\eta_1 = p, \quad \eta_2 = q, \quad \eta_3 = r.$$

Hence

$$(X, Y)f = X(Y, f) - Y(X, f) = \sum \left(\frac{\partial f}{\partial \beta_\alpha^1} \beta_\alpha^2 - \frac{\partial f}{\partial \beta_\alpha^2} \beta_\alpha^1 \right) = Zf,$$

and, similarly,

$$(Y, Z)f = Xf, \quad (Z, X)f = Yf.$$

Consequently, the infinitesimal operators X, Y, Z form a Lie group of virtual displacements. The nonzero structural constants of the Lie group of virtual displacements are

$$c_{123} = 1, \quad c_{231} = 1, \quad c_{312} = 1, \quad c_{213} = -1, \quad c_{321} = -1, \quad c_{132} = -1.$$

In this case the operator $X_0 f = \dfrac{\partial f}{\partial t}$ is commutative with the Lie group X, Y, Z of virtual displacements, and this means that $c_{0\alpha\beta} = 0$ (α, $\beta = 1, 2, 3$).

With the use of the notation $X_1 = X$, $X_2 = Y$, $X_3 = Z$ Poincaré's equations have the form

$$\frac{d}{dt} \frac{\partial T}{\partial \eta_i} = \sum c_{\alpha i \beta} \eta_\alpha \frac{\partial T}{\partial q_\beta} + X_i(T + U) \quad (i = 1, 2, 3).$$

We adopt as the fixed axes the principal axes of an ellipsoid of inertia of a rigid body constructed for a fixed point. In these axes

$$2T = Ap^2 + Bq^2 + Cr^2,$$

where A, B, C are constant moments of inertia of the rigid body about the x, y, z axes respectively. Hence ($i = 1$)

$$\frac{d}{dt} Ap = c_{213} qCr + c_{312} rBq + XU,$$

$$\frac{d}{dt} Ap = (B - C)qr + XU$$

and so on. These are Euler's well-known equations.

9. Example. *Similarity transformation of a body* (D. N. Zeiliger) (Fig. 164). Assume that β_i^k is the cosine of the angle between a fixed axis $\bar{x}_i$ and a moving axis x_k, and that a_1, a_2, a_3 are the coordinates of O which is the origin of the moving coordinate system and is the centre of gravity of the body. The coordinates of the particles m of the body are

$$\bar{x}_i = a_i + \sum_{k=1}^{3} \beta_i^k x_k \quad (i = 1, 2, 3).$$

We assume a_i, β_i^k, x_k to be the defining coordinates, and then

$$\frac{da_i}{dt} = u_i \quad (i = 1, 2, 3), \quad \frac{d\beta_r^1}{dt} = r\beta_r^2 - q\beta_r^3, \quad \frac{d\beta_r^2}{dt} = p\beta_r^3 - r\beta_r^1,$$

$$\frac{d\beta_r^3}{dt} = q\beta_r^1 - p\beta_r^2 \quad (r = 1, 2, 3).$$

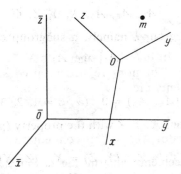

Figure 164

Ray dilatation. If η is the velocity of the ray dilatation, then

$$\frac{dx_k}{dt} = \eta x_k \quad (k = 1, 2, 3).$$

Let $f(t, a_i, \beta_i^k, x_k)$. Then

$$\frac{df}{dt} = \frac{\partial f}{\partial t} + \sum_\alpha u_\alpha \frac{\partial f}{\partial a_\alpha} + p \sum_\alpha \left(\frac{\partial f}{\partial \beta_\alpha^2} \beta_\alpha^3 - \frac{\partial f}{\partial \beta_\alpha^3} \beta_\alpha^2 \right)$$

$$+ q \sum_\alpha \left(\frac{\partial f}{\partial \beta_\alpha^3} \beta_\alpha^1 - \frac{\partial f}{\partial \beta_\alpha^1} \beta_\alpha^3 \right) + r \sum_\alpha \left(\frac{\partial f}{\partial \beta_\alpha^1} \beta_\alpha^2 - \frac{\partial f}{\partial \beta_\alpha^2} \beta_\alpha^1 \right)$$

$$+ \eta \sum \left(\frac{\partial f}{\partial x} x + \frac{\partial f}{\partial y} y + \frac{\partial f}{\partial z} z \right).$$

Consequently, Lie's infinitesimal group of the true displacements of a body under similarity transformation consists of operators

$$X_0 = \frac{\partial}{\partial t}, \quad A_1 = \frac{\partial}{\partial a_1}, \quad A_2 = \frac{\partial}{\partial a_2}, \quad A_3 = \frac{\partial}{\partial a_3},$$

$$X = \sum_\alpha \left(\frac{\partial}{\partial \beta_\alpha^2} \beta_\alpha^3 - \frac{\partial}{\partial \beta_\alpha^3} \beta_\alpha^2 \right), \quad Y = \sum_\alpha \left(\frac{\partial}{\partial \beta_\alpha^3} \beta_\alpha^1 - \frac{\partial}{\partial \beta_\alpha^1} \beta_\alpha^3 \right),$$

$$Z = \sum_\alpha \left(\frac{\partial}{\partial \beta_\alpha^1} \beta_\alpha^2 - \frac{\partial}{\partial \beta_\alpha^2} \beta_\alpha^1 \right), \quad \Phi = \sum \left(x \frac{\partial f}{\partial x} + y \frac{\partial f}{\partial y} + z \frac{\partial f}{\partial z} \right).$$

The subgroup of virtual displacements

$$A_1, \; A_2, \; A_3, \; X, \; Y, \; Z, \; \Phi$$

consists of three subgroups, namely, a subgroup of translations

$$A_1, \; A_2, \; A_3$$

with the property

$$(A_\alpha, \; A_\beta) = 0 \quad (\alpha, \; \beta = 1, \; 2, \; 3),$$

a subgroup of rotations X, Y, Z with the property (proved in the preceding example)

$$(X, \; Y) = Z, \quad (Y, \; Z) = X, \quad (Z, \; X) = Y,$$

a subgroup of ray dilatation consisting of only one transformation Φ.
These subgroups are commutable in the sense that

$$(A_\alpha, \; X_\beta) = 0, \quad (A_\alpha, \; \Phi) = 0, \quad (X_\alpha, \; \Phi) = 0 \quad (\alpha, \; \beta = 1, \; 2, \; 3).$$

Let us calculate the kinetic energy

$$T = \sum \frac{m}{2} \left(\bar{x}'^2 + \bar{y}'^2 + \bar{z}'^2 \right).$$

We have

$$\bar{x}_i' = a_i' + \sum x_k \frac{d\beta_i^k}{dt} + \sum \beta_i^k \frac{dx^k}{dt}$$

$$= a_i' + x_1(r\beta_i^2 - q\beta_i^3) + x_2(p\beta_i^3 - r\beta_i^1) + x_3(q\beta_i^1 - p\beta_i^2) + \eta \sum \beta_i^k x_k$$

$$= a_i' + p(x_2\beta_i^3 - x_3\beta_i^2) + q(x_3\beta_i^1 - x_1\beta_i^3) + r(x_1\beta_i^2 - x_2\beta_i^1) + \eta \sum \beta_i^k x_k$$

and hence

$$T = \sum\sum \frac{m}{2} \bar{x}_i'^2 = \sum \frac{m}{2} \sum \left(a_i' + p(x_2\beta_i^3 - x_3\beta_i^2) \right.$$
$$\left. + q(x_3\beta_i^1 - x_1\beta_i^3) + r(x_1\beta_i^2 - x_2\beta_i^1) + \eta \sum \beta_i^k x_k \right)^2.$$

Since O is the centre of mass of the system, it follows that

$$\sum m x_\alpha = 0 \quad (\alpha = 1, \; 2, \; 3),$$

$$\sum m \sum_i \left(\sum \beta_i^k x_k \right)^2 = \sum m(x^2 + y^2 + z^2) = \Pi,$$

where Π is the polar moment of inertia of the body about the centre of mass O. Furthermore,

$$\sum m \sum [p(x_2\beta_i^3 - x_3\beta_i^2) + q(x_3\beta_i^1 - x_1\beta_i^3) + r(x_1\beta_i^2 - x_2\beta_i^1)]^2$$
$$= Ap^2 + Bq^2 + Cr^2 - 2Dqr - 2Erp - 2Fpq$$

where

$$A = \sum m(y^2 + z^2), \quad B = \sum m(z^2 + x^2), \quad C = \sum m(x^2 + y^2),$$

$$D = \sum myz, \quad E = \sum mzx, \quad F = \sum mxy.$$

Next we have

$$\sum m \sum_i [p(x_2\beta_i^3 - x_3\beta_i^2) + \ldots] \sum \beta_i^k x_k$$

$$= \sum m \left(p \sum_k x_2 x_k \underbrace{\sum_i \beta_i^3 \beta_i^k}_{\delta_{3k}} - p \sum_k x_3 x_k \underbrace{\sum_i \beta_i^2 \beta_i^k}_{\delta_{2k}} + \ldots \right)$$

$$= \sum m[p(x_2 x_3 - x_3 x_2) + \ldots] = 0.$$

The final result is

$$T = \frac{M}{2}(a_1'^2 + a_2'^2 + a_3'^2)$$
$$+ \frac{1}{2}(Ap^2 + Bq^2 + Cr^2 - 2Dqr - 2Erp - 2Fpq) + \Pi\frac{\eta^2}{2}.$$

If we assume the principal axes of the central ellipsoid of inertia to be the coordinate axes x, y, z, then $D = E = F = 0$ and the kinetic energy assumes a simple form

$$T = \frac{M}{2}(a_1'^2 + a_2'^2 + a_3'^2) + \frac{1}{2}(Ap^2 + Bq^2 + Cr^2) + \Pi\frac{\eta^2}{2}.$$

Then Poincaré's equations of motion become

$$M\frac{d^2 a_1}{dt^2} = A_1 U, \quad M\frac{d^2 a_2}{dt^2} = A_2 U, \quad M\frac{d^2 a_3}{dt^2} = A_3 U,$$

$$\frac{d}{dt} Ap = c_{213}qCr + c_{312}rBq + XU,$$

or

$$\frac{d}{dt} Ap = (B - C)qr + XU,$$

$$\ldots\ldots\ldots\ldots\ldots\ldots\ldots\ldots$$

and, furthermore,

$$\frac{d\Pi\eta}{dt} = \Phi(T + U),$$

but

$$\Phi A = \sum \left(x\frac{\partial}{\partial x} + y\frac{\partial}{\partial y} + z\frac{\partial}{\partial z} \right) \sum m(y^2 + z^2) = \sum m(2y^2 + 2z^2) = 2A,$$

$$\Phi B = 2B, \quad \Phi C = 2C, \quad \Phi\Pi = 2\Pi.$$

Consequently,

$$\Phi(T) = Ap^2 + Bq^2 + Cr^2 + \Pi\eta^2, \quad \Phi U \text{ is a virial.}$$

Thus, the system of equations of motion of a body under a similarity transformation is

$$\frac{d}{dt}\Pi\eta = Ap^2 + Bq^2 + Cr^2 + \Pi\eta^2 + \Phi U, \quad \frac{d}{dt}Ap = (B - C)qr + XU,$$

$$\frac{d}{dt}Bq = (C - A)rp + YU, \quad \frac{d}{dt}Cr = (A - B)pq + ZU,$$

$$M\frac{d^2a_1}{dt^2} = A_1 U, \quad M\frac{d^2a_2}{dt^2} = A_2 U, \quad M\frac{d^2a_3}{dt^2} = A_3 U.$$

These equations were derived by Professor D. N. Zeiliger.

Zeiliger's first equation has the form

$$\frac{d^2\Pi}{dt^2} = 2(Ap^2 + Bq^2 + Cr^2 + \Pi\eta^2) + 2\Phi U$$

since

$$\Pi\eta = \eta \sum m\varrho^2 = \sum m\varrho\frac{d\varrho}{dt} = \frac{1}{2}\frac{d}{dt}\sum m\varrho^2 = \frac{1}{2}\frac{d\Pi}{dt}.$$

10. Example. *A body under an affine transformation.* The space x, y, z is fixed. A transformation under which the zero values of the parameters a_i, p, q, r, ε_1, ε_2, ε_3, σ_1, σ_2, σ_3 are associated with an identity transformation is taken in the form

$$x' = x + a_1 + qz - ry + \varepsilon_1 x + \sigma_3 y + \sigma_2 z,$$
$$y' = y + a_2 + rx - pz + \sigma_3 x + \varepsilon_2 y + \sigma_1 z,$$
$$z' = z + a_3 + py - qx + \sigma_2 x + \sigma_1 y + \varepsilon_3 z.$$

Hence the group of virtual displacements of a body under an affine transformation consists of the operators for translation

$$X_1 = \frac{\partial}{\partial x}, \quad X_2 = \frac{\partial}{\partial y}, \quad X_3 = \frac{\partial}{\partial z},$$

rotation

$$X_p = y\frac{\partial}{\partial z} - z\frac{\partial}{\partial y}, \quad X_q = z\frac{\partial}{\partial x} - x\frac{\partial}{\partial z}, \quad X_r = x\frac{\partial}{\partial y} - y\frac{\partial}{\partial x}$$

and deformation

$$X_{\varepsilon_1} = x\frac{\partial}{\partial x}, \quad X_{\varepsilon_2} = y\frac{\partial}{\partial y}, \quad X_{\varepsilon_3} = z\frac{\partial}{\partial z},$$

$$X_{\sigma_1} = y\frac{\partial}{\partial z} + z\frac{\partial}{\partial y}, \quad X_{\sigma_2} = z\frac{\partial}{\partial x} + x\frac{\partial}{\partial z}, \quad X_{\sigma_3} = x\frac{\partial}{\partial y} + y\frac{\partial}{\partial x}.$$

Example 1. *A plane body under an affine transformation.* The general affine group in the plane has the form

$$X_1 = \sum \frac{\partial}{\partial x}, \quad X_2 = \sum \frac{\partial}{\partial y}, \quad X_3 = \sum x\frac{\partial}{\partial x},$$

$$X_4 = \sum x\frac{\partial}{\partial y}, \quad X_5 = \sum y\frac{\partial}{\partial x}, \quad X_6 = \sum y\frac{\partial}{\partial y}.$$

Example 2. An affine unimodular group in the plane is

$$X_1 = \sum \frac{\partial}{\partial x}, \quad X_2 = \sum \frac{\partial}{\partial y}, \quad X_3 = \sum x\frac{\partial}{\partial y},$$

$$X_4 = \sum y\frac{\partial}{\partial x}, \quad X_5 = \sum \left(x\frac{\partial}{\partial x} - y\frac{\partial}{\partial y} \right).$$

11. If there is no force function and a particle of mass m is acted on by a force with projections U, V, W on fixed axes, then the Euler-Lagrange principle has the form

$$\sum m\left(\frac{d^2u}{dt^2}\delta u + \frac{d^2v}{dt^2}\delta v + \frac{d^2w}{dt^2}\delta w \right) = \sum (U\delta u + V\delta v + W\delta w).$$

Repeating all the calculations we have carried out when deriving Poincaré's equations, we get equations of motion in the form

$$\frac{d}{dt}\frac{\partial T}{\partial \eta_i} - XT = \sum c_{\alpha i\beta}\eta_\alpha \frac{\partial T}{\partial \eta_\beta} + \sum c_{0i\beta}\frac{\partial T}{\partial \eta_\beta} + Q_i,$$

where $Q_i = \sum (UX_i u + VX_i v + WX_i w)$ is a generalized force. The mechanical meaning of the generalized force follows from the relation

$$A = \sum (U\delta u + V\delta v + W\delta w) = \sum \omega_i Q_i,$$

whence Q_i is the virtual work A_i of the forces, when one parameter ω_i of the virtual displacement is nonzero and the other parameters ω_j $(j \neq i)$ are zero, divided by ω_i, i.e.

$$Q_i = \frac{A_i}{\omega_i}.$$

12. Cyclic displacements. Poincaré's equations are

$$\frac{d}{dt}\left(\frac{\partial T}{\partial \eta_i}\right) = \sum c_{\alpha i \beta} \eta_\alpha \frac{\partial T}{\partial \eta_\beta} + \sum c_{0i\beta} \frac{\partial T}{\partial \eta_\beta} + X_i(T + U)$$

$$(i = 1, \ldots, k).$$

Assume that for $i = s + 1, \ldots, k$ the following equalities hold true:

1°. $c_{\alpha i \beta} = 0$ $(\alpha = 1, 2, \ldots, k)$, $(X_\alpha, X_i) = 0$;
2°. $c_{0i\beta} = 0$ $(X_0, X_i) = 0$;
3°. $X_i(T + U) = 0$ $(i = s + 1, \ldots, k)$.

The displacements X_i with the indicated properties are called cyclic.

Cyclic displacements form an Abel subgroup of the group of virtual displacements, which is commutable with all X_α $(\alpha = 1, \ldots, k)$. Poincaré's equations for cyclic displacements yield

$$\frac{d}{dt} \frac{\partial T}{\partial \eta_\alpha} = 0 \quad (\alpha = s + 1, \ldots, k),$$

or, in other words, for cyclic displacements we have integrals

$$\frac{\partial T}{\partial \eta_\alpha} = \beta_\alpha \quad (\alpha = s + 1, \ldots, k),$$

where β_α are integration constants.

The parameters $\eta_{s+1}, \ldots, \eta_k$ appearing in the last system can be expressed in terms of β_α and the defining variables $t, x_1, \ldots, x_n$ and $\eta_1, \ldots, \eta_s$. Assuming this substitution to be fulfilled, we consider a function

$$R(t, x_1, \ldots, x_n, \eta_1, \ldots, \eta_s, \beta_{s+1}, \ldots, \beta_k) = T + U - \sum_{\alpha = s+1}^{k} \frac{\partial T}{\partial \eta_a} \eta_\alpha.$$

22*

From this we get

$$\delta R = \sum_{j=1}^{k} \omega_j X_j R + \sum_{1}^{s} \frac{\partial R}{\partial \eta_r} \delta \eta_r + \sum_{s+1}^{k} \frac{\partial R}{\partial \beta_\alpha} \delta \beta_\alpha$$

$$= \sum_{1}^{k} \omega_j X_j (T + U) + \sum_{1}^{k} \frac{\partial T}{\partial \eta_j} \delta \eta_j - \sum_{s+1}^{k} \eta_\alpha \delta \beta_\alpha - \sum_{s+1}^{k} \beta_\alpha \delta \eta_\alpha$$

$$= \sum_{1}^{k} \omega_j X_j (T + U) + \sum_{1}^{s} \frac{\partial T}{\partial \eta_r} \delta \eta_r - \sum_{s+1}^{k} \eta_\alpha \delta \beta_\alpha.$$

Comparing the coefficients in the same variations and taking into account that $X_\alpha(T + U) = 0$ $(\alpha = s + 1, \ldots, k)$, we obtain

$$X_r R = X_r(T + U), \quad X_\alpha R = 0,$$

$$\frac{\partial R}{\partial \eta_r} = \frac{\partial T}{\partial \eta_r}, \quad \frac{\partial R}{\partial \beta_\alpha} = -\eta_\alpha \quad (r = 1, \ldots, s, \; \alpha = s + 1, \ldots, k).$$

In accordance with the relations obtained, Poincaré's equations for noncyclic displacements assume the form

$$\frac{d}{dt} \frac{\partial R}{\partial \eta_r} = \sum c_{\alpha r \beta} \eta_\alpha \frac{\partial R}{\partial \eta_\beta} + \sum c_{\alpha r \gamma} \eta_\alpha \beta_\gamma + \sum c_{0 r \beta} \frac{\partial R}{\partial \eta_\beta} + \sum c_{0 r \gamma} \beta_\gamma$$

$$+ X_r R \quad (r, \alpha, \beta = 1, 2, \ldots, s, \; \gamma = s + 1, \ldots, k).$$

When the last equations are integrated, the values of Poincaré's parameters η_α $(\alpha = s + 1, \ldots, k)$ are defined by the relations:

$$\eta_\alpha = - \frac{\partial R}{\partial \beta_\alpha}.$$

Example. The first elementary integrals of this kind were given in Chaplygin's paper *On Some Possible Generalization of the Area Theorem as Applied to the Rolling of Balls* written in 1897.

Remark. Integrals associated with cyclic coordinates or cyclic displacements are linear with respect to velocities or linear with respect to the Poincaré parameters η_α of true displacements.

13. Canonical Equations. We introduce new variables

$$y_i = \frac{\partial T}{\partial \eta_i} \quad (i = 1, \ldots, k).$$

This system can be solved for η_i. As a result we get a unique solution $\eta_i = \eta_i(t, x_1, x_2, \ldots, x_n, y_1, \ldots, y_k)$ since $\det \left\| \dfrac{\partial^2 T}{\partial \eta_\alpha \partial \eta_\beta} \right\| > 0$, because T is a positive function of $\eta_1, \ldots, \eta_k$ by definition.

Consider a function

$$H(t, x_1, \ldots, x_n, y_1, \ldots, y_k) = \sum y_\alpha \eta_\alpha - T - U.$$

It follows that

$$\delta H = \sum \omega_j X_j H + \sum \frac{\partial H}{\partial y_\alpha} \delta y_\alpha$$

$$= \sum y_\alpha \delta \eta_\alpha + \sum \eta_\alpha \delta y_\alpha - \sum \omega_j X_j (T + U) - \sum \frac{\partial T}{\partial \eta_\alpha} \delta \eta_\alpha$$

(the sum of the first and the last term on the right-hand side is zero). Comparing the coefficients in the same variations, we get

$$X_j H = -X_j (T + U), \quad \eta_\alpha = \frac{\partial H}{\partial y_\alpha}.$$

Consequently, the canonical equations have the form

$$\frac{dy_i}{dt} = \sum c_{\alpha i \beta} \frac{\partial H}{\partial y_\alpha} y_\beta + \sum c_{0 i \beta} y_\beta - X_i H, \quad \eta_\alpha = \frac{\partial H}{\partial y_\alpha}.$$

We can impart another form to these equations:

$$\frac{dy_i}{dt} = \sum c_{\alpha i \beta} \frac{\partial H}{\partial y_\alpha} y_\beta + \sum c_{0 i \beta} y_\beta - X_i H, \tag{14}$$

$$\frac{dx_j}{dt} = X_0 x_j + \sum \frac{\partial H}{\partial y_\alpha} X_\alpha x_j \quad (i, \alpha, \beta = 1, \ldots, k; \quad j = 1, \ldots, n).$$

14. A partial differential equation. Consider the following theorem. *If we know the complete integral*

$$V(t, x_1, \ldots, x_n, a_1, \ldots, a_n) + a_{n+1}$$

of the first-order partial differential equation

$$X_0 V + H(t, x_1, \ldots, x_n, X_1 V, \ldots, X_k V) = 0,$$

then the solution of the canonical equations (14) *is defined by the system*

$$\frac{\partial V}{\partial a_j} = b_j, \quad y_\alpha = X_\alpha V \quad (j = 1, \ldots, n, \ \alpha = 1, \ldots, k). \tag{15}$$

Indeed, for the complete integral the functional determinant

$$\left\| \frac{\partial^2 V}{\partial x_j \partial a_r} \right\|$$

is not zero since otherwise we would have a ratio $\partial V / \partial x_j$, i.e. the function V would be an integral of one more equation, which does not include $\partial V / \partial t$, and this contradicts the definition of a complete integral.

We differentiate (15),

$$0 = \frac{\partial^2 V}{\partial t \, \partial a_j} + \sum \frac{\partial^2 V}{\partial x_r \partial a_j} \frac{dx_r}{dt}.$$

Then substitute the complete integral into the result and differentiate the obtained identity with respect to a_j.

$$0 = X_0 \frac{\partial V}{\partial a_j} + \sum \frac{\partial H}{\partial y_\alpha} X_\alpha \frac{\partial V}{\partial a_j}$$

$$= \frac{\partial^2 V}{\partial a_j \partial t} + \sum \frac{\partial^2 V}{\partial a_j \partial x_r} X_0 x_r + \sum \frac{\partial H}{\partial y_\alpha} \frac{\partial^2 V}{\partial a_j \partial x_r} X_\alpha x_r$$

$$= \frac{\partial^2 V}{\partial a_j \partial t} + \sum \frac{\partial^2 V}{\partial a_j \partial x_r} \left(X_0 x_r + \sum \frac{\partial H}{\partial y_\alpha} X_\alpha x_r \right).$$

Since

$$\left\| \frac{\partial^2 V}{\partial a_j \partial x_r} \right\| \neq 0,$$

the last two systems have a unique solution. Therefore

$$\frac{dx_r}{dt} = X_0 x_r + \sum \frac{\partial H}{\partial y_\alpha} X_\alpha x_r.$$

Thus x_r, found from relations (15), satisfy the second system of the canonical equations. Furthermore,

$$\frac{dy_i}{dt} = \frac{d}{dt} X_i V = X_0(X_i V) + \sum \eta_\alpha X_\alpha(X_i V) = X_0(X_i V)$$

$$+ \sum \frac{\partial H}{\partial y_\alpha} X_\alpha(X_i V) = \sum c_{0i\beta} X_\beta V + X_i(X_0 V)$$

$$+ \sum \frac{\partial H}{\partial y_\alpha} c_{\alpha i \beta} X_\beta V + \sum \frac{\partial H}{\partial y_\alpha} X_i(X_\alpha V) + X_i H - X_i H$$

$$= \sum c_{\alpha i \beta} \frac{\partial H}{\partial y_\alpha} y_\beta + \sum c_{0i\beta} y_\beta - X_i H + X_i(X_0 V$$

$$+ H(t, x_1, \ldots, x_n, X_1 V, \ldots, X_k V)).$$

And this is what we wished to prove.

15. The kinetic energy integral. We multiply Poincaré's equations $\left(\text{the case } c_{0i\beta} = 0, \quad X_0 = \dfrac{\partial}{\partial t}\right)$

$$\frac{d}{dt}\frac{\partial T}{\partial \eta_i} = \sum c_{\alpha i\beta}\eta_\alpha \frac{\partial T}{\partial \eta_\beta} + X_i(T + U)$$

by η_i and add the results together:

$$\sum \eta_i \frac{d}{dt}\frac{\partial T}{\partial \eta_i} = \sum \frac{\partial T}{\partial \eta_\beta}\sum c_{\alpha i\beta}\eta_\alpha\eta_i + \sum \eta_i X_i(T + U)$$

(the first sum on the right-hand side is zero since $c_{\alpha i\beta} = -c_{i\alpha\beta}$);

$$\frac{d}{dt}\left(\sum \frac{\partial T}{\partial \eta_i}\eta_i\right) = \sum \frac{\partial T}{\partial \eta_i}\frac{d\eta_i}{dt} + \sum \eta_i X_i(T + U).$$

If $\dfrac{\partial}{\partial t}(T + U) = 0$, then

$$\frac{d}{dt}\left(\sum \frac{\partial T}{\partial \eta_i}\eta_i\right) = \frac{d}{dt}(T + U),$$

and, consequently,

$$\sum \frac{\partial T}{\partial \eta_i}\eta_i = T + U + h.$$

If $T = T_2 + T_1 + T_0$, then

$$2T_2 + T_1 = T_2 + T_1 + T_0 + U + h, \quad T_2 = T_0 + U + h.$$

This is the generalization of the kinetic energy integral. Its special cases are well-known, they are (1) Jacobi's integral and (2) the ordinary kinetic energy integral.

16. The case of integrability. Generalization of cyclic variables. Let $c_{0\alpha\beta} = 0$,

$$\frac{d}{dt}\frac{\partial T}{\partial \eta_i} = \sum c_{\alpha i\beta}\eta_\alpha \frac{\partial T}{\partial \eta_\beta} + X_i(T + U),$$

and i is fixed. Let

$$T = \frac{1}{2}\sum a_{\alpha\beta}\eta_\alpha\eta_\beta \quad (a_{\alpha\beta} = a_{\beta\alpha}).$$

Hence

$$\sum c_{\alpha i\beta}\eta_\alpha a_{\gamma\beta}\eta_\beta = \sum \eta_\alpha\eta_\gamma \sum_{\gamma=1}^{k} c_{\alpha i\beta}a_{\gamma\beta}.$$

It may so happen that

1°. $\displaystyle\sum_{\beta=1}^{k} c_{\alpha i\beta} a_{\gamma\beta} = 0$

for all α, $\beta = 1, \ldots, k$ and a certain i. And if

2°. $X_i(T + U) = 0$,

then Poincaré's equation can be reduced to the form

$$\frac{d}{dt}\frac{\partial T}{\partial \eta_i} = 0,$$

whence follows an integral

$$\frac{\partial T}{\partial \eta_i} = \beta_i.$$

Example. This case is encountered when a heavy rigid body with one fixed point is moving, i.e. Lagrange's case. If the defining variables are taken to be Euler's angles, which define the position of the principal axes of an ellipsoid of inertia of the body constructed for the fixed point relative to the fixed axes $x_1 y_1 z_1$, where z_1 is directed vertically upwards, then

$$p = \psi' \sin\theta \sin\varphi + \theta' \cos\varphi,$$

$$q = \psi' \sin\theta \cos\varphi - \theta' \sin\varphi, \quad r = \psi' \cos\theta + \varphi'.$$

Hence

$$\theta' = p\cos\varphi - q\sin\varphi, \quad \psi' = \frac{1}{\sin\theta}(p\sin\varphi + q\cos\varphi),$$

$$\varphi' = r - \cot\theta(p\sin\varphi + q\cos\varphi),$$

and then

$$\frac{df}{dt} = \frac{\partial f}{\partial t} + pXf + qYf + rZf,$$

$$X = -\cot\theta\sin\varphi\,\frac{\partial}{\partial\varphi} + \frac{\sin\varphi}{\sin\theta}\,\frac{\partial}{\partial\psi} + \cos\varphi\,\frac{\partial}{\partial\theta},$$

$$Y = -\cot\theta\cos\varphi\,\frac{\partial}{\partial\varphi} + \frac{\cos\varphi}{\sin\theta}\,\frac{\partial}{\partial\psi} - \sin\varphi\,\frac{\partial}{\partial\theta},$$

$$Z = \frac{\partial}{\partial\varphi}$$

is a transitive group. From this we have

$$(X, Y) = Z, \quad (Y, Z) = X, \quad (Z, X) = Y.$$

The kinetic energy is

$$T = \frac{1}{2}(Ap^2 + Bq^2 + Cr^2).$$

Let $i = 3$. The conditions for integrability are

$1°$. $X_3 U = 0$, $X_3 = Z = \dfrac{\partial}{\partial\varphi}$,

$2°$. $\sum c_{\alpha i\beta} a_{\gamma\beta} = 0$;

and in our case we have

$$c_{132}a_{22} + c_{231}a_{11} = -1 \cdot B + 1 \cdot A = A - B = 0.$$

Thus, for this case of integrability we must have $A = B$, $\dfrac{\partial U}{\partial\varphi} = 0$. This condition is fulfilled in Lagrange's case $U = -mgz_0 \cos\theta$. The integral is known, it is $r = r_0$.

Generalization:

$$\frac{d}{dt}\frac{\partial T}{\partial\eta_i} = \sum c_{\alpha i\beta}\eta_\alpha \frac{\partial T}{\partial\eta_\beta} + \sum c_{0i\beta}\frac{\partial T}{\partial\eta_\beta} + X_i(T + U),$$

here i is fixed. Let

$$T = \frac{1}{2}\sum a_{\alpha\beta}\eta_\alpha\eta_\beta + \sum b_\beta\eta_\beta + T_0,$$

and then

$$\sum c_{\alpha i\beta}\eta_\alpha \frac{\partial T}{\partial\eta_\beta} + \sum c_{0i\beta}\frac{\partial T}{\partial\eta_\beta}$$

$$= \sum c_{\alpha i\beta}\eta_\alpha a_{\gamma\beta}\eta_\gamma + \sum c_{\alpha i\beta}\eta_\alpha b_\beta + \sum c_{0i\beta}a_{\gamma\beta}\eta_\gamma + \sum c_{0i\beta}b_\beta.$$

The conditions for the generalized cyclic displacements are

$1°$. $\displaystyle\sum_{\beta=1}^{k} c_{\alpha i\beta}a_{\gamma\beta} = 0 \quad (\alpha, \gamma = 1, \ldots, k),$

$2°$. $\displaystyle\sum_{\beta=1}^{k} c_{\alpha i\beta}b_\beta + \sum_{\beta=1}^{k} c_{0i\beta}a_{\alpha\beta} = 0,$

$3°$. $\displaystyle\sum_{\beta=1}^{k} c_{0i\beta}b_\beta = 0,$

$4°$. $X_i(T + U) = 0.$

The integral is
$$\frac{\partial T}{\partial \eta_i} = \text{const.}$$

The integration method presented in item 12 is valid for these cyclic displacements.

17. A complete integral. We must now return to the group of true displacements. The constraints imposed on the mechanical system are expressed by the integrable equations

$$\pi_j \equiv \sum_{s=1}^{n} a_{js} dx_s + a_j dt = 0 \quad (j = k + 1, \ldots, n).$$

Since this system is integrable, it can be reduced to the form

$$d\Phi_j = 0 \quad (j = k + 1, \ldots, n).$$

To construct an infinitesimal group of true displacements, we complete the equations by independent forms

$$\pi_\alpha \equiv \sum_{s=1}^{n} a_{\alpha s} dx_s \quad (\alpha = 1, \ldots, k), \quad \pi_0 \equiv dt$$

under the condition that all the forms π_0, π_1, $\ldots$, π_n are independent

$$|a_{rs}| \neq 0 \quad (r, s = 1, \ldots, n).$$

The differential of the function $f(t, x_1, \ldots, x_n)$ can be expressed in terms of the operators

$$df = \pi_0 X_0 f + \sum_{\alpha=1}^{n} \pi_\alpha X_\alpha f.$$

For the true displacements $\pi_{k+1} = 0$, $\ldots$, $\pi_n = 0$ and, consequently,

$$df = \pi_0 X_0 f + \sum_{r=1}^{k} \pi_r X_r f.$$

Now we have the following from $d\Phi_j = 0$:

$$X_0 \Phi_j = 0, \quad X_r \Phi_j = 0 \quad (r = 1, \ldots, k, \, j = k + 1, \ldots, n)$$

for the functions Φ_j $(j = k + 1, \ldots, n)$. Therefore, we can add terms linear with respect to Φ_j and with arbitrary constant coefficients

$$V + \sum_{j=k+1}^{n} c_j \Phi_j$$

to the complete integral V of the equation

$$X_0 V + H(t, x_1, \ldots, x_n, X_1 V, \ldots, X_k V) = 0. \tag{16}$$

Consequently, of the n essential constants $a_1, \ldots, a_n$, none of which is additive, $n - k$ constants are $c_{k+1}, \ldots, c_n$, and these arbitrary constants enter the last formula in the indicated form. However, there cannot be more than n constants, and, consequently, the complete integral has the form

$$W = V(t, x_1, \ldots, x_n, a_1, \ldots, a_k) + \sum_{j=k+1}^{n} a_j \Phi_j + a_{n+1}.$$

Thus, $n - k$ of the relations in the theorem on a complete integral (item 14) refer to the definition of constant holonomic constraints

$$\frac{\partial W}{\partial a} = b_j = \Phi_j \quad (j = k + 1, \ldots, n).$$

How can we find the principal part of a complete integral when the Lie group of true displacements is intransitive?

We call the function V the principal part of the complete integral

$$X_\alpha W = X_\alpha V \quad (\alpha = 0, 1, \ldots, k),$$

but $X_\alpha W = X_\alpha V = \varphi_\alpha (t, x_1, \ldots, x_n, a_1, \ldots, a_k)$. Excluding $a_1, \ldots, a_k$ form these $k + 1$ relations, we get (16). In other words, we must have

$$\frac{\partial(X_1 V, \ldots, X_k V)}{\partial(a_1, \ldots, a_k)} \neq 0.$$

Thus, using the principal part $V(t, x_1, \ldots, x_n, a_1, \ldots, a_k)$, we get the solution of (16) in the form

$$y_\alpha = X_\alpha V, \quad b_\alpha = \frac{\partial W}{\partial a_\alpha} \quad (\alpha = 1, \ldots, k). \tag{17}$$

When the constraints $\Phi_j = b_j (j = k + 1, \ldots, n)$ are added, the solution must be completely determined from the last relation $b_\alpha = \partial W / \partial a_\alpha$ ($\alpha = 1, \ldots, k$):

$$x_s = x_s(t, a_1, \ldots, a_k, b_1, \ldots, b_k, b_{k+1}, \ldots, b_n).$$

The complete integral W can be found by the ordinary routine methods. How can we find the principal part V? This question has sense when n is large as compared to k.

18. Hamilton's principle. In order to discuss the properties of the complete integral of the Hamilton-Jacobi partial differential equation in more detail, we must consider the action integral. We shall first derive Hamilton's principle from the Euler-Lagrange principle (item 8). We have

$$\int_{t_0}^{t_1} \left(\sum m \left(\frac{du'}{dt} \delta u + \frac{dv'}{dt} \delta v + \frac{dw'}{dt} \delta w \right) - \delta U \right) dt = 0.$$

Under the conditions

$$\omega_\alpha = 0 \quad (\alpha = 1, \ldots, k) \quad \text{for} \quad t = t_0, t_1,$$

(the endpoints are fixed) we have the true motion. We consider

$$\int_{t_0}^{t_1} m \frac{du'}{dt} \delta u \, dt = [mu' \delta u]_{t_0}^{t_1} - \int_{t_0}^{t_1} mu' \frac{d\delta u}{dt} \, dt,$$

but

$$[\delta u]_{t_0}^{t_1} = \left[\sum \omega_\alpha X_\alpha u \right]_{t_0}^{t_1} = 0$$

and

$$\frac{d\delta u}{dt} = \delta \frac{du}{dt}, \quad \ldots$$

by virtue of the relations of commutativity (item 7). Furthermore,

$$-\int_{t_0}^{t_1} mu' \frac{d\delta u}{dt} \, dt = -\int_{t_0}^{t_1} mu' \delta u' \, dt = -\delta \int_{t_0}^{t_1} m \frac{u'^2}{2} \, dt_1.$$

This means that the expression in question assumes the form

$$\delta \int_{t_0}^{t_1} (T + U) dt = 0$$

provided that $\omega_\alpha = 0$ ($\alpha = 1, \ldots, k$) for $t = t_0, t$. (To practice the process of calculation, I have made this conclusion by calculating $\frac{d}{dt} \delta u = \frac{d}{dt} \sum \omega_\alpha X_\alpha$ and using the commutativity relations.) Poincaré first derived his equations from Hamilton's principle. In my papers I also started from Hamilton's principle.

19. Hamilton's principal function. The principal function, or the action integral, is defined by the relation

$$V(t, x_1, \ldots, x_n, t_0, x_1^0, \ldots, x_n^0) = \int_{t_0}^{t} L \, dt = \int_{t_0}^{t} \left(\sum y_\alpha \eta_\alpha = H \right) dt,$$

where, in accordance with what we said in item 13, we have

$$H(t, x_1, \ldots, x_n, y_1, \ldots, y_k) = \sum y_\alpha \eta_\alpha - L$$

and we integrate along the true trajectory of the mechanical system. We have

$$\delta V = \sum \omega_\alpha X_\alpha V + \sum \omega_\alpha^0 X_\alpha^0 V$$

$$= \int_{t_0}^{t} \left(\sum \eta_\alpha \delta y_\alpha + \sum y_\alpha \delta\eta_\alpha - \sum \omega_\alpha X_\alpha H - \sum \frac{\partial H}{\partial y_\alpha} \delta y_\alpha \right) dt$$

$$= \int_{t_0}^{t} \left[\sum \delta y_\alpha \left(\eta_\alpha - \frac{\partial H}{\partial y_\alpha} \right) + \sum y_i \left(\frac{d\omega_i}{dt} + \sum c_{\beta\alpha i}\eta_\beta\omega_\alpha \right. \right.$$

$$\left. + \sum c_{0\alpha i}\omega_\alpha \right) - \sum \omega_\alpha X_\alpha H \right] dt$$

$$= \left[\sum y_i\omega_i \right]_{t_0}^{t_1} + \int_{t_0}^{t} \left[\sum \delta y_\alpha \left(\eta_\alpha - \frac{\partial H}{\partial y_\alpha} \right) \right.$$

$$\left. + \sum \omega_\alpha \left(-\frac{dy_\alpha}{dt} + \sum c_{\beta\alpha i}\eta_\beta y_i + \sum c_{0\alpha i}y_i - X_\alpha H \right) \right] dt$$

$$= \sum y_i\omega_i - \sum y_i^0\omega_i^0$$

by virtue of the canonical equations of motion. Here ω_α^0 are the parameters of the initial variation of the trajectory, X_α^0 denote the infinitesimal operators X_α when the variables t_0 and x_s^0 are substituted into the latter.

A comparison of the coefficients in the arbitrary similar ω_α^0 and ω_α yields

$$y_\alpha = X_\alpha V, \quad -y_\alpha^0 = X_\alpha^0 V \quad (\alpha = 1, \ldots, k).$$

The second group of equations defines the law of motion when the principal function is known.

Furthermore,

$$\frac{dV}{dt} = X_0 V + \sum \eta_\alpha X_\alpha V = L,$$

or

$$X_0 V + H(t, x_1, \ldots, x_n, y_1, \ldots, y_k) = 0,$$

where

$$H(t, x_1, \ldots, x_n, y_1, \ldots, y_k) = \sum y_\alpha \eta_\alpha - L.$$

Consequently, V is a complete integral of the equation

$$X_0 V + H(t, x_1, \ldots, x_n, X_1 V, \ldots, X_k V) = 0.$$

We see that the essential ones of the constants b_α (item 17) of the complete integral, in the case when V is an action, are

$$-y_\alpha^0 = X_\alpha^0 V$$

and are defined by some operators X_α^0 of the complete integral.

20. Different forms of Hamilton's principle of least action. We must discuss this principle in more detail. Hamilton's principle has the form

$$\delta \int_{t_0}^{t_1} (T + U)dt = 0$$

provided that $\omega_\alpha = 0$ $(\alpha = 1, \ldots, k)$ and $t = t_0, t_1$ at the endpoints. Within the interval (t_0, t_1) the parameters ω_α are quite arbitrary.

We shall prove that we can get the equations of motion in canonical form from the condition

$$\delta \int_{t_0}^{t_1} \left(\sum y_\alpha \eta_\alpha - H \right) dt = 0, \tag{18}$$

where H is Hamilton's function

$$H(t, x_1, \ldots, x_n, y_1, \ldots, y_k),$$

when $\omega_\alpha = 0$ $(\alpha = 1, \ldots, k)$ and $t = t_0, t_1$ at the endpoints and y_α in (18) are independent variables, i.e. when ω_α and δy_α are arbitrary and independent within the interval (t_0, t_1). Indeed,

$$\delta \int_{t_0}^{t_1} \left(\sum y_\alpha \eta_\alpha - H \right) dt$$

$$= \int_{t_0}^{t_1} \left(\sum y_\alpha \delta \eta_\alpha + \sum \eta_\alpha \delta y_\alpha - \sum \frac{\partial H}{\partial y_\alpha} \delta y_\alpha - \sum \omega_\alpha X_\alpha H \right) dt$$

$$= \int_{t_0}^{t_1} \left\{ \sum \delta y_\alpha \left(\eta_\alpha - \frac{\partial H}{\partial y_\alpha} \right) + \sum y_i \left(\frac{d\omega_i}{dt} + \sum c_{\beta \alpha i} \eta_\beta \omega_\alpha + \sum c_{0\alpha i} \omega_\alpha \right) \right.$$

$$\left. - \sum \omega_\alpha X_\alpha H \right\} dt = \left[\sum y_i \omega_i \right]_{t_0}^{t_1} + \int_{t_0}^{t_1} \left\{ \sum \delta y_\alpha \left(\eta_\alpha - \frac{\partial H}{\partial y_\alpha} \right) \right.$$

$$+ \sum \omega_\alpha \left(- \frac{dy_\alpha}{dt} + \sum c_{\beta\alpha i}\eta_\beta y_i + \sum c_{0\alpha i}y_i - X_\alpha H \right) \Big\} dt = 0.$$

The expression outside the integral is zero by virtue of the conditions at the endpoints. Because of the assumption that δy_α and ω_α are independent and arbitrary, we obtain

$$\frac{dy_\alpha}{dt} = \sum c_{\beta\alpha i}\eta_\beta y_i + \sum c_{0\alpha i}y_i - X_\alpha H, \quad \eta_\alpha = \frac{\partial H}{\partial y_\alpha}.$$

These are the canonical equations derived in item 13.

Equation (18) has a significance of its own because of the assumption that the functions y_α are independent, or, to be more precise, the variations δy_α of the variational parameters ω_α are arbitrary and independent within the interval $[t_0, t_1]$.

This principle proved for the form (18) can also be stated as follows: the relation

$$\delta \int_{t_0}^{t_1} \Big(\sum y_\alpha \eta_\alpha - H \Big) dt = 0 \tag{19}$$

is valid for true motions provided that $\omega_\alpha = 0$ and $\delta y_\alpha = 0$ and $t = t_0, t_1$ at the endpoints and that δy_α and ω_α are arbitrary within the interval $[t_0, t_1]$.

We can immediately prove this hypothesis by noting that when integration is carried out by parts outside the interval, the expression is linearly dependent on ω_α for $t = t_0, t_1$ and is independent of δy_α.

Remark. Equations (18) and (19) can be proved on the basis of the canonical differential equations of motion of the mechanical system we proved earlier.

The canonical equations (14) can be derived from Hamilton's principle. For definiteness, I present the derivation here.

Thus Hamilton's principle of least action yields

$$\delta \int_{t_0}^{t_1} (T + U) dt = 0$$

provided that $\omega_\alpha = 0$ $(\alpha = 1, \ldots, k)$ and $t = t_0, t_1$ at the endpoints. Within the interval ω_α are assumed to be arbitrary and mutually independent. The variations of the parameters of the true displacements $\delta\eta_\alpha$ are related to ω_α as

$$\delta\eta_i = \frac{d\omega_i}{dt} + \sum c_{\beta\alpha i}\eta_\beta\omega_\alpha + \sum c_{0\alpha i}\omega_\alpha,$$

and, consequently, the canonical variables (item 13)

$$y_\alpha = \frac{\partial T}{\partial \eta_\alpha} \quad (\alpha = 1, \ldots, k), \tag{20}$$

being dependent on $t, x_1, \ldots, x_n, \eta_1, \ldots, \eta_k$, have variations δy_α dependent on $\omega_1, \ldots, \omega_k$. We must remember this when discussing Hamilton's principle. We can impart to this principle the form

$$\delta \int_{t_0}^{t_1} \left(\sum y_\alpha \eta_\alpha - H \right) dt = 0$$

when $\omega_\alpha = 0$ ($\alpha = 1, \ldots, k$) for $t = t_0, t_1$, where

$$H(t, x_1, \ldots, x_n, y_1, \ldots, y_k) = \sum y_\alpha \eta_\alpha - T - U \tag{21}$$

and consider y_α in Hamilton's principle to be independent of η_α according to formulas (20). Then we have

$$\delta \int_{t_0}^{t_1} \left(\sum y_\alpha \eta_\alpha - H \right) dt$$

$$= \left[\sum y \omega_i \right]_{t_0}^{t_1} + \int_{t_0}^{t_1} \left[\sum \omega_\alpha \left(-\frac{dy_\alpha}{dt} + \sum c_{\beta\alpha i} \eta_\beta y_i + \sum c_{0\alpha i} y_i \right. \right.$$

$$\left. \left. - X_\alpha H \right) + \sum \delta y_\alpha \left(\eta_\alpha - \frac{\partial H}{\partial y_\alpha} \right) \right] dt = 0.$$

The expression $\left[\sum y_i \omega_i \right]_{t_0}^{t_1}$ is zero by virtue of the conditions at the endpoints. From relation (21), which was obtained with the use of the transformation formulas (20), we have

$$\delta H = \sum \omega_\alpha X_\alpha H + \sum \delta y_\alpha \frac{\partial H}{\partial y_\alpha}$$

$$= \sum \delta y_\alpha \eta_\alpha + \sum y_\alpha \delta \eta_\alpha - \sum \frac{\partial T}{\partial \eta_\alpha} \delta \eta_\alpha - \sum \omega_\alpha X_\alpha (T + U).$$

The sum of the second and the last term on the right-hand side is zero. Since the initial values of the independent coordinates and velocities are independent for the initial moment, which we can assume to be the moment we consider, the coefficients of the corresponding ω_α and δy_α must be equal, i.e.

$$\eta_\alpha = \frac{\partial H}{\partial y_\alpha}.$$

The following relation remains from Hamilton's principle

$$\int_{t_0}^{t_1} \sum \omega_\alpha \left(-\frac{dy_\alpha}{dt} + \sum c_{\beta\alpha i}\eta_\beta y_i + \sum c_{0\alpha i}y_i - X_\alpha H \right) dt = 0,$$

whence, because of the arbitrariness of ω_α within the interval $[t_0, t_1]$, we have

$$-\frac{dy_\alpha}{dt} + \sum c_{\beta\alpha i}\eta_\beta y_i + \sum c_{0\alpha i}y_i - X_\alpha H = 0 \quad (\alpha = 1, \ldots, k).$$

21. Contact transformations. Let us consider new variables $\alpha_1, \ldots,$ α_n which define the position of the system and a Lie group of true displacements A_α $(\alpha = 0, 1, \ldots, k)$ with a subgroup of virtual displacements $A_1,$ $\ldots, A_k$, which is dependent on t.

Let $\gamma_{\alpha\beta i}$ be structural constants of the group

$$(A_\alpha, A_\beta) = \sum \gamma_{\alpha\beta i}A_i$$

and assume that $\gamma_{0\beta i} = 0$. To put it otherwise, let $A_0 = \partial/\partial t$. Let the parameters of the virtual displacement be designated as π_α and the parameters of the true displacement as θ_α,

$$\delta f(t, \alpha_1, \ldots, \alpha_n) = \sum \pi_\alpha A_\alpha f, \quad \frac{df}{dt} = \frac{\partial f}{\partial t} + \sum \theta_\alpha A_\alpha f.$$

We express the transformation of variables in terms of the characteristic function

$$V(t, x_1, \ldots, x_n, \alpha_1, \ldots, \alpha_n), \quad \frac{\partial \left(\dfrac{\partial V}{\partial x_1}, \ldots, \dfrac{\partial V}{\partial x_n} \right)}{\partial(\alpha_1, \ldots, \alpha_n)} \neq 0$$

using the formulas

$$y_s = X_s V, \quad -\beta_s = A_s V \quad (s = 1, \ldots, k). \tag{22}$$

Formulas (22) can be used for a complete definition of the transformation. We must, first of all, add to them the constraint equations (item 17)

$$\Phi_j = b_j \quad (j = n - k, \ldots, n) \tag{23}$$

in the variables $x_1, \ldots, x_n$ and the constraint relations written in the variables $\alpha_1, \ldots, \alpha_n$. As we saw in item 17, the latter are defined by the solutions of the system of equations

$$A_\alpha Z = 0 \quad (\alpha = 0, 1, \ldots, k).$$

Assume that the system of independent solutions is designated as

$$Z = Z_j \quad (j = n - k, \ldots, k),$$

and, consequently, the constraint equations are

$$Z_j = Z_j^0(t_0, \alpha_1^0, \ldots, \alpha_n^0). \tag{24}$$

Systems (23) and (24) make it possible for us to isolate the independent variables $x_1, \ldots, x_n$ and the independent variables $\alpha_1, \ldots, \alpha_n$ and $2k$ equations of (22) will be sufficient to define contact transformations.

Next we multiply the relations by η_s and θ_s respectively and add the results together:

$$\sum y_s\eta_s - \sum \beta_s\theta_s = \sum \eta_s X_s V + \sum \theta_s A_s V + X_0 V - X_0 V$$

$$= \frac{dV}{dt} - X_0 V.$$

Remark. We must have $\gamma_{0\beta i} = 0$ for the last transformation since otherwise the equalities

$$\frac{dV}{dt} = X_0 V + \sum \eta_\alpha X_\alpha V + \sum \theta_\alpha A_\alpha V$$

would not exist. In this case Hamilton's principle of least action

$$\delta \int_{t_0}^{t_1} \left(\sum \eta_s y_s - H \right) dt = 0,$$

under Poincaré's conditions

$$\omega_\alpha = 0, \quad \delta y_\alpha = 0 \quad \text{for} \quad t = t_0, \, t_1$$

and arbitrary ω_α, δy_α within the interval of integration, reduces to the form

$$\delta \int_{t_0}^{t_1} \left(\sum \theta_s\beta_s - H^* \right) dt + \delta \int_{t_0}^{t_1} \frac{dV}{dt} \, dt = 0. \tag{25}$$

The variation of the last integral is evidently zero. The conditions of variation in the new variables reduce to the form

$$\pi_\alpha = 0, \quad \delta\beta_\alpha = 0 \quad \text{for} \quad t = t_0, \, t_1$$

and π_α, $\delta\beta_\alpha$ are arbitrary and independent within the integration interval $(t_0, \, t_1)$,

$$H^*(t, \alpha_1, \ldots, \alpha_n, \beta_1, \ldots, \beta_k) = X_0 V + H(t, x_1, \ldots, x_n, X_1 V, \ldots, X_k V).$$

In accordance with Hamilton's principle (25), we have the following canonical equations of motion for the new variables:

$$\frac{d\beta_i}{dt} = \sum \gamma_{\alpha i\beta}\theta_\alpha\beta_\beta - A_i H^*, \quad \theta_\alpha = \frac{\partial H^*}{\partial \beta_\alpha}.$$

If the characteristic function of transformation

$$V(t, x_1, \ldots, x_n, \alpha_1, \ldots, \alpha_n)$$

is a complete integral of the equation

$$H^* = 0,$$

then the canonical equations in the new variables are

$$\frac{d\beta_i}{dt} = 0, \quad \theta_\alpha = 0$$

and, consequently,

$$\frac{d\alpha_i}{dt} = \sum \theta_\alpha A_\alpha \alpha_i = 0.$$

The integration of the equations

$$\frac{d\alpha_i}{dt} = 0, \quad \frac{d\beta_i}{dt} = 0 \quad (i = 1, \ldots, n)$$

yields

$$\alpha_i = \text{const}, \quad \beta_i = \text{const}.$$

This is a generalization of the Hamilton-Jacobi theorem.

Thus, if $V(t, x_1, \ldots, x_n, \alpha_1, \ldots, \alpha_n)$ is a complete integral of the equation

$$X_0 V + H(t, x_1, \ldots, x_n, X_1 V, \ldots, X_k V) = 0,$$

then the solution of the canonical equations of motion

$$\frac{dy_i}{dt} = \sum c_{\alpha i\beta}\eta_\alpha y_\beta + \sum c_{0i\beta}y_\beta - X_i H, \quad \eta_\alpha = \frac{\partial H}{\partial y_\alpha}$$

is defined by the formulas

$$y_\alpha = X_\alpha V, \quad -\beta_\alpha = A_\alpha V,$$

where α_i and β_i are constants.

A.3 The Special Theory of Relativity

A.3.1

1. The theory of ether at rest was accepted in electrodynamics after Lorentz.

Given this theory, it seemed possible to measure the velocity of the earth relative to the ether. To do this, Michelson performed an experiment in 1881 which indicated that when the earth rotates, the ether moves with it. The experiment was very accurate. Physicists then recalled Fizeau's experiments and the phenomenon of aberration, the explanations of which seemed artificial.

Fitzgerald and Lorentz hypothesized that solid bodies contract in the direction of motion (1895).

2. In his article entitled "Electromagnetic phenomena in a system moving with any velocity lower than that of light" Lorentz,[*] in 1904, discussed transformations of Maxwell's equations which would not change their form. This is the fundamental problem of the theory of relativity.

In Newtonian mechanics, the equations of motion

$$m \frac{d^2x}{dt^2} = X, \quad m \frac{d^2y}{dt^2} = Y, \quad m \frac{d^2z}{dt^2} = Z$$

do not change if a new system of coordinates is considered which moves uniformly and in a straight line relative to the first system. In other words, the equations of motion are invariant under the transformation

$$x' = x + at, \quad y' = y + bt, \quad z' = z + ct, \quad t' = t.$$

This is the Newtonian relativity principle.

The equations of electromagnetics are the main part of the theory of the phenomena of optics. Although Lorentz began the discussion on the relativity principle for the electromagnetics equations, he did not solve the problem in its entirety.

3. Poincaré was the first to solve Lorentz' problem in his article on the dynamics of an electron[**] by postulating that it was impossible, in principle, to determine the absolute motion of the earth.

In his article on the electrodynamics of moving bodies written in 1905 Einstein[***] considered the kinematics of the Lorentz transformations and

---

[*] *Proc. Acad. Sci. Amsterdam* 6: 809 (1904).
[**] J.H. Poincaré, "Sur la dynamique de l'electron", *Comp. Rend.* 140: 1504-1508 (1905).
[***] A. Einstein, "Zur Elektrodynamik bewegter körper", *Ann. d. phys.* 17: 891 (1905).

thus opened the way to Minkowski's idea of a four-dimensional space, i.e. space plus time.

Minkowski's essay* on space and time greatly influenced further research. Einstein's ideas were widely disseminated thanks to their simplicity. Poincaré, with his usual depth and thoroughness, remained in the background.

"... Either everything that exists in the world is of an electromagnetic origin, or this property, which is general, so to say, for all physical phenomena, is nothing other than an outward appearance, something connected with the methods of our measurements.**

4. Let us designate the components of the electric vector as X, Y, Z and the components of the magnetic vector as L, M, N.***

Maxwell's equations for a vacuum are

$$\frac{1}{c}\frac{\partial X}{\partial t} = \frac{\partial N}{\partial y} - \frac{\partial M}{\partial z}, \quad \frac{1}{c}\frac{\partial L}{\partial t} = -\left(\frac{\partial Z}{\partial y} - \frac{\partial Y}{\partial z}\right),$$

$$\frac{1}{c}\frac{\partial Y}{\partial t} = \frac{\partial L}{\partial z} - \frac{\partial N}{\partial x}, \quad \frac{1}{c}\frac{\partial M}{\partial t} = -\left(\frac{\partial X}{\partial z} - \frac{\partial Z}{\partial x}\right),$$

$$\frac{1}{c}\frac{\partial Z}{\partial t} = \frac{\partial M}{\partial x} - \frac{\partial L}{\partial y}, \quad \frac{1}{c}\frac{\partial N}{\partial t} = -\left(\frac{\partial Y}{\partial x} - \frac{\partial X}{\partial y}\right),$$

$$\frac{\partial X}{\partial x} + \frac{\partial Y}{\partial y} + \frac{\partial Z}{\partial z} = 0, \quad \frac{\partial L}{\partial x} + \frac{\partial M}{\partial y} + \frac{\partial N}{\partial z} = 0.$$

Consider a new system which translates with velocity $v < c$ along the x-axis. For the equations to retain the same form in the new system, it is necessary also to change the time reckoning:

$$t' = \beta\left(t - \frac{v}{c^2}\right)x, \quad \beta = \frac{1}{\sqrt{1 - \dfrac{v^2}{c^2}}},$$

$$x' = \beta(x - vt), \quad y' = y, \quad z' = z,$$

---

* H. Minkowski, "Raum and Zeit", *Phys. Zs.* **10**: 104 (1909).
** J.H. Poincaré, *Comp. Rend.* **140**: 1504-1508 (1905).
*** The vectors of the electric and magnetic strength. — *Ed.*

$$\frac{1}{c} \frac{\partial X}{\partial t'} = \frac{1}{c} \frac{\partial X}{\partial t} \beta + \frac{1}{c} \frac{\partial X}{\partial x} \beta v$$

$$= \left(\frac{\partial N}{\partial y} - \frac{\partial M}{\partial z} \right) \beta + \beta \frac{v}{c} \left(- \frac{\partial Y}{\partial y} - \frac{\partial Z}{\partial z} \right)$$

$$= \frac{\partial \beta \left(N - \frac{v}{c} Y \right)}{\partial y'} - \frac{\partial \beta \left(M + \frac{v}{c} Z \right)}{\partial z'}.$$

Similarly,

$$\frac{1}{c} \frac{\partial L}{\partial t'} = - \left(\frac{\partial \beta \left(Z + \frac{v}{c} M \right)}{\partial y'} - \frac{\partial \beta \left(Y - \frac{v}{c} N \right)}{\partial z'} \right).$$

Then we have

$$\frac{1}{c} \frac{\partial Y}{\partial t'} = \frac{1}{c} \frac{\partial Y}{\partial t} \beta + \frac{1}{c} \frac{\partial Y}{\partial x} \beta v = \beta \left(\frac{\partial L}{\partial z} - \frac{\partial N}{\partial x} \right) + \frac{v}{c} \beta \frac{\partial Y}{\partial x}$$

$$= \beta \frac{\partial L}{\partial z'} - \beta \frac{\partial N}{\partial x'} \beta - \beta \frac{\partial N}{\partial t'} \left(- \beta \frac{v}{c^2} \right)$$

$$+ \beta \frac{v}{c} \left(\frac{\partial Y}{\partial x'} \beta + \frac{\partial Y}{\partial t'} \left(- \beta \frac{v}{c^2} \right) \right),$$

whence it follows that

$$\frac{1}{c} \frac{\partial}{\partial t'} \left\{ Y \left(1 + \beta^2 \frac{v^2}{c^2} \right) - \beta^2 \frac{v}{c} N \right\} = \beta \frac{\partial L}{\partial z'} - \beta \frac{\partial \beta \left(N - \frac{v}{c} Y \right)}{\partial x'}.$$

Consequently,

$$\frac{1}{c} \frac{\partial \beta \left(Y - \frac{v}{c} N \right)}{\partial t'} = \frac{\partial L}{\partial z'} - \frac{\partial \beta \left(N - \frac{v}{c} Y \right)}{\partial x'}.$$

It follows that the components of the electric and magnetic vectors in a moving system must be proportional to the corresponding expressions in Maxwell's transformed equations.

The transformation factor obtained from symmetry and transformation equals unity (as a matter of fact, this factor which exists in Maxwell's equations as well, is more due to the choice of units). We obtain

$$X' = X, \quad Y' = \beta \left(Y - \frac{v}{c} N \right), \quad Z' = \beta \left(Z + \frac{v}{c} M \right),$$

$$L' = L, \quad M' = \beta \left(M + \frac{v}{c} Z \right), \quad N' = \beta \left(N - \frac{v}{c} Y \right). \tag{1}$$

In both systems Maxwell's equations have the same form.

In the transformations only time changes. The principle of relativity of electrodynamics introduces the notion that time is related to the reference system. The concept of absolute time is lost. Furthermore, electric and magnetic forces do not exist independently of the state of motion of the coordinate system.

5. Let us find the algebraic invariant of the Lorentz transformation (an interval) which we shall need. We have

$$x'^2 + y'^2 + z'^2 - c^2 t'^2$$

$$= \beta^2 (x^2 - 2vxt + v^2 t^2) + y^2 + z^2 - c^2 \beta^2 \left(t^2 - 2 \frac{v}{c^2} xt + \frac{v^2}{c^4} x^2 \right)$$

$$= x^2 \left(\beta^2 - \frac{v^2}{c^2} \beta^2 \right) + y^2 + z^2 + \beta^2 t^2 (v^2 - c^2) = x^2 + y^2 + z^2 - c^2 t^2.$$

The inverse transformation (Poincaré's) is

$$t = \beta \left(t' + \frac{v}{c^2} x' \right), \quad x = \beta (x' + vt'), \quad y = y', \quad z = z'.$$

The Lorentz transformations (Poincaré's) form a group.

If our physical ideas are based on Maxwell's equations, then the Lorentz transformations form the basis of our concepts of time and length (i.e. the transformations rid us of the necessity to give a thorough definition of these concepts).

As soon as we believe the definitions and accept them as postulates, the interesting world depicted by Einstein and Minkowski opens up before us.

A.3.2

1. Thus if, in our constructions, we approach the realization of phenomena via Maxwell's equations, which are invariant with respect to uniform translations, then, with such a hypothetic approach, the space and time, which actually exist objectively will be understood in a different way. Minkowski brilliantly explained how differently.

Now let us study Minkowski's universe.

2. Space and time. In 1908 Minkowski developed the following geometric idea of space and time. Assume for simplicity that $c = 1$. In the system of coordinates of a stationary observer (t, x) (say, Cartesian coordinates) any event is associated with a point in space (t, x). The evolution of this event in space (t, x) traces a curve. The position of an observer moving with a constant velocity v traces a straight line t' such that

$$\tan \alpha = v.$$

Let us consider, in a plane world (t, x), a hyperbola (Fig. 165)

$$t^2 - x^2 = 1.$$

If we take the units of length to be the segments OA' and $OC' = OA'$ in the (t', x') system and the segments OA and $OC = OA$ in the (t, x) system, then the equation of this hyperbola in the new axes t', x' will be

$$t'^2 - x'^2 = 1,$$

where x' is a ray conjugate to t'.

And if this is the case, then, since the interval $t^2 - x^2$ is invariant, the passage to the new axes must be a Lorentz transformation.

Indeed, it follows from the figure that

$$t \cdot OA = t' \cdot OA' \cos \alpha + x' \cdot OC' \sin \alpha,$$
$$x \cdot OC = t' \cdot OA' \sin \alpha + x' \cdot OC' \cos \alpha.$$

But

$$\frac{AD}{OA} = v = \tan \alpha.$$

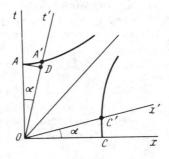

Figure 165

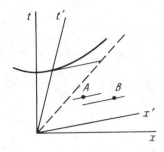

Figure 166

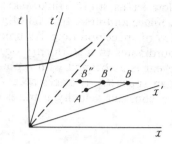

Figure 167

From the equation of the hyperbola $OA'^2 (\cos^2 \alpha - \sin^2 \alpha) = OA^2$ we have

$$\frac{OA'}{OA} = \frac{1}{\cos \alpha \sqrt{1 - \tan^2 \alpha}},$$

and this means that

$$t = \beta(t' + vx'), \quad x = \beta(x' + vt'), \quad \beta = \frac{1}{\sqrt{1 - \tan^2 \alpha}}.$$

The time vector t' completely defines the system of the moving observer.

Different observers carry out different measurements of the geometric coordinates and time at the same space-time point.

3. Simultaneity of events. Causality. Two events A and B defined by space-time points are simultaneous in the (t, x) system if the curve AB is parallel to the x-axis (Fig. 166).

Let us consider another reference frame, defined by the time vector t' (Fig. 167). In this new system, the events A and B are no longer simultaneous. Figure 167 illustrates the following case: in the (t, x) system event A precedes events B, B', B'' and in the (t', x') system, event B precedes event A, while event A is synchronous with event B' and precedes B''. This means that if these events are functionally related, then the observers in the systems (t, x) and (t', x') will differ in their understanding of the cause and effect. We do not lose causality here, but its formulation becomes more general and it expresses a strict relationship between the events. Event A can only be the cause of event B if event B does not precede event A in any moving coordinate system. This in only possible when point B is inside the angle bAb' (Fig. 168). All the events which can cause event A fill up the angle $a'Aa$. By contrast the points which are inside the angles $a'Ab'$ and bAa correspond to the events, which, from the viewpoint of

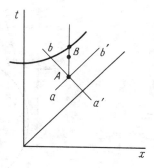

Figure 168

the moving observer, succeed or precede event A, depending on the velocity with which the observer is moving.

A.3.3 Mapping Minkowski's Space-Time

1. Let the speed of light be $c = 1$. We shall consider a three-dimensional world (flat space plus time). The ends of the unit (scale) vectors of time form the upper sheet (for $t > 0$) of a two-sheet hyperboloid. Every t-axis on the hyperboloid is associated with a point which we shall agree to call by the same letter as before, i.e. A.

2. Poincaré* proved that if we agree to call a plane section across the diameter of a two-sheet hyperboloid a *straight line*, to call a plane section not across the diameter of such a surface a *circle*, to define the *angle* between two plane sections ("straight lines") across the diameter (which pass through a point of the surface of a two-sheet hyperboloid) as the logarithm, divided by $\sqrt{-1}$, of an anharmonic ratio of a pair of imaginary rectilinear generatrices and a pair of tangents to these two sections, and to define the *length* of a segment of a section across the diameter as the logarithm of the anharmonic ratio of the two ends of the segment and two infinitely distant points on the conic section, then we get the system of terms of Lobachevskian geometry.

3. For convenience, we use a conic transformation to map a sheet of a hyperboloid ($t > 0$) onto a plane π which is orthogonal to the t-axis (Fig. 169).

---

* J.H. Poincaré "Sur les hypothéses fondamentales de la géométrie", *Oeuvres* **XI**: 79-81; *Bull. Soc. Math., Paris, 1887,* **15**: 203-216 (1887).

P.E. Appell, *Traité de méchanique rationnelle*, V. 1, 2 (Cautier-Villars, Paris: 1953).

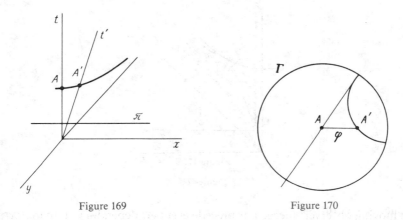

Figure 169   Figure 170

The Lobachevsky plane will be mapped onto the interior of a circle (the mapping of an asymptotic circular cone).

We shall denote the images of the points by the same letters. The "straight lines" which pass through a point A, will be Euclidean straight lines. The other lines will be straight lines "orthogonal" to the main circle (which will be called the absolute and denoted by the letter Γ) (Fig. 170). We shall get a peculiar image of Lobachevsky's plane. In his *Acta Mathematica* (Vol. 1) Poincaré gave a more interesting image. Lobachevsky proved the theorem that the geometry of a hyperbolic plane coincides with the geometry on a sphere of an imaginary radius. We shall widely use this theorem.

4. Without presenting the derivation of the fundamental formulas of Lobachevskian geometry, which I could give in the same way as for Poincaré's mapping, I shall directly use, for simplicity, a theorem proved by Lobachevsky, namely, that the formulas of Lobachevskian geometry coincide with those of the spherical geometry on a sphere of an imaginary radius i:

$$i \sin \frac{x}{i} = \sinh x, \quad \cos \frac{x}{i} = \cosh x.$$

In what follows I shall use the formulas of the spherical geometry assuming them to be known.

5. Consider a parametric equation of a hyperbola

$$t = \cosh \varphi, \quad x = \sinh \varphi, \quad t^2 - x^2 = \cosh^2 \varphi - \sinh^2 \varphi = 1,$$

where φ is the length in Lobachevsky's sense. In Fig. 171 φ is the doubled

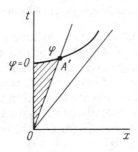

Figure 171

hatched area. Indeed,

$$\int_0^\varphi \left(-x\frac{dt}{d\varphi} + t\frac{dx}{d\varphi} \right) d\varphi = \int_0^\varphi (-\sinh^2\varphi + \cosh^2\varphi)d\varphi = +\varphi.$$

Assume that for a stationary observer $\varphi = 0$

$$t = \cosh\varphi, \quad x = \sinh\varphi.$$

For a moving observer $-\varphi = \varphi_0$. We have

$$t' = \cosh(\varphi - \varphi_0), \quad x' = \sinh(\varphi - \varphi_0),$$
$$t'^2 - x'^2 = 1,$$
$$t' = \cosh(\varphi - \varphi_0) = \cosh\varphi\cosh\varphi_0 - \sinh\varphi\sinh\varphi_0$$
$$= t\cosh\varphi_0 - x\sinh\varphi_0,$$
$$x' = \sinh(\varphi - \varphi_0) = \sinh\varphi\cosh\varphi_0 - \cosh\varphi\sinh\varphi_0$$
$$= x\cosh\varphi_0 - t\sinh\varphi_0.$$

(2)

The coordinates of the point A' are

$$x^0 = \sinh\varphi_0, \quad t^0 = \cosh\varphi_0.$$

Then

$$\frac{x^0}{t^0} = \tanh\varphi_0 = v.$$

Hence formulas (2) assume the form of a Lorentz transformation

$$t' = \beta(t - xv), \quad x' = \beta(x - tv).$$

A.3.4 Applications to Physics

1. The law of addition of velocities. Assume that A is a fixed system, B is a system which moves in A in an indicated direction, C is a system which moves in B at an angle B to the motion of A relative to B.

In Lobachevsky's plane a triangle ABC results with sides a, b, c (Fig. 172). Since tanh c is the velocity of motion of the system B relative to the system A, tanh a is the velocity of the system C relative to the system B, tanh b is the velocity of the system C relative to the system A, it is evident that

$$\cosh b = \cosh a \cdot \cosh c - \sinh a \cdot \sinh c \cdot \cos B.$$

This formula is Einstein's law of addition of velocities. (You can find it in Poincaré's article (see the footnote on p. 356).)

2. Consider the case when the directions of motions of systems B and C coincide. The angle B is zero if the motion of B relative to A and the motion of C relative to B are different in direction. It is equal to π if B moves in the same direction as A and C moves in the same direction as B (Fig. 173). We have

$$\tanh b = \tanh(c \pm a) = \frac{\tanh c \pm \tanh a}{1 \pm \tanh a \tanh c}.$$

In the first case we must take the plus sign and in the second the minus sign.

3. Propagation of plane harmonic waves. Assume that a plane harmonic wave is propagated in space-time along the x-axis with a constant velocity v. In the system of a stationary observer A, points with the same phase are a system of parallel planes in space-time (Fig. 174). The plane S of the phase passing through O yields a section across the diameter of the sheet ($t > 0$) of a two-sheet hyperboloid if the speed of propagation of the wave

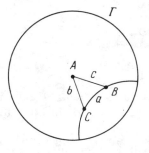

Figure 172

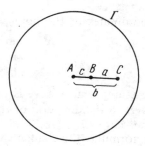

Figure 173

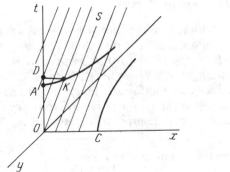

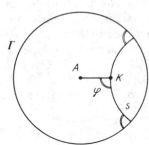

Figure 174   Figure 175

is slower than that of light. In Lobachevsky's plane a straight line s corresponds to the section S (Fig. 175).

Let φ be the non-Euclidean length of the perpendicular drawn from A to s.

The speed v of propagation of a wave is the ratio of the path traversed $DK = \sinh \varphi$ to the elapsed time $OD = \cosh \varphi$ (Fig. 174)

$$v = \tanh \varphi.$$

The direction of propagation of the wave in the system A is along the perpendicular AK drawn from A to the straight line s.

The straight line s is the locus of the time axes for which the wave is a standing wave ($\tanh 0 = 0$). Assume that μ waves can fill in OC (the quantity $1/\mu$ is the wave length for the observer A) and that the ν waves can fill in OA (ν is the frequency of the waves for the observer A). In the triangle ODK the segment DK cuts the $\mu \sinh \varphi$ waves and the segment OD cuts the $\nu \cosh \varphi$ waves. But the segment DK cuts as many waves as the segment OD and hence

$$\mu \sinh \varphi = \nu \cosh \varphi,$$

or

$$\frac{\mu}{\cosh \varphi} = \frac{\nu}{\sinh \varphi} \ (= \varkappa).$$

4. Assume that a wave is observed from two systems A and A' and A' moves in the direction of the wave in the system A. In Lobachevsky's plane we have (Fig. 176)

$$\tanh \varphi' \equiv \tanh(\varphi - u) = \frac{\tanh \varphi - \tanh u}{1 - \tanh \varphi \tanh u}.$$

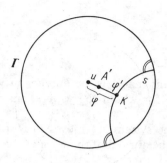

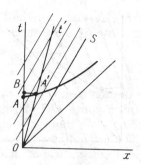

Figure 176   Figure 177

The Fizeau experiment. A wave passes through a liquid A' with the velocity tanh φ. The liquid A' translates in A in the direction of the wave with the velocity tanh u. What is the velocity of the wave relative to the system A? We have

$$\tanh \varphi = \tanh(\varphi' + u) = \frac{\tanh \varphi' + \tanh u}{1 + \tanh \varphi' \tanh u}$$

$$= \tanh \varphi' + \tanh u(1 - \tanh^2 \varphi') + \ldots$$

The coefficient $1 - \tanh^2 \varphi'$ is the Fizeau-Fresnel drag coefficient. Let us consider the change in frequency. The number of waves cutting the segment OA' (ν' is the frequency of a wave in the system A') is equal to the number of waves cutting the segment OB ($\nu \cosh u$) minus the number of waves cutting the segment BA' ($\mu \sinh u$, i.e. $\nu' = \nu \cosh u - \mu \sinh u$, Fig. 177). This means that

$$\nu' = \varkappa(\sinh \varphi \cosh u - \cosh \varphi \sinh u) = \varkappa \sinh(\varphi - u) = \varkappa \sinh \varphi'.$$

Hence

$$\frac{\nu'}{\sinh \varphi'} = \varkappa = \frac{\nu}{\sinh \varphi} \text{ is the invariant.}$$

The Doppler effect. According to the invariant obtained, the change in frequency is given by

$$\nu' = \nu \cosh u - \mu \sinh u = \nu \left(\cosh u - \frac{\sinh u}{\tanh \varphi} \right).$$

If s is a light wave in vacuum, then $\tanh \varphi = 1$ and

$$\nu' = \nu(\cosh u - \sinh u) = \nu \frac{1 - v}{\sqrt{1 - v^2}} = \nu \sqrt{\frac{1 - v}{1 + v}} \quad (v = \tanh u).$$

This is the formula for the Doppler effect.

The sense of the invariant $\varkappa$ is

$$\varkappa = \frac{\tilde{\mu}}{\cosh \tilde{\varphi}}.$$

If $\tilde{\varphi} = 0$, then $\cosh \tilde{\varphi} = 1$, and this means that $\varkappa$ is the number of waves on a unit segment OC for the observer for whom the wave seems to be a standing wave ($\tanh \tilde{\varphi} = 0$).

5. The general case. Assume that the system A' moves in a straight line in a fixed system A at an angle α to the normal of the wave s (Fig. 178). It follows from Lobachevskian geometry that

$$\sinh a = \sinh b \cosh c - \sinh c \cosh b \cos \alpha.$$

Directly from the figure we have

$$\sinh \varphi' = \sinh \varphi \cosh u - \sinh u \cosh \varphi \cos \alpha.$$

This formula explains a more complicated Fizeau experiment.

Assume that a liquid moves in a system A with a velocity $\tanh u$ and a wave s moves in the liquid with a velocity $\tanh \varphi$ at an angle α to the direction of motion of A' relative to A. Using the formula obtained we can get a relation for the velocity $\tanh \varphi'$.

Assume now that the liquid moves with the velocity $\tanh u$. A wave passes through the liquid with a velocity $\tanh \varphi'$. If α is the angle between the velocity of the wave and the velocity of the observer, then this formula yields the velocity of the waves for the stationary observer A. It is as easy to derive equation for the change in frequency as before (Fig. 179). The segment OA' cuts as many waves ν' (ν' is the frequency in the system A') as the segment OB ($\nu \cosh u$) minus the number of waves which cut

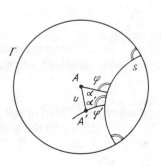

Figure 178

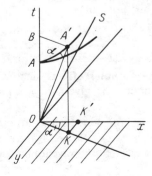

Figure 179

the segment BA', but BA' cuts as many waves as the segment OK, i.e. μ (sinh u cosh α) waves. Hence

$$\nu' = \nu \cosh u - \mu \sinh u \cos \alpha,$$

or

$$\nu' = \varkappa(\sinh \varphi \cosh u - \cosh \varphi \sinh u \cos \alpha) = \varkappa \sinh \varphi'.$$

Hence $\varkappa = \dfrac{\nu'}{\sinh \varphi'}$ is the invariant. Consequently,

$$\frac{\nu'}{\nu} = \frac{\sinh \varphi'}{\sinh \varphi} = \cosh u - \sinh u \frac{\cos \alpha}{\tanh \varphi} = \cosh u \left(1 - \frac{\tanh u}{\tanh \varphi} \cos \alpha\right).$$

This is the Doppler effect.

6. Assume that a stationary observer A and a moving observer A' see the same light wave moving in a vacuum at the speed of light, which is unity. The wave s in Lobachevsky's plane will degenerate to a point on the absolute (Fig. 180). The triangle $AA's$ has an angle s equal to zero. We use Napier's analogies*

$$\tan \frac{A+B}{2} = \cot \frac{C}{2} \frac{\cosh \dfrac{a-b}{2}}{\cosh \dfrac{a+b}{2}}, \quad \tan \frac{A-B}{2} = \cot \frac{C}{2} \frac{\sinh \dfrac{a-b}{2}}{\sinh \dfrac{a+b}{2}}.$$

We use the second formula setting $A = \alpha$, $B = 0$, $C = \alpha'$,

$$a = \varphi', \quad b = u, \quad c = \varphi, \quad \tan \frac{\alpha}{2} = \cot \frac{\alpha'}{2} \frac{\sinh \dfrac{\varphi'-u}{2}}{\sinh \dfrac{\varphi'+u}{2}}.$$

But $\tanh \dfrac{\varphi'}{2} = \tanh \dfrac{\varphi}{2} = 1$. Then

$$\tan \frac{\alpha}{2} \tan \frac{\alpha'}{2} = \frac{\sinh \dfrac{\varphi'}{2} \cosh \dfrac{u}{2} - \cosh \dfrac{\varphi'}{2} \sinh \dfrac{u}{2}}{\sinh \dfrac{\varphi'}{2} \cosh \dfrac{u}{2} + \cosh \dfrac{\varphi'}{2} \sinh \dfrac{u}{2}} = \frac{\tanh \dfrac{\varphi'}{2} - \tanh \dfrac{u}{2}}{\tanh \dfrac{\varphi'}{2} + \tanh \dfrac{u}{2}}$$

$$= \frac{1 - \tanh \dfrac{u}{2}}{1 + \tanh \dfrac{u}{2}} = \frac{e^{\frac{u}{2}} + e^{-\frac{u}{2}} - \left(e^{\frac{u}{2}} - e^{-\frac{u}{2}}\right)}{e^{\frac{u}{2}} + e^{-\frac{u}{2}} + e^{\frac{u}{2}} - e^{-\frac{u}{2}}} = e^{-u},$$

* A, B, C are the angles of the triangle and a, b, c are its sides. — Ed.

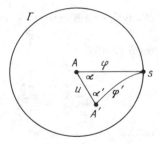

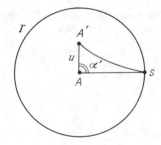

Figure 180   Figure 181

i.e. $\tan \dfrac{\alpha}{2} \tan \dfrac{\alpha'}{2} = e^{-u}$.

This formula explains aberration.

7. Consider an interesting case $\alpha = \pi/2$, $\alpha' = \Pi(u)$ (Fig. 181). The angle α' is equal to the angle of Lobachevsky's parallelism. The preceding formula passes into the fundamental formula of Lobachevskian geometry

$$\tan \frac{\Pi(u)}{2} = e^{-u}.$$

8. The reflection of a light wave from a moving mirror. Assume that a mirror B is moving in a straight line in a fixed system A at a velocity $\tanh u$ (Fig. 182). The normal of the mirror coincides with the direction of the velocity of motion. A plane light wave s falls on the mirror and is reflected as a wave s'; s, s' lie on the absolute Γ. We designate the angles A and B of the triangles ABs and ABs' as α, β and α', β' respectively. The ordinary law of reflection is valid in the system B of the moving mirror,

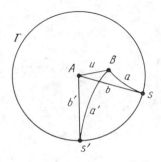

Figure 182

i.e. the angle of incidence is equal to the angle of reflection

$$\beta - \frac{\pi}{2} = \frac{\pi}{2} - \beta', \quad \cos \beta = -\cos \beta',$$

or

$$\frac{\beta}{2} = \frac{\pi}{2} - \frac{\beta'}{2}, \quad \tan \frac{\beta}{2} = \cot \frac{\beta'}{2}, \quad \tan \frac{\beta}{2} \tan \frac{\beta'}{2} = 1.$$

The aberration formulas for the rays s and s' are

$$\tan \frac{\alpha}{2} \tan \frac{\beta}{2} = e^{-u}, \quad \tan \frac{\alpha'}{2} \tan \frac{\beta'}{2} = e^{-u}.$$

If we equate the left-hand sides, we obtain

$$\tan \frac{\alpha}{2} \tan \frac{\beta}{2} = \tan \frac{\alpha'}{2} \tan \frac{\beta'}{2},$$

or

$$\tan \frac{\alpha}{2} \tan^2 \frac{\beta}{2} = \tan \frac{\alpha'}{2}.$$

After multiplication we have

$$\tan \frac{\alpha}{2} \tan \frac{\beta}{2} \tan \frac{\alpha'}{2} \tan \frac{\beta'}{2} = e^{-2u} = \frac{e^u - e^{-u} - (e^u - e^{-u})}{e^u - e^{-u} + (e^u - e^{-u})}$$

$$= \frac{1 - \tanh u}{1 + \tanh u}.$$

The final result is

$$\tan \frac{\alpha}{2} \tan \frac{\alpha'}{2} = \frac{1 - \tanh u}{1 + \tanh u}.$$

For an observer B moving with the mirror the length of the incident wave is equal to that of the reflected wave, i.e. $\mu = \mu'$,

$$\varkappa = \frac{\mu}{\cosh a} = \frac{\mu'}{\cosh a'}, \quad \frac{\cosh a}{\cosh a'} = 1.$$

For the stationary observer

$$\frac{v_{\text{ref}}}{v_{\text{inc}}} = \frac{\sinh b'}{\sinh b}$$

$$= \frac{\sinh a' \cosh u - \cosh a' \sinh u \cos \beta'}{\sinh a \cosh u - \cosh a \sinh u \cos \beta}$$

$$= \frac{\tanh a' \cosh u - \sinh u \cos \beta' \cosh a'}{\tanh a \cosh u - \sinh u \cos \beta \cosh a}$$

$$= \frac{1 - \tanh u \cos \alpha + \tanh u \cos \alpha - \tanh^2 u}{1 - \tanh u \cos \alpha - \tanh u \cos \alpha + \tanh^2 u}$$

$$= \frac{1 - \tanh^2 u}{1 - 2\tanh u \cos \alpha + \tanh^2 u}.$$

Hence

$$\frac{v_{\text{inc}}}{v_{\text{ref}}} = \cosh^2 u (1 - 2\tanh u \cos \alpha + \tanh^2 u).$$

9. You are invited to work out the situation when the normal to the plane of a mirror makes an angle $\varphi \neq 0$ with the direction of its motion.

We can also consider the laws of refraction of light at the boundary of a moving media. Since in refraction problems the speed of the wave must be smaller than unity, they are not very simple to calculate. We can make simplification if we consider the ideal domain of Lobachevsky's plane which lies beyond the absolute and, according to Poincaré, corresponds to a one-sheet hyperboloid. In this domain the wave s can be represented by a single point which corresponds to the normal to the wave in the system where the wave seems to be standing.

A.3.5

1. Consider Minkowski's four-dimensional universe (the three-dimensional space plus time). The locus of a pencil of unit vectors of time is a sheet of a two-sheet hyperboloid. According to Poincaré, the metric of this pencil of vectors will coincide with the metric of Lobachevsky's three-dimensional space.

Thus every point of Lobachevsky's space is associated with a time direction in Minkowski's space-time, and then the angles at a point A are measured as the angles between the purely spatial directions in the reference system A.

We can map Lobachevsky's space into the interior of a sphere of unit radius as it was done by Poincaré, or by a generalization of the technique suggested in Sec. A.3.4.

I will use the last method although in this method the angles cannot obviously be measured by ordinary techniques at any point except for the centre. But in this representation the "straight lines" will remain straight whereas in Poincaré's method the "angles" remain invariant.

2. We are interested in electromagnetic phenomena. Let us consider an electromagnetic field in a vacuum.

Let A be a stationary observer. Assume that an electromagnetic wave (light) moves along the x-axis of the stationary observer (Fig. 183).

In Lobachevsky's space, the wave is associated with a point s on the absolute since the speed of the wave is 1. The direction m of the magnetic vector and the direction f of the electric vector in Minkowski's space-time must be orthogonal to the direction of the wave. We assume the direction f to be the y- and the direction m to be the z-axis.

We draw planes through the time axis t and the y- and z-axes. They cut the superhyperboloid along the curves Af and Am.

In Lobachevsky's space, we draw planes containing As through the y- and z-axes. These planes cut the π-plane, which is tangent to the absolute at the point s along the (f) and (m) directions.

3. If we recall the property of screws in Lobachevsky's space, we note that the directions of the electric and magnetic vectors in a plane electromagnetic wave, propagating in a vacuum, form the directions of the screw whose principal axes (f) and (m) are perpendicular and touch the absolute at a point s. The normalizing plane (i.e. the plane containing both f and m) for the point A touches a limit surface at A which passes through A and the centre s of the principal axes of the screw.

4. We shall show that the Lorentz' transformations for an electromagnetic field are the normalization formulas of the screw for a new origin.

Assume that for the point A the screw on the perpendicular axes x_1, x_2, x_3 has components $\xi_i = f_i + m_i\sqrt{-1}$ $(i = 1, 2, 3)$. If, for the point A', the coordinate axes x_i' form *complex angles*, whose cosines are α_{ki}, with the x_i-axes, then the components of the screw for the system A' are

$$\xi_k' = \sum \alpha_{ki}\xi_i.$$

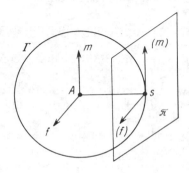

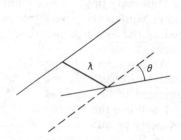

Figure 183   Figure 184

The complex angle between two straight lines in Lobachevsky's space is the number

$$\theta = \lambda\sqrt{-1},$$

where θ is the angle between the straight lines and λ is the shortest distance between the lines (Fig. 184).

We write out the matrix $\|\alpha_{ki}\|$ of the cosines of the angles between the axes x'_k and x_i for the case shown in Fig. 185. Measuring the angles from the old axes to the new ones, we obtain

$$\angle(x', x) = 0, \qquad\qquad \alpha_{11} = \cos 0 = 1,$$

$$\angle(x', y) = -\frac{\pi}{2}, \qquad\qquad \alpha_{12} = \cos\left(-\frac{\pi}{2}\right) = 0,$$

$$\angle(x', z) = \frac{\pi}{2}, \qquad\qquad \alpha_{13} = \cos\frac{\pi}{2} = 0,$$

$$\angle(y', x) = \frac{\pi}{2}, \qquad\qquad \alpha_{21} = \cos\frac{\pi}{2} = 0,$$

$$\angle(y', y) = \varphi\sqrt{-1}, \qquad\qquad \alpha_{22} = \cos\varphi\sqrt{-1} = \cosh\varphi,$$

$$\angle(y', z) = -\frac{\pi}{2} + \varphi\sqrt{-1}, \quad \alpha_{23} = \cos\left(-\frac{\pi}{2} + \varphi\sqrt{-1}\right) = \sqrt{-1}\,\sinh\varphi,$$

$$\angle(z', x) = -\frac{\pi}{2}, \qquad\qquad \alpha_{31} = \cos\left(-\frac{\pi}{2}\right) = 0,$$

$$\angle(z', y) = \frac{\pi}{2} + \varphi\sqrt{-1}, \quad \alpha_{32} = \cos\left(\frac{\pi}{2} + \varphi\sqrt{-1}\right) = -\sqrt{-1}\,\sinh\varphi,$$

$$\angle(z', z) = \varphi\sqrt{-1}, \qquad\qquad \alpha_{33} = \cos\varphi\sqrt{-1} = \cosh\varphi,$$

$$-\sin i\varphi = -\frac{(e^{i\varphi} - e^{i\varphi})}{2i} = \frac{e^{i\varphi} + e^{-\varphi}}{2i} = -i\,\sinh\varphi,$$

$$\sinh i\varphi = \frac{e^{-i\varphi} - e^{i\varphi}}{2i} = i\,\frac{e^{i\varphi} - e^{-i\varphi}}{2i} = i\,\sin\varphi.$$

The matrix of the direction cosines assumes the form

$$
\begin{array}{c}
 \\
x' \\
y' \\
z'
\end{array}
\begin{array}{ccc}
x & y & z
\end{array}
\left\|
\begin{array}{ccc}
1 & 0 & 0 \\
0 & \cosh\varphi & i\,\sinh\varphi \\
0 & -i\,\sinh\varphi & \cosh\varphi
\end{array}
\right\|.
$$

If in the reference system A the screw has coordinates

$$f_j + m_j\sqrt{-1},$$

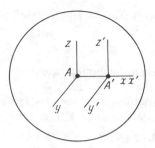

Figure 185

then in the system A' its coordinates are

$$\xi_1' = f_1 + m_1 \sqrt{-1},$$

$$\xi_2' = \sum \alpha_{2j} \xi_j = \cosh \varphi (f_2 + m_2 \sqrt{-1}) + i \sinh \varphi (f_3 + m_3 \sqrt{-1})$$
$$= (f_2 \cosh \varphi - m_3 \sinh \varphi) + \sqrt{-1} (m_2 \cosh \varphi + f_3 \sinh \varphi),$$

$$\xi_3' = \sum a_{3j} \xi_j = -\sqrt{-1} \sinh \varphi (f_2 + m_2 \sqrt{-1}) + \cosh \varphi (f_3 + m_3 \sqrt{-1})$$
$$= (f_3 \cosh \varphi + m_2 \sinh \varphi) + \sqrt{-1} (m_3 \cosh \varphi - f_2 \sinh \varphi).$$

A comparison between these formulas and those of Poincaré (1) proves that the transformation of the electric and the magnetic vector coincides with the transformation of the screw in Lobachevsky's space.

A.3.6

1. The preceding usage of Lobachevskian geometry is marvellous. In Poincaré's mapping the main factors of the relativity theory became obvious.

The Lorentz transformations resulted from the assumption that Maxwell's electromagnetic equations are the same both for a stationary observer and for all the other observers who move in a straight line relative to the stationary one. However, the observers do not only move uniformly in a straight line. In order to generalize, there must be analytical methods for investigating the relativity postulate.

2. Let the speed of light in a vacuum again be c. The Lorentz transformations are

$$t' = \beta \left(t - \frac{v}{c^2} x \right), \quad x' = \beta(x - vt), \quad y' = y, \quad z' = z$$

and have, in addition to Maxwell's equations, the following algebraic invariant:

$$c^2 t'^2 - x'^2 - y'^2 - z'^2 = c^2 t^2 - x^2 - y^2 - z^2.$$

This invariant yields the space-time metric. Minkowski was the first to find this metric.

In the world of the stationary observer (A) the vectors are defined by the projections on the axes t, x, y, z (Fig. 186). For instance

$$\overrightarrow{OA'}\ (t,\ x,\ y,\ z),\qquad \overrightarrow{OC'}\ (t_1,\ x_1,\ y_1,\ z_1).$$

If the vector $\overrightarrow{OC'}$ is parallel to a tangent plane D drawn to the hyperboloid at a point A', then the moving observer (A') will perceive the direction of $\overrightarrow{OC'}$ purely spatially, as *orthogonal* to the $\overrightarrow{OA'}$ axis.

This property yields the principal metric property of the orthogonality of the two vectors.

If T, X, Y, Z are the running coordinates of the point in the plane D, then we obtain the following from the equation of the tangent plane $c^2 tT - xX - yY - zZ = 0$:

$$c^2 tt_1 - xx_1 - yy_1 - zz_1 = 0.$$

The scale units are given by the algebraic invariant.

Thanks to Poincaré's marvellous results we can study this space-time metric mapped into Lobachevsky's space, in the large.

When studying natural phenomena, the exact natural sciences use the successful method of infinitesimals, i.e. the technique of representing phenomena in the small.

3. Consider the world line shown in Fig. 187. If this line represents the course of a physical phenomenon, then the velocity of the particle is not

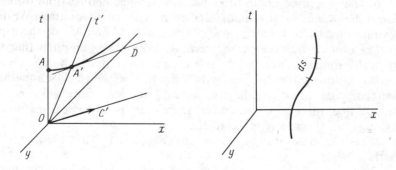

Figure 186   Figure 187

higher than the speed of light, and, hence, a tangent to the world line is always steeper than the generatrices of an asymptotic cone.

Consider an element ds on the line. For a stationary observer its projections are dt, dx, dy, dz. For a moving observer we get, from the Lorentz transformations, the following projections:

$$dt' = \beta \left(dt - \frac{v}{c^2} dx \right), \quad dx' = \beta(dx - vdt), \quad dy' = dy, \quad dz' = dz.$$

From this we get the following invariant:

$$c^2 dt'^2 - dx'^2 - dy'^2 - dz'^2 = c^2 dt^2 - dx^2 - dy^2 - dz^2 = ds^2 = c^2 d\tau,$$

where τ is the inherent time of the particle in question.

We have thus found an expression for an element of length in the space-time. This expression defines the metric of the space-time. What is this space?

4. The expression

$$dx^2 = dx^2 + dy^2 + dz^2 - c^2 dt^2$$

for the differential element for the directed element in space (which is invariant under the Lorentz transformations) yields an expression for the length of the segment ds in systems for which ds seems to be at rest at some moment.

We can see from the expression for ds^2 that the metric of the space-time is non-Euclidean. Let us find the space the element of whose arc is expressed in this way. We begin with a simple case.

We consider a sphere of radius R with centre at O (Fig. 188): π is its tangent plane, x, y, z are its Cartesian coordinates reckoned from the point O. We map the sphere onto the π-plane by rays emanating from the centre. The straight lines are associated with the arcs of the large circles. We define the non-Euclidean distance between the points M and M' of the π-plane by the distance* between the corresponding points of the sphere (measured along the sphere).

Let us use another technique to define the distance. The equation of the sphere is

$$x^2 + y^2 + z^2 = R^2.$$

---

* The distance between points on a sphere is measured along the arc of the large circle which passes through these points. — *Ed.*

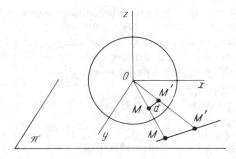

Figure 188

Let x, y, z and x', y', z' be the coordinates of the points M and M' of the sphere. From this it follows that

$$R^2 \cos \frac{d}{R} = xx' + yy' + zz'.$$

The isotropic cone

$$x^2 + y^2 + z^2 = c^2 t^2$$

cuts the π-plane along the conic section P (called the *absolute*). The coordinates x, y, z of the points of the sphere are the projective coordinates of the points M of the π-plane.

Let N_1 and N_2 be two complex-conjugate points at which the straight line MM' cuts the absolute. The coordinates of the running point on the line MM' can be written as

$$x + \lambda x', \quad y + \lambda y', \quad z + \lambda z'.$$

Assume that the point M is associated with $\lambda = 0$, the point M' with $\lambda = \infty$, the point N_1 with $\lambda = \lambda_1$, and the point N_2 with $\lambda = \lambda_2$. The values λ_1 and λ_2 can be found from the condition that N_1 and N_2 lie on the absolute:

$$\lambda^2(x'^2 + y'^2 + z'^2) + 2\lambda(xx' + yy' + zz') + x^2 + y^2 + z^2 = 0.$$

Hence

$$\lambda^2 + 2\lambda \cos \frac{d}{R} + 1 = 0.$$

This means that

$$\lambda_1 + \lambda_2 = -2 \cos \frac{d}{R}, \quad \lambda_1 \lambda_2 = 1.$$

Let $\lambda_1 = ke^{i\varphi}$, $\lambda_2 = ke^{-i\varphi}$. From the preceding relations we get

$$k^2 = 1 \quad \text{and} \quad k \cdot 2 \cos \varphi = -2 \cos \frac{d}{R},$$

$$k = -1 \quad \text{and} \quad \varphi = \frac{d}{R}.$$

The anharmonic ratio is

$$(MM'N_1N_2) = (0, \infty, \lambda_1, \lambda_2) = \frac{0 - \lambda_1}{0 - \lambda_2} \frac{\infty - \lambda_2}{\infty - \lambda_1} = \frac{\lambda_1}{\lambda_2} = e^{2i\varphi} = e^{2i\frac{d}{R}}$$

and, hence,

$$d = \frac{R}{2i} \ln(MM'N_1N_2).$$

Consequently, the distance d can be defined (Cayley) by the product of $R/2i$ multiplied by the logarithm of the anharmonic ratio of the four points M, M', N_1, N_2.

We define the angle between the straight lines of the π-plane as the angle between the corresponding large circles of the sphere R. We can find the angle by means of projections.

At the point of tangency of the π-plane the angles can be measured as usual. Assume that we are given an angle at the point M. The angle is defined by the tangents to the corresponding large circles. We transfer the tangents to the centre O and circumscribe a sphere of unit radius. The angles will be measured by the arcs of the large circles of the unit sphere. Using what we have proved, we find the angle

$$\varphi = \frac{1}{2i} \ln(m_1 m_2 n_1 n_2),$$

where n_1 and n_2 are tangents to the absolute which pass through M, and m_1 and m_2 are the sides of the angle with the vertex at the point M (on the π-plane). Let us find the required expression for the differential of the arc in the π-plane.

We consider a point A at which the sphere touches the plane (Fig. 189). We define the point A by the vector $\overrightarrow{OA}$ and take orthogonal unit vectors $\mathbf{e}_1$ and $\mathbf{e}_2$ in the π-plane. The vector $\overrightarrow{OM}$, which defines the point M of the sphere, can be determined as follows:

$$\overrightarrow{OM} = t \, \overrightarrow{OA} + x\mathbf{e}_1 + y\mathbf{e}_2,$$

where t, x, y are scalars. From this and from the orthogonality of $\overrightarrow{OA}$, $\mathbf{e}_1$ and $\mathbf{e}_2$ it follows that

$$\overrightarrow{OM}^2 = R^2 = R^2 t^2 + x^2 + y^2.$$

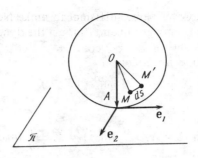

Figure 189

We take a point M' infinitely close to M:
$$\overrightarrow{OM'} = (t + dt)\, \overrightarrow{OA} + (x + dx)\mathbf{e}_1 + (y + dy)\mathbf{e}_2.$$

Let $\mathbf{ds}$ be MM'. Then
$$\mathbf{ds} = dt\, \overrightarrow{OA} + dx\, \mathbf{e}_1 + dy\, \mathbf{e}_2,$$

and this means that
$$ds^2 = (dx)^2 + (dy)^2 + R^2 dt^2.$$

This differs from the expression for ds^2 of the space-time from the relativity theory only by the sign of the last term. Hence, if we assume that $R = c\sqrt{-1}$, we find that the expression for ds^2 of the three-dimensional world coincides with the expression for the element of the arc of the sphere of imaginary radius.

We mentioned earlier that the geometry on a sphere of imaginary radius is Lobachevskian geometry.

A.3.7 Dynamics of a Particle

1. We designate the inherent time of a moving particle as
$$c^2 d\tau^2 = c^2 dt^2 - dx^2 - dy^2 - dz^2.$$

Setting $f \equiv df/d\tau$, we express the impulse in terms of the components
$$m_0\dot{x}, \quad m_0\dot{y}, \quad m_0\dot{z}, \quad m_0\dot{t},$$

where m_0 is the rest mass.

The components of the impulse vector thus defined can be transformed in the same way as the coordinates x, y, z, t with the use of the formulas for the Lorentz transformation.

This definition proves to be sufficient to make Newton's law covariant with respect to the Lorentz group and thus get the dynamics of the relativity principle.

Indeed, let us add an equation

$$m_0 \ddot{t} = \dot{t}T$$

to Newton's law

$$\frac{d}{dt}(m_0 \dot{x}) = X, \quad \frac{d}{dt}(m_0 \dot{y}) = Y, \quad \frac{d}{dt}(m_0 \dot{z}) = Z,$$

or, what is the same, to the equations

$$m_0 \ddot{x} = \dot{t}X, \quad m_0 \ddot{y} = \dot{t}Y, \quad m_0 \ddot{z} = \dot{t}Z.$$

On the left-hand sides we get the complete four-vector

$$m_0 \ddot{x}, \quad m_0 \ddot{y}, \quad m_0 \ddot{z}, \quad m_0 \ddot{t}$$

which can be transformed in the same way as the coordinates x, y, z, t with the use of the Lorentz transformation formulas since m_0 is a constant and $d\tau$ is an invariant.

We shall now find T. We have

$$c^2 d\tau^2 = c^2 dt^2 - dx^2 - dy^2 - dz^2,$$

whence it follows that $c^2 = c^2 \dot{t}^2 - \dot{x}^2 - \dot{y}^2 - \dot{z}^2$, or, after the differentiation with respect to τ,

$$0 = c^2 \dot{t}\ddot{t} - \ddot{x}\dot{x} - \ddot{y}\dot{y} - \ddot{z}\dot{z}.$$

Multiplying both sides of this relation by m_0 we obtain

$$c^2 \dot{t}T = X\dot{x} + Y\dot{y} + Z\dot{z}.$$

Hence

$$T = \frac{1}{c^2}\left(X\frac{\dot{x}}{\dot{t}} + Y\frac{\dot{y}}{\dot{t}} + Z\frac{\dot{z}}{\dot{t}} \right),$$

or

$$T = \frac{1}{c^2}\left(X\frac{dx}{dt} + Y\frac{dy}{dt} + Z\frac{dz}{dt} \right).$$

The equations of dynamics

$$m_0 \ddot{x} = \dot{t}X, \quad m_0 \ddot{y} = \dot{t}Y, \quad m_0 \ddot{z} = \dot{t}Z, \quad m_0 \ddot{t} = \dot{t}T$$

show that $X\dot{t}$, $Y\dot{t}$, $Z\dot{t}$, $T\dot{t}$ can be transformed according to the Lorentz formulas in the same way as the coordinates x, y, z, t. This means that the equations of dynamics we have written are covariant with respect to the Lorentz group.

2. A stationary observer defines the momentum as a vector with components

$$m \frac{dx}{dt}, \quad m \frac{dy}{dt}, \quad m \frac{dz}{dt}.$$

Comparing this expression to the momentum we have introduced, we get

$$m \frac{dx}{dt} = m_0 \dot{x} = m_0 \frac{dx}{dt} \frac{dt}{d\tau}, \quad \ldots,$$

whence we can see that the relative mass m is connected with the rest mass m_0 by the formula

$$m = m_0 \frac{dt}{d\tau}.$$

It is clear that m_0 is a mass relative to the system in which the moving particle seems to be at rest.

From the formula

$$c^2 d\tau^2 = c^2 dt^2 - dx^2 - dy^2 - dz^2$$

we have

$$c^2 \left(\frac{d\tau}{dt} \right)^2 = c^2 - \left[\left(\frac{dx}{dt} \right)^2 + \left(\frac{dy}{dt} \right)^2 + \left(\frac{dz}{dt} \right)^2 \right],$$

or

$$\frac{d\tau}{dt} = \sqrt{1 - \frac{v^2}{c^2}},$$

where v is the velocity of motion of the particle relative to the stationary observer and, hence, the relative mass

$$m = \frac{m_0}{\sqrt{1 - v^2/c^2}}.$$

The fourth component of the momentum vector

$$m_0 \dot{t} = m$$

is equal to the relative mass.

For a low velocity of motion v (small ratio v/c)

$$m = \frac{m_0}{\sqrt{1 - v^2/c^2}} = m_0 + \frac{1}{2} m_0 \frac{v^2}{c^2} + O \left(\left(\frac{v}{c} \right)^4 \right).$$

$$m - m_0 = \frac{1/2 m_0 v^2}{c^2} + O\left(\left(\frac{v}{c}\right)^4\right).$$

Hence, the variation of the mass upon a change in the velocity is equal to the kinetic energy divided by c^2. This demonstrates that mass and energy are identical for the stationary observer, i.e. (mc^2 is the internal energy of the body) $m = E/c^2$. The principle of conservation of mass has turned into the principle of the conservation of energy.

3. Assume that a particle moves along the x-axis with velocity v. The momentum is

$$\frac{mv}{\sqrt{1 - v^2/c^2}}, \quad 0.$$

Assume now that the velocity incremented by δv in the direction of the x-axis and δw in the direction of the y-axis. The momentum is

$$\frac{m(v + \delta v)}{\sqrt{1 - \dfrac{(v + \delta v)^2 + \delta w^2}{c^2}}}, \quad \frac{m\delta w}{\sqrt{1 - \dfrac{(v + \delta v)^2 + \delta w^2}{c^2}}}.$$

To within second-order terms, the variation of the impulses is

$$\frac{m\delta v}{(\sqrt{1 - v^2/c^2})^3}, \quad \frac{m\delta w}{\sqrt{1 - v^2/c^2}},$$

or

$$m\beta^3 \delta v, \quad m\beta \delta w, \quad \beta = \frac{1}{\sqrt{1 - v^2/c^2}}.$$

The first component of the impulse $m\beta^2 \delta v$ is greater than the second component $m\beta \delta w$.

A.3.8 Analytical Dynamics of a Particle

1. We have introduced the impulses via the relations

$$p_1 = m_0 \dot{x} = m_0 \frac{dt}{d\tau} \frac{dx}{dt}, \quad p_2 = m_0 \frac{dt}{d\tau} \frac{dy}{dt}, \quad p_3 = m_0 \frac{dt}{d\tau} \frac{dz}{dt}.$$

But

$$\frac{dt}{d\tau} = \frac{1}{\sqrt{1 - v^2/c^2}},$$

and, consequently, the impulses have the form

$$p_1 = \frac{m_0 \dfrac{dx}{dt}}{\sqrt{1 - v^2/c^2}}, \quad p_2 = \frac{m_0 \dfrac{dy}{dt}}{\sqrt{1 - v^2/c^2}}, \quad p_3 = \frac{m_0 \dfrac{dz}{dt}}{\sqrt{1 - v^2/c^2}}.$$

We define the function T^* so that $p_\nu = \dfrac{\partial T^*}{\partial x_\nu'}$. This gives

$$T^* = -m_0 c^2 \sqrt{1 - v^2/c^2}.$$

2. In the case of conservative forces, the differential equations of motion are

$$\frac{d}{dt}\left(\frac{m_0 x'}{\sqrt{1 - v^2/c^2}}\right) - \frac{\partial U}{\partial x} = 0,$$

$$\frac{d}{dt}\left(\frac{m_0 y'}{\sqrt{1 - v^2/c^2}}\right) - \frac{\partial U}{\partial y} = 0,$$

$$\frac{d}{dt}\left(\frac{m_0 z'}{\sqrt{1 - v^2/c^2}}\right) - \frac{\partial U}{\partial z} = 0.$$

They have the form of Lagrange's equations

$$\frac{d}{dt}\left(\frac{\partial L}{\partial q_i'}\right) - \frac{\partial L}{\partial q_i} = 0,$$

if we take the orthogonal Cartesian coordinates x, y, z as q_i and

$$L = -mc^2 \sqrt{1 - \frac{v^2}{c^2}} + U,$$

or

$$L = T^* + U.$$

T^* is not the kinetic energy of the particle. Together with the inherent energy, the kinetic energy is

$$m_0 c^2 \frac{1}{\sqrt{1 - \dfrac{v^2}{c^2}}}.$$

3. In accordance with the Lagrangian form of the equations of motion, Hamilton's principle of least action in the dynamics of a particle from relativity theory has the form

$$\delta \int_{t_0}^{t_1} L\, dt = 0.$$

Let the endpoints be fixed. Then Hamilton's principal function is defined by the integral

$$V = \int_{t_0}^{t} L \, dt$$

taken along the true trajectory. If we assume that V is a function of t, x, y, z, x^0, y^0, z^0, then

$$\frac{\partial V}{\partial x} = p_1, \quad \frac{\partial V}{\partial y} = p_2, \quad \frac{\partial V}{\partial z} = p_3,$$

$$\frac{\partial V}{\partial x_0} = -p_1^0, \quad \frac{\partial V}{\partial y_0} = -p_2^0, \quad \frac{\partial V}{\partial z_0} = -p_3^0.$$

The proof does not differ from the ordinary one.

4. To derive the canonical equations and obtain the Hamiltonian H, we use the method suggested by Poincaré;

$$\delta \int_{t_0}^{t_1} \left[p_1 \frac{dx}{dt} + p_2 \frac{dy}{dt} + p_3 \frac{dz}{dt} - \left(p_1 \frac{dx}{dt} + p_2 \frac{dy}{dt} + p_3 \frac{dz}{dt} - L \right) \right] dt = 0.$$

Hence

$$H = p_1 \frac{dx}{dt} + v_2 \frac{dy}{dt} + p_3 \frac{dz}{dt} - L.$$

The explicit form of the function H is

$$H = \frac{m_0 v^2}{\sqrt{1 - v^2/c^2}} + m_0 c^2 \sqrt{1 - \frac{v^2}{c^2}} - U$$

$$= m_0 c^2 \left[\frac{1}{\sqrt{1 - v^2/c^2}} \right] - U = T - U.$$

This means that H is the total energy. We have

$$p_1^2 + p_2^2 + p_3^2 = m_0^2 c^2 \frac{v^2/c^2}{(1 - v^2/c^2)} = m_0^2 c^2 \left(\frac{1}{1 - v^2/c^2} - 1 \right),$$

and, hence

$$\frac{1}{\sqrt{1 - v^2/c^2}} = \sqrt{m_0^2 c^2 + p_1^2 + p_2^2 + p_3^2} \cdot \frac{1}{m_0 c}$$

$$= \sqrt{1 + \frac{1}{m_0^2 c^2} (p_1^2 + p_2^2 + p_3^2)},$$

and this means that the function H in the impulses has the form

$$H = m_0 c^2 \sqrt{1 + \frac{1}{m_0^2 c^2} (p_1^2 + p_2^2 + p_3^2)} - U.$$

5. An interesting feature of the special theory of relativity is the form of the Hamilton-Jacobi equations

$$\frac{\partial V}{\partial t} + mc^2 \sqrt{1 + \frac{1}{m^2 c^2} \left[\left(\frac{\partial V}{\partial x} \right)^2 + \left(\frac{\partial V}{\partial y} \right)^2 + \left(\frac{\partial V}{\partial z} \right)^2 \right]} - U = 0,$$

or

$$\left(\frac{\partial V}{\partial t} - U \right)^2 - m^2 c^4 \left\{ 1 + \frac{1}{m^2 c^2} \left[\left(\frac{\partial V}{\partial x} \right)^2 + \left(\frac{\partial V}{\partial y} \right)^2 + \left(\frac{\partial V}{\partial z} \right)^2 \right] \right\} = 0,$$

i.e. in this case Jacobi's equation includes the square of the derivative $\partial V/\partial t$. This led Dirac to the well-known equations of quantum mechanics.

A.3.9

1. Consider the general case of the quadratic metric

$$ds^2 = \sum g_{\alpha\beta} dx^\alpha dx^\beta,$$

here dx^α are the components of the (contravariant) vector and $g_{\alpha\beta}$ is a fundamental tensor.

We set

$$dx^1 = dx, \quad dx^2 = dy, \quad dx^3 = dz, \quad dx^4 = cdt.$$

In the theory of relativity we have

$$\{g_{\alpha\beta}\} = \begin{pmatrix} -1 & 0 & 0 & 0 \\ 0 & -1 & 0 & 0 \\ 0 & 0 & -1 & 0 \\ 0 & 0 & 0 & 1 \end{pmatrix}.$$

After transforming the *coordinate x of the particle* we have

$$dx'^\mu = \sum \frac{\partial x'^\mu}{\partial x^\nu} dx^\nu.$$

2. We call the quantities A^α, which can be transformed according to the formulas

$$A'^\mu = \sum \frac{\partial x'_\mu}{\partial x_\nu} A^\nu,$$

the *contravariant* components of the vector **A**.

We term A_α the *covariant* components of the vector **A** if $\sum A_\nu B^\nu$ is an invariant of the arbitrary vector **B**. Hence follows the transformation of the covariant components

$$A'_\mu = \sum \frac{\partial x_\nu}{\partial x'_\mu} A_\nu.$$

We agree to delete the sign of the sum.

3. Tensors. Forming tensors by multiplying vectors.

1°. $A^{\alpha\beta} = A^\alpha A^\beta$ is a contravariant tensor.

The transformation formulas are

$$A'^{\alpha\beta} = \frac{\partial x'_\alpha}{\partial x_\nu} \frac{\partial x'_\beta}{\partial x_\mu} A^{\nu\mu}.$$

2°. $A_{\alpha\beta} = A_\alpha A_\beta$ is a covariant tensor.

3°. $A^\alpha_\beta = A^\alpha A_\beta$ is a mixed tensor.

The symmetric tensors are

$$A^{\alpha\beta} = A^{\beta\alpha}, \quad A_{\alpha\beta} = A_{\beta\alpha}.$$

The skew-symmetric tensors are

$$A^{\alpha\beta} = -A^{\beta\alpha}, \quad A_{\alpha\beta} = -A_{\beta\alpha}.$$

If $\sum A_{\alpha\beta} B^{\alpha\beta}$ is an invariant for any tensor $B^{\alpha\beta}$, then $A_{\alpha\beta}$ is a tensor.

4°. The inner contraction of a tensor: $A^{\alpha \cdot \beta}_{\cdot \alpha} = A^\beta$, $B_{\cdot \nu} = B$.

4. Transposition of an index.

$$ds^2 = g_{\alpha\beta} dx^\alpha dx^\beta.$$

Here $dx^\alpha dx^\beta$ is an arbitrary tensor and ds^2 is an invariant. Hence $g_{\alpha\beta}$ is a covariant tensor. It follows that $g_{\alpha\beta} A^\beta$ is a vector in covariant components. We write this vector (dependent on A) as $A_\alpha = g_{\alpha\beta} A^\beta$. This is the definition of covariant components and the transposition of an index. We have $T^\alpha_{\cdot\beta\gamma} = g_{\beta\mu} g_{\gamma\nu} T^{\alpha\mu\nu}$ and so on.

5. Consider a determinant

$$g = |g_{\alpha\beta}|.$$

Let $g^{\alpha\beta}$ be a minor (with a sign) for the element $g_{\alpha\beta}$ divided by g. We have

$$g_{\mu\sigma} g^{\nu\sigma} = \delta^\nu_\mu = \begin{cases} 1, & \text{if } \mu = \nu, \\ 0, & \text{if } \mu \neq \nu. \end{cases}$$

Hence

$$ds^2 = g_{\alpha\beta} dx^\beta dx^\alpha = g_{\beta\alpha} \delta^\beta_\nu dx^\nu dx^\alpha = g_{\alpha\beta} g_{\nu\sigma} g^{\beta\sigma} dx^\nu dx^\alpha$$
$$= g^{\beta\sigma}(g_{\alpha\beta} dx^\alpha)(g_{\nu\sigma} dx^\nu) = g^{\beta\sigma} dx_\beta dx_\sigma,$$

where $dx_\beta dx_\sigma$ is a covariant tensor. This means that $g^{\beta\sigma}$ is a contravariant tensor.

A.3.10

1. The equation of a geodesic line is

$$\delta \int_{P_0}^{P_1} ds = 0.$$

We introduce a parameter u. Assume that on comparable curves it has values u_1 and u_0 for P_1 and P_0. It follows that

$$\delta \int_{P_0}^{P_1} ds = \delta \int_{u_0}^{u_1} \frac{ds}{du} \, du = \int_{u_0}^{u_1} \delta \left(\frac{ds}{du} \right) du = 0.$$

But

$$\left(\frac{ds}{du} \right)^2 = g_{\alpha\beta} \frac{dx^\alpha}{du} \frac{dx^\beta}{du} \,,$$

$$2 \frac{ds}{du} \delta \left(\frac{ds}{du} \right) = \frac{\partial g_{\alpha\beta}}{\partial x^\nu} \frac{dx^\alpha}{du} \frac{dx^\beta}{du} \delta x^\nu + 2 g_{\alpha\beta} \frac{dx^\alpha}{du} \delta \frac{dx^\beta}{du}$$

and

$$\delta \frac{dx^\beta}{du} = \frac{d}{du} (\delta x^\beta),$$

$$0 = \int_{u_0}^{u_1} du \left[\frac{1}{2} \frac{\partial g_{\alpha\beta}}{\partial x^\nu} \frac{dx^\alpha}{du} \frac{dx^\beta}{du} \delta x^\nu + g_{\alpha\nu} \frac{dx^\alpha}{du} \delta \frac{dx^\nu}{du} \right] \frac{1}{(ds/du)} \,,$$

$$\int_{u_0}^{u_1} \frac{1}{(ds/du)} g_{\alpha\nu} \frac{dx^\alpha}{du} \frac{d\delta x^\nu}{du} \, du = - \int_{u_0}^{u_1} \frac{d}{du} \left[\frac{g_{\alpha\nu}}{(ds/du)} \frac{dx^\alpha}{du} \right] \delta x^\nu du,$$

$$0 = \int_{u_0}^{u_1} \delta x^\nu \left[\frac{1}{2(ds/du)} \frac{\partial g_{\alpha\beta}}{\partial x^\nu} \frac{dx^\alpha}{du} \frac{dx^\beta}{du} - \frac{d}{du} \left(\frac{g_{\alpha\nu}}{(ds/du)} \frac{dx^\alpha}{du} \right) \right] du.$$

From this we get the equation of a geodesic line

$$\frac{d}{du} \left(\frac{g_{\alpha\nu}}{(ds/du)} \frac{dx^\alpha}{du} \right) - \frac{1}{2(ds/du)} \frac{\partial g_{\alpha\beta}}{\partial x^\nu} \frac{dx^\alpha}{du} \frac{dx^\beta}{du} = 0.$$

We can take $s = u$ as a parameter if s is nonzero. Then

$$\frac{d}{ds}\left(g_{\alpha\nu}\frac{dx^{\alpha}}{ds}\right) - \frac{1}{2}\frac{\partial g_{\alpha\beta}}{\partial x^{\nu}}\frac{dx^{\alpha}}{ds}\frac{dx^{\beta}}{ds} = 0,$$

or

$$g_{\alpha\nu}\frac{d^2x^{\alpha}}{ds^2} + \frac{\partial g_{\alpha\nu}}{\partial x^{\beta}}\frac{dx^{\alpha}}{ds}\frac{dx^{\beta}}{ds} - \frac{1}{2}\frac{\partial g_{\alpha\beta}}{\partial x^{\nu}}\frac{dx^{\alpha}}{ds}\frac{dx^{\beta}}{ds} = 0,$$

$$g_{\alpha\nu}\frac{d^2x^{\nu}}{ds^2} + \frac{1}{2}\left(\frac{\partial g_{\alpha\nu}}{\partial x^{\beta}} + \frac{\partial g_{\beta\nu}}{\partial x^{\alpha}} - \frac{\partial g_{\alpha\beta}}{\partial x^{\nu}}\right)\frac{dx^{\alpha}}{ds}\frac{dx^{\beta}}{ds} = 0.$$

We introduce the notation

$$\Gamma_{\alpha\nu\beta} = \Gamma_{\beta\nu\alpha} = \frac{1}{2}\left(\frac{\partial g_{\alpha\nu}}{\partial x^{\beta}} + \frac{\partial g_{\beta\nu}}{\partial x^{\alpha}} - \frac{\partial g_{\alpha\beta}}{\partial x^{\nu}}\right).$$

In this notation

$$g^{\varkappa\nu}\left|g_{\alpha\nu}\frac{d^2x^{\alpha}}{ds^2} + \Gamma_{\alpha\nu\beta}\frac{dx^{\alpha}}{ds}\frac{dx^{\beta}}{ds} = 0,\right.$$

$$g^{\varkappa\nu}g_{\alpha\nu}\frac{d^2x^{\alpha}}{ds^2} = \delta^{\varkappa}_{\alpha}\frac{d^2x^{\alpha}}{ds^2} = \frac{d^2x^{\varkappa}}{ds^2}$$

and hence

$$\frac{d^2x^{\varkappa}}{ds^2} + \Gamma^{\varkappa}_{\alpha\beta}\frac{dx^{\alpha}}{ds}\frac{dx^{\beta}}{ds} = 0,$$

where

$$\Gamma^{\varkappa}_{\alpha\beta} = g^{\varkappa\nu}\Gamma_{\alpha\nu\beta};$$

$\Gamma_{\alpha\nu\beta}$ and $\Gamma^{\varkappa}_{\alpha\beta}$ are Christoffel symbols.

Remark. The surfaces $u = \text{const}$ cut the field of comparable curves (Fig. 190), $u = u_0$ passes through P_0 and $u = u_1$ passes through P_1.

Taking u as a parameter, we have

$$\delta\int\limits_{u_0}^{u_1}\frac{ds}{du}\,du = 0.$$

We isolate the geodesic Γ. Its arc $s = f(u)$, and then, instead of the parameter u, we can take a parameter

$$v = f(u),$$

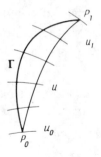

Figure 190

v has the sense of s only on Γ. For comparable curves v has some other sense. For this choice of v, we have the following only on the geodesic line

$$\frac{ds}{dv} = 1.$$

We use this condition when deriving the equation

$$\frac{d^2x^\alpha}{ds^2} + \Gamma^\alpha_{\mu\nu} \frac{dx^\mu}{ds} \frac{dx^\nu}{ds} = 0.$$

This means that it is senseless to choose both $u = s$ and $\frac{ds}{dv} = 1$ for comparable curves (Einstein).

2. Forming tensors by differentiation. Assume that φ is an invariant function. Then $d\varphi/ds$ is also invariant, $d\varphi$ and ds are invariants. It follows from $d\varphi = \frac{\partial\varphi}{\partial x_\nu} dx^\nu$ that $A_\nu = \frac{\partial\varphi}{\partial x_\nu}$ is a covariant four-vector (the gradient of φ).

Furthermore,

$$\frac{d\varphi}{ds} = \frac{\partial\varphi}{\partial x_\mu} \frac{dx^\mu}{ds},$$

$$\frac{d}{ds}\left(\frac{d\varphi}{ds}\right) = \frac{\partial^2\varphi}{\partial x_\mu \partial x_\nu} \frac{dx^\mu}{ds} \frac{dx^\nu}{ds} + \frac{\partial\varphi}{\partial x_\mu} \frac{d^2x^\mu}{ds^2}.$$

Assume that we are differentiating along a geodesic line. Then

$$\frac{d^2\varphi}{ds^2} = \frac{\partial^2\varphi}{\partial x_\mu \partial x_\nu} \frac{dx^\mu}{ds} \frac{dx^\nu}{ds} - \Gamma^\alpha_{\mu\nu} \frac{\partial\varphi}{\partial x_\alpha} \frac{dx^\mu}{ds} \frac{dx^\nu}{ds}$$

$$= \left(\frac{\partial^2\varphi}{\partial x_\mu \partial x_\nu} - \Gamma^\alpha_{\mu\nu} \frac{\partial\varphi}{\partial x_\alpha}\right) \frac{dx^\mu}{ds} \frac{dx^\nu}{ds}.$$

But if $\dfrac{d^2\varphi}{ds^2}$ is an invariant and $\dfrac{dx^\mu}{ds}\dfrac{dx^\nu}{ds}$ is a tensor, then

$$D_{\mu\nu} = \frac{\partial^2 \varphi}{\partial x_\mu \partial x_\nu} - \Gamma_{\mu\nu}^\alpha \frac{\partial \varphi}{\partial x_\alpha}$$

is a covariant tensor. Or, if we assume that $A_\mu = \dfrac{\partial \varphi}{\partial x_\mu}$, we have

$$D_\nu A_\mu = A_{\mu\nu} = \frac{\partial A_\mu}{\partial x_\nu} - \Gamma_{\mu\nu}^\alpha A_\alpha.$$

We term this tensor a *tensorial derivative of the tensor* A_μ. This rule is valid for any tensor A_μ. Indeed,

$$\varphi = A_\mu \frac{dx^\mu}{ds}$$

is an invariant and

$$\frac{d\varphi}{ds} = \frac{\partial A_\mu}{\partial x_\nu} \frac{dx^\mu}{ds} \frac{dx^\nu}{ds} + A_\mu \frac{d^2 x^\mu}{ds^2} = \frac{dx^\mu}{ds} \frac{dx^\nu}{ds} \left(\frac{\partial A_\mu}{\partial x_\nu} - \Gamma_{\mu\nu}^\alpha A_\alpha \right)$$

as well. This means that in the general case

$$A_{\mu\nu} = \frac{\partial A_\mu}{\partial x^\nu} - \Gamma_{\mu\nu}^\alpha A_\alpha, \qquad A_{\mu\nu\sigma} = \frac{\partial A_{\mu\nu}}{\partial x^\sigma} - \Gamma_{\sigma\nu}^\alpha A_{\mu\alpha} - \Gamma_{\mu\sigma}^\alpha A_{\alpha\nu}.$$

3. According to the rule of differentiating determinants*

$$dg = g^{\mu\nu} g \, dg_{\mu\nu} = -g_{\mu\nu} g \, dg^{\mu\nu}$$

and hence (instead of $\sqrt{g}$ we introduce a real value $\sqrt{-g}$),

$$\frac{1}{\sqrt{-g}} \frac{\partial \sqrt{-g}}{\partial x_\sigma} = \frac{1}{2} \frac{\partial \ln(-g)}{\partial x^\sigma} = \frac{1}{2} g^{\mu\nu} \frac{\partial g_{\mu\nu}}{\partial x_\sigma} = \frac{1}{2} g_{\mu\nu} \frac{\partial g^{\mu\nu}}{\partial x_\sigma}. \tag{3}$$

From $g_{\mu\sigma} g^{\nu\sigma} = \delta_\mu^\nu$ we have

$$g_{\mu\sigma} dg^{\nu\sigma} = -g^{\nu\sigma} dg_{\mu\sigma}.$$

Consequently,

$$dg^{\mu\nu} = -g^{\mu\alpha} g^{\nu\beta} dg_{\alpha\beta}, \qquad dg_{\mu\nu} = -g_{\mu\alpha} g_{\nu\beta} dg^{\alpha\beta},$$

$$\frac{\partial g^{\mu\nu}}{\partial x_\sigma} = -g^{\mu\alpha} g^{\nu\beta} \frac{\partial g_{\alpha\beta}}{\partial x_\sigma}; \qquad \frac{\partial g_{\mu\nu}}{\partial x_\sigma} = -g_{\mu\alpha} g_{\nu\beta} \frac{\partial g^{\alpha\beta}}{\partial x_\sigma}.$$

---

* For your information: $g^{\mu\nu} g_{\mu\nu} = 1 = \delta_\nu^\nu$, $g_{\mu\nu} dg^{\mu\nu} = -g^{\mu\nu} dg_{\mu\nu}$. — *Ed.*

But by virtue of the fundamental formula

$$\frac{\partial g_{\alpha\beta}}{\partial x_\sigma} = \Gamma_{\alpha\beta\sigma} + \Gamma_{\beta\alpha\sigma} = \frac{1}{2}\left(\frac{\partial g_{\alpha\beta}}{\partial x_\sigma} + \frac{\partial g_{\beta\alpha}}{\partial x_\sigma}\right).$$

Hence

$$\frac{\partial g^{\mu\nu}}{\partial x_\sigma} = -(g^{\mu\tau}\Gamma^\nu_{\tau\sigma} + g^{\nu\tau}\Gamma^\mu_{\tau\sigma}).$$

And, hence, if we substitute the last expression into (3), we obtain

$$\frac{1}{\sqrt{-g}} = \frac{\partial\sqrt{-g}}{\partial x_\sigma} = \Gamma^\mu_{\mu\sigma}.$$

4. Some principal tensors.

$1°.\ A_{\mu\nu} = \dfrac{\partial A_\mu}{\partial x_\nu} - \Gamma^\alpha_{\mu\nu}A_\alpha.$

From this it follows that

$$g^{\mu\nu}A_{\mu\nu} = g^{\mu\nu}\frac{\partial A_\mu}{\partial x_\nu} - g^{\mu\nu}\Gamma^\alpha_{\mu\nu}A_\alpha$$

$$= \frac{\partial}{\partial x_\nu}(g^{\mu\nu}A_\mu) - A_\mu\frac{\partial g^{\mu\nu}}{\partial x^\nu} - g^{\mu\nu}\Gamma^\alpha_{\mu\nu}A_\alpha \quad .$$

$$= \frac{\partial A^\nu}{\partial x_\nu} + A_\mu g^{\mu\alpha}\Gamma^\nu_{\alpha\nu} = \frac{\partial A^\nu}{\partial x_\nu} + A^\alpha\Gamma^\nu_{\alpha\nu}$$

$$= \frac{\partial A^\nu}{\partial x^\nu} + A^\nu\Gamma^\alpha_{\nu\alpha} = \frac{\partial A^\nu}{\partial x^\nu} + \frac{1}{\sqrt{-g}}\frac{\partial\sqrt{-g}}{\partial x_\nu}A^\nu = \frac{1}{\sqrt{-g}}\frac{\partial}{\partial x_\nu}(\sqrt{-g}A^\nu).$$

This is a *divergence* of a contravariant vector.

$2°.$ **Curl.** $A_{\mu\nu} - A_{\nu\mu} = \dfrac{\partial A_\mu}{\partial x_\nu} - \dfrac{\partial A_\nu}{\partial x_\mu}$ is a skew-symmetric tensor. The tensorial derivative of a (skew-symmetric) six-vector

$$A_{\mu\nu\sigma} = \frac{\partial A_{\mu\nu}}{\partial x_\sigma} - \Gamma^\alpha_{\sigma\nu}A_{\mu\alpha} - \Gamma^\alpha_{\mu\sigma}A_{\alpha\nu},$$

$$B_{\mu\nu\sigma} = A_{\mu\nu\sigma} + A_{\nu\sigma\mu} + A_{\sigma\mu\nu} = \frac{\partial A_{\mu\nu}}{\partial x_\sigma} + \frac{\partial A_{\nu\sigma}}{\partial x_\mu} + \frac{\partial A_{\sigma\mu}}{\partial x_\nu}.$$

$3°.$ The divergence of a six-vector.

$$A^{\alpha\beta}_\sigma = \frac{\partial A^{\alpha\beta}}{\partial x_\sigma} + \Gamma^\alpha_{\sigma\varkappa}A^{\varkappa\beta} + \Gamma^\beta_{\sigma\varkappa}A^{\alpha\varkappa} \quad (A^{\alpha\beta} = -A^{\beta\alpha}),$$

whence

$$A^{\alpha} = A^{\alpha\beta}_{\beta} = \frac{\partial A^{\alpha\beta}}{\partial x_{\beta}} + \Gamma^{\beta}_{\beta x} A^{\alpha x} + \Gamma^{\alpha}_{\beta x} A^{x\beta},$$

$$A^{\alpha} = \frac{1}{\sqrt{-g}} \frac{\partial(\sqrt{-g} A^{\alpha\beta})}{\partial x^{\beta}}.$$

A.3.11

1. Let us return to Maxwell's equations. Putting them in tensor notation is the beginning of further generalization. It is interesting to show that we can impart a tensor form to Maxwell's equations. We write out Maxwell's equations in their ordinary form, i.e.

$$\frac{\partial N}{\partial y} - \frac{\partial M}{\partial z} - \frac{1}{c}\frac{\partial X}{\partial t} = \frac{4\pi}{c} j_x, \qquad \frac{\partial Y}{\partial z} - \frac{\partial Z}{\partial y} - \frac{1}{c}\frac{\partial L}{\partial t} = 0,$$

$$\frac{\partial L}{\partial z} - \frac{\partial N}{\partial x} - \frac{1}{c}\frac{\partial Y}{\partial t} = \frac{4\pi}{c} j_y, \qquad \frac{\partial Z}{\partial x} - \frac{\partial X}{\partial z} - \frac{1}{c}\frac{\partial M}{\partial t} = 0,$$

$$\frac{\partial M}{\partial x} - \frac{\partial L}{\partial y} - \frac{1}{c}\frac{\partial Z}{\partial t} = \frac{4\pi}{c} j_z, \qquad \frac{\partial X}{\partial y} - \frac{\partial Y}{\partial x} - \frac{1}{c}\frac{\partial N}{\partial t} = 0,$$

$$\frac{\partial X}{\partial x} + \frac{\partial Y}{\partial y} + \frac{\partial Z}{\partial z} = 4\pi\varrho, \qquad \frac{\partial L}{\partial x} + \frac{\partial M}{\partial y} + \frac{\partial N}{\partial z} = 0,$$

where ϱ is the density of the charges and j_x, j_y, j_z are the components of the density of current **j**.

2. We designate the coordinates ct, x, y, z, as x_0, x_1, x_2, x_3 and $4\pi\varrho$, $\frac{4\pi}{c} j_x$, $\frac{4\pi}{c} j_y$, $\frac{4\pi}{c} j_z$ as S^0, S^1, S^2, S^3. We introduce into consideration a second-order skew-symmetric contravariant tensor

$$(F^{\mu\nu}) = \begin{pmatrix} 0 & X & Y & Z \\ -X & 0 & N & -M \\ -Y & -N & 0 & L \\ -Z & M & -L & 0 \end{pmatrix}.$$

Passing to covariant components,

$$(F_{\mu\nu}) = \begin{pmatrix} 0 & -X & -Y & -Z \\ X & 0 & N & -M \\ Y & -N & 0 & L \\ Z & M & -L & 0 \end{pmatrix}.$$

In this notation, we rewrite Maxwell's equations as

$$0 + \frac{\partial F^{01}}{\partial x_1} + \frac{\partial F^{02}}{\partial x_2} + \frac{\partial F^{03}}{\partial x_3} = S^0,$$

$$\frac{\partial F^{10}}{\partial x_0} + 0 + \frac{\partial F^{12}}{\partial x_2} + \frac{\partial F^{13}}{\partial x_3} = S^1,$$

$$\frac{\partial F^{20}}{\partial x_0} + \frac{\partial F^{21}}{\partial x_1} + 0 + \frac{\partial F^{23}}{\partial x_3} = S^2,$$

$$\frac{\partial F^{30}}{\partial x_0} + \frac{\partial F^{31}}{\partial x_1} + \frac{\partial F^{32}}{\partial x_2} + 0 = S^3,$$

$$0 + \frac{\partial F_{32}}{\partial x_1} + \frac{\partial F_{13}}{\partial x_2} + \frac{\partial F_{21}}{\partial x_3} = 0,$$

$$\frac{\partial F_{32}}{\partial x_0} + 0 + \frac{\partial F_{03}}{\partial x_2} + \frac{\partial F_{20}}{\partial x_3} = 0,$$

$$\frac{\partial F_{13}}{\partial x_0} + \frac{\partial F_{30}}{\partial x_1} + 0 + \frac{\partial F_{01}}{\partial x_3} = 0,$$

$$\frac{\partial F_{21}}{\partial x_0} + \frac{\partial F_{02}}{\partial x_1} + \frac{\partial F_{10}}{\partial x_2} + 0 = 0.$$

The first group is the divergence of the six-vector $F^{\mu\nu}$

$$\frac{\partial F^{\mu\nu}}{\partial x^\nu} = S^\mu.$$

The second group is a tensorial derivative

$$F_{\mu\nu,\sigma} + F_{\nu\sigma,\mu} + F_{\sigma\mu,\nu} = 0$$

of the covariant skew-symmetric tensor $F_{\mu\nu}$. The tensor form of the equations of an electromagnetic field immediately shows that Maxwell's equations are invariant with respect to the Lorentz transformations.

Changing to a new observer, the densities of the charge and the current S^0, S^1, S^2, S^3 are transformed as a contravariant vector. The electric and magnetic vectors are transformed according to the rules for a covariant skew-symmetric tensor ($F_{\mu\nu}$).

3. To see the tensor nature of the quantities $F_{\mu\nu}$ more clearly, let us consider another definition of electromagnetic vectors.

Assume that φ is a scalar potential and A_x, A_y, A_z are vector potentials. In terms of φ and **A** the electromagnetic field strength can be expressed as follows:

$$L = \frac{\partial A_3}{\partial y} - \frac{\partial A_2}{\partial z}, \quad X = -\frac{\partial \varphi}{\partial x} - \frac{1}{c}\frac{\partial A_1}{\partial t},$$

$$M = \frac{\partial A_1}{\partial z} - \frac{\partial A_3}{\partial x}, \quad Y = -\frac{\partial \varphi}{\partial y} - \frac{1}{c}\frac{\partial A_2}{\partial t},$$

$$N = \frac{\partial A_2}{\partial x} - \frac{\partial A_1}{\partial y}, \quad Z = -\frac{\partial \varphi}{\partial z} - \frac{1}{c}\frac{\partial A_3}{\partial t}.$$

In Minkowski's space-time we introduce a vector with covariant components $\varphi_0 = \varphi$, $\varphi_1 = -A_1$, $\varphi_2 = -A_2$, $\varphi_3 = -A_3$. The curl of this vector yields a skew-symmetric tensor

$$F_{\mu\nu} = \frac{\partial \varphi_\mu}{\partial x_\nu} - \frac{\partial \varphi_\nu}{\partial x_\mu}.$$

We have thus proved the tensor character of the quantities $F_{\mu\nu}$.

A.3.12

1. In the electromagnetic theory of light, Maxwell's equations (a system of first-order equations) yield a second-order equation

$$\frac{1}{c^2}\frac{\partial^2 \psi}{\partial t^2} - \Delta\psi = 0.$$

It is interesting to see whether the Hamilton-Jacobi equation in the mechanics of a particle (when there are no forces, $U = 0$)

$$\left(\frac{1}{c}\frac{\partial V}{\partial t}\right)^2 - \left[\left(\frac{\partial V}{\partial x}\right)^2 + \left(\frac{\partial V}{\partial y}\right)^2 + \left(\frac{\partial V}{\partial z}\right)^2\right] = m^2 c^2$$

is replaced by a system of first-order partial differential equations.

2. We consider a complex vector space. The metric is given by the bilinear form

$$\xi^1 \eta^2 - \xi^2 \eta^1 = \sum g_{\mu\nu}\xi^\mu \eta^\nu.$$

The metric matrix is skew-symmetric, i.e.

$$(g_{\mu\nu}) = \begin{pmatrix} 0 & 1 \\ -1 & 0 \end{pmatrix}.$$

Let us consider a group of linear transformations which leave the bilinear form invariant:

$$\xi^{1'} = \alpha\xi^1 + \beta\xi^2, \quad \xi^{2'} = \gamma\xi^1 + \delta\xi^2.$$

This yields

$$\begin{vmatrix} \alpha & \beta \\ \gamma & \delta \end{vmatrix} = 1.$$

The vectors (ξ^1, ξ^2), (η^1, η^2) of this space are known as *spinors*.

3. The metric tensor $g_{\mu\nu}$ makes it possible to introduce covariant components of the spinor by means of the formulas

$$\eta_\mu = \sum g_{\mu\nu} \eta^\nu.$$

We can now write the bilinear fundamental form of the space as

$$\xi^1 \eta^2 - \xi^2 \eta^1 = \sum g_{\mu\nu} \xi^\mu \eta^\nu = \xi^\mu \eta_\mu = -\xi_\nu \eta^\nu.$$

This notation establishes a rule for contracting tensors, i.e. the rule for obtaining *invariants*.

Hence the linear transformation, which is contragradient with respect to the principal transformation, is

$$\eta_1' = \delta\eta_1 - \gamma\eta_2, \quad \eta_2' = -\beta\eta_1 + \alpha\eta_2.$$

4. We call a *spin tensor* a quantity $a^\nu_{\lambda\mu\sigma}$, which changes when the bilinear group is transformed in the same way as the quantities $\xi_\lambda \eta_\mu$, $\zeta^\nu \chi_\sigma$.

Spinor algebra can be understood by analogy with tensor algebra. There is a correspondence between the spinor ξ^1, ξ^2 and the conjugate spinor $\bar{\xi}^1$, $\bar{\xi}^2$ $\left(\bar{\xi}_\mu = \sum g_{\mu\nu} \bar{\xi}^\nu \right)$. Spin tensors can be transformed as conjugate spinors with respect to some indices. We shall mark this fact by a bar over the indices. For instance,

$$a^{\bar\lambda}_{\mu\bar\nu} \sim \xi_\mu \bar{\eta}_\nu \bar{\zeta}^\lambda.$$

5. In a conservative system with one degree of freedom a bilinear form can be understood as Poincaré's invariant and the spinor ξ^1, ξ^2 as a particular solution of Poincaré's equations in variations for some perturbed motion. It is well known that the group of transformations of motion is a binary group.

This means that we can interpret the quantities ξ^1, ξ^2 as deviations of canonical variables of the conservative system.

A.3.13

1. Consider a covariant spin tensor of the second rank with four components

$$a_{1\bar1}, \; a_{1\bar2}, \; a_{2\bar1}, \; a_{2\bar2},$$

which are transformed as

$$\xi_1\bar{\xi}_1, \quad \xi_1\bar{\xi}_2, \quad \xi_2\bar{\xi}_1, \quad \xi_2\bar{\xi}_2$$

under a binary transformation. Let us prove that the determinant is invariant:

$$\begin{vmatrix} a_{1\bar{1}} & a_{1\bar{2}} \\ a_{2\bar{1}} & a_{2\bar{2}} \end{vmatrix} = a_{1\bar{1}}a_{2\bar{2}} - a_{2\bar{1}}a_{1\bar{2}}$$

$$= \frac{1}{2}\left(a_{1\bar{1}}a^{1\bar{1}} + a_{2\bar{2}}a^{2\bar{2}} + a_{2\bar{1}}a^{2\bar{1}} + a_{1\bar{2}}a^{1\bar{2}}\right) = \frac{1}{2}\,a_{\mu\bar{\nu}}a^{\bar{\mu}\nu}.$$

2. The quantities $a_{1\bar{1}}, a_{2\bar{2}}$ are real and $a_{1\bar{2}}$ and $a_{2\bar{1}}$ are complex-conjugate. Therefore we can set

$$a_{1\bar{1}} = ct + z, \quad a_{1\bar{2}} = x + iy, \quad a_{2\bar{2}} = ct - z, \quad a_{2\bar{1}} = x - iy.$$

We have

$$a_{1\bar{1}} = a^{2\bar{2}}, \quad a_{2\bar{2}} = a^{1\bar{1}}, \quad a_{1\bar{2}} = -a^{2\bar{1}}, \quad a_{2\bar{1}} = -a^{1\bar{2}}.$$

In the new notation we get

$$\frac{1}{2}\,a_{\mu\bar{\nu}}a^{\mu\bar{\nu}} = \begin{vmatrix} ct + z & x + iy \\ x - iy & ct - z \end{vmatrix} = c^2t^2 - x^2 - y^2 - z^2$$

for the invariant $\frac{1}{2}\,a_{\mu\bar{\nu}}a^{\mu\bar{\nu}}$. The main invariant of Minkowski's space-time will be the invariant of the binary group if ct, x, y, z have the *mechanical* sense of coordinates,

$$x^0 = ct = \frac{1}{2}\,(a_{1\bar{1}} + a_{2\bar{2}}), \quad x^1 = x = \frac{1}{2}\,(a_{1\bar{2}} + a_{2\bar{1}}),$$

$$x^2 = y = -\frac{i}{2}\,(a_{1\bar{2}} - a_{2\bar{1}}), \quad x^3 = z = \frac{1}{2}\,(a_{1\bar{1}} - a_{2\bar{2}}),$$

$$x^0 = \frac{1}{2}\,(a^{2\bar{2}} + a^{1\bar{1}}), \quad x^1 = -\frac{1}{2}\,(a^{2\bar{1}} + a^{1\bar{2}}),$$

$$x^2 = \frac{i}{2}\,(a^{2\bar{1}} - a^{1\bar{2}}), \quad x^3 = \frac{1}{2}\,(a^{2\bar{2}} - a^{1\bar{1}}).$$

3. This correspondence allows us to construct a spinor analysis. Let us consider an invariant function φ. We have

$$\text{inv} \equiv d\varphi = \frac{\partial\varphi}{\partial x^0}\,dx^0 + \frac{\partial\varphi}{\partial x_1}\,dx^1 + \frac{\partial\varphi}{\partial x_2}\,dx^2 + \frac{\partial\varphi}{\partial x_3}\,dx^3$$

$$= \frac{\partial\varphi}{\partial x^0}\,\frac{1}{2}\,(a^{2\bar{2}} + a^{1\bar{1}}) - \frac{\partial\varphi}{\partial x_1}\,\frac{1}{2}\,(a^{2\bar{1}} + a^{1\bar{2}})$$

$$+ \frac{\partial \varphi}{\partial x_2} \frac{i}{2} (a^{2\bar{1}} - a^{1\bar{2}}) + \frac{\partial \varphi}{\partial x_3} \frac{1}{2} (a^{2\bar{2}} - a^{1\bar{1}})$$

$$= \frac{1}{2} \left[a^{1\bar{1}} \left(\frac{\partial \varphi}{\partial x_0} - \frac{\partial \varphi}{\partial x_3} \right) + a^{1\bar{2}} \left(- \frac{\partial \varphi}{\partial x_1} - i \frac{\partial \varphi}{\partial x_2} \right) \right.$$

$$\left. + a^{2\bar{1}} \left(- \frac{\partial \varphi}{\partial x_1} + i \frac{\partial \varphi}{\partial x_2} \right) + a^{2\bar{2}} \left(\frac{\partial \varphi}{\partial x_0} + \frac{\partial \varphi}{\partial x_3} \right) \right].$$

Since this expression is an invariant for the arbitrary $a^{\mu\bar{\nu}}$ we have a spin tensor $\partial_{\mu\bar{\nu}}$

$$\partial_{1\bar{1}} \varphi = \frac{\partial \varphi}{\partial x_0} - \frac{\partial \varphi}{\partial x_3}, \quad \partial_{1\bar{2}} \varphi = - \frac{\partial \varphi}{\partial x_1} - i \frac{\partial \varphi}{\partial x_2},$$

$$\partial_{2\bar{1}} \varphi = - \frac{\partial \varphi}{\partial x_1} + i \frac{\partial \varphi}{\partial x_2}, \quad \partial_{2\bar{2}} \varphi = \frac{\partial \varphi}{\partial x_0} + \frac{\partial \varphi}{\partial x_3}.$$

Problem. Find directly and from the transformation formulas the spin tensor $\partial^{\mu\bar{\nu}}$

$$\partial_{\mu\bar{\nu}} = g_{\mu\alpha} g_{\mu\beta} \partial^{\alpha\bar{\beta}}.$$

4. Putting the vector a^i into correspondence with the spin tensor $a_{\mu\bar{\nu}}$ and the vector $\frac{\partial}{\partial x_i}$ into correspondence with the spin tensor $\partial_{\lambda\bar{\mu}}$, we can prove that

$$\frac{1}{2} \partial_{\mu\bar{\nu}} a^{\mu\bar{\nu}} = \frac{\partial a^i}{\partial x^i},$$

$$\frac{1}{2} \partial_{\mu\bar{\nu}} \partial^{\mu\bar{\nu}} = \frac{\partial^2}{\partial x_0^2} - \frac{\partial^2}{\partial x_1^2} - \frac{\partial^2}{\partial x_2^2} - \frac{\partial^2}{\partial x_3^2},$$

$$\frac{1}{2} \partial_{\mu\bar{\nu}} \varphi \partial^{\mu\bar{\nu}} \varphi = \left(\frac{\partial \varphi}{c \partial t} \right)^2 - \left(\frac{\partial \varphi}{\partial x} \right)^2 - \left(\frac{\partial \varphi}{\partial y} \right)^2 - \left(\frac{\partial \varphi}{\partial z} \right)^2.$$

5. In this respect the main problem (not yet solved) is to find the mechanical meaning of spinors with the ordinary interpretation of the coordinates ct, x, y, z.

A.3.14

1. The group of binary transformations include true perturbed motions, and, hence, the invariant of the binary transformations of *motion* is an integral of Poincaré's equations in variations.

Stability obtains when there is a positive definite invariant (the corollary of Lyapynov's theorem on stability).

Consider an example

$$\frac{dx}{dt} = y, \quad \frac{dy}{dt} = -x.$$

From this we have $x = -\lambda y$, $y = \lambda x$, or

$$\xi^1 = -\lambda \xi^2 = -\lambda \xi_1, \quad \xi^2 = \lambda \xi^1 = -\lambda \xi_2.$$

This means that

$$\eta^\nu = \overline{\xi^\nu} = -\overline{\lambda \xi_\nu},$$

and, consequently, the invariant

$$\xi_\nu \eta^\nu = -\lambda \sum_\nu \xi_\nu \overline{\xi}_\nu$$

is a function of fixed sign. We have stability here.

2. We shall write the condition of existence for a quadratic invariant $\xi_\nu \eta^\nu$ of fixed sign as

$$\eta^{\overline{\nu}} = \xi_\nu.$$

How can we write this condition in the form of equations?

We can get a new spinor from the spinor ξ_μ using the formula

$$\partial^{\mu \overline{\nu}} \xi_\mu + \frac{i}{\lambda_0} \eta^{\overline{\nu}} = 0. \tag{4}$$

If the spinor η^ν yields an invariant $\xi_\nu \eta^\nu$ of fixed sign, we have, in accordance with the main condition (4), equations

$$\sum_\mu \partial^{\mu \overline{\nu}} \eta^{\overline{\mu}} + \frac{i}{\lambda_0} \eta^{\overline{\nu}} = 0.$$

We write out the sum as follows:

$$\nu = 1, \quad \partial^{1\overline{1}} \eta^{\overline{1}} + \partial^{2\overline{1}} \eta^{\overline{2}},$$
$$\nu = 2, \quad \partial^{1\overline{2}} \eta^{\overline{1}} + \partial^{2\overline{2}} \eta^{\overline{2}}.$$

3. Let us study the operator $\partial_{\mu \overline{\nu}}$. We have determined it as a function of ξ_ν, i.e.

$$a_{\mu \overline{\nu}} = \xi_\mu \xi_{\overline{\nu}}.$$

As a function of η^ν we have

$$a_{\mu \overline{\nu}} = \xi_\mu \xi_{\overline{\nu}} = \eta^{\overline{\mu}} \eta^\nu = c^{\overline{\mu} \nu}.$$

Naturally, this gives us a distinction between the (ξ)-operators and the (η)-operators:

$$\partial_{\mu\bar{\nu}} = \Delta^{\mu\bar{\nu}}.$$

Hence

$$\sum \partial^{\mu\bar{\nu}}\eta^{\bar{\mu}} = \sum \Delta_{\bar{\mu}\nu}\eta^{\bar{\mu}},$$

and this means that the last condition can be simply written in the (η)-operators as

$$\Delta_{\nu\bar{\mu}}\eta^{\bar{\mu}} + \frac{i}{\lambda_0}\,\xi_\nu = 0. \tag{5}$$

Equations (4) and (5) yield Dirac's equations.

NAME INDEX

SUBJECT INDEX